Ecology and Landscape Development: A History of the Mersey Basin

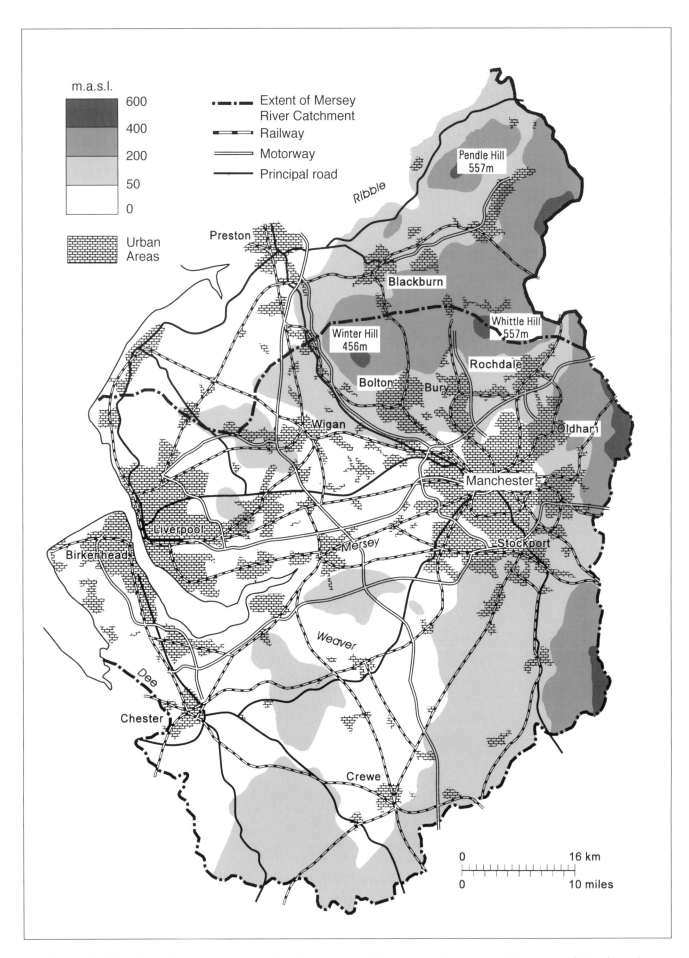

Figure 0.1. The Mersey Basin (map prepared by Department of Planning and Landscape, University of Manchester).

Ecology and Landscape Development: A History of the Mersey Basin

PROCEEDINGS OF A CONFERENCE
HELD AT MERSEYSIDE MARITIME MUSEUM,
LIVERPOOL, 5–6 JULY 1996

Edited by E.F. Greenwood

LIVERPOOL UNIVERSITY PRESS

NATIONAL MUSEUMS & GALLERIES ON MERSEYSIDE

First published 1999 by Liverpool University Press
copyright © 1999 The Board of Trustees of
the National Museums & Galleries on Merseyside

The right of E.F. Greenwood
to be identified as the editor of this work
has been asserted by him in accordance with
the Copyright, Design and Patents Act, 1988

British Library Cataloguing-in-Publication Data
A British Library CIP record is available

ISBN 0-85323-653-4

Design and production: Janet Allan

Typeset in 10/12pt Palatino by
XL Publishing Services, Lurley, Tiverton
Printed by Redwood Books, Trowbridge

Contents

Forewords

As Chairman of the Mersey Basin Campaign I was honoured to be invited to give the opening address to the conference which gave rise to this publication.

The Mersey Basin Campaign is, like the rest of north-west England, a consequence of the history that the conference traced. We are dealing with the basin's geography, geology and hydrology as well as the effects of its economic past and its hopes for the future. The conference and the papers published here amply illustrate the scale and variety of those topics.

There have been so many influences on the nature and landscape and its flora and fauna. Few of the influences have been planned and, until relatively recently, almost none sought to shape the future environment positively. Those influences, both natural and man-made are explored in great detail through these conference papers. Fascinating as that is in its own right, perhaps the true value of the papers lies in the pointers they can give about how we move on.

Understanding the ingredients that have brought us to where we are today is an essential part of the process of taking us towards tomorrow. That future is in the hands of a wide range of individuals and organisations with many different roles. The biggest question is perhaps, not so much what they will do and what they will decide, but how will they work together? The Mersey Basin Campaign is one organisation which invites partners, whose objectives overlap, to co-operate – I believe that the conference in 1996, and the chapters published here, also encourage that process.

Brian Alexander
Chairman, Mersey Basin Campaign
December 1997

One of the main aims in 1986 of the newly created National Museums & Galleries on Merseyside was the promotion of its scholarly and scientific programmes. The Trustees quickly established a Scholarship Committee under the Chairmanship of Sir David Wilson. It seemed particularly important to the Trustees that they should procure a study of man's impact on the animals, plants and natural features of the Merseyside landscape.

Eric Greenwood, as Liverpool Museum's leading natural historian, was asked to co-ordinate the preparation of this study and to find a way of funding it. A tall order indeed! Nevertheless Eric achieved all this through meticulous preparation of the conference held in July 1996.

On behalf of NMGM's Trustees and staff I would like to acknowledge both the importance and practical value of this benchmark survey. The result is a tribute to Eric Greenwood's determination and tenacity as the organiser and editor of the many distinguished papers presented by the experts who attended and gave papers to this important gathering.

Richard Foster
Director, National Museums & Galleries on Merseyside
October 1998

Stratigraphic correlation chart — Sefton Coast

Thousands of Years B.P.	14 13 12 11 10 9 8 7 6 5 4 3 2 1
Stage	PLEISTOCENE — Devensian (Glacial) (115,000 to 10,000 B.P., peak 17,000 B.P.) → HOLOCENE — Flandrian (Interglacial); Early / Mid / Late
Chronozone (Godwin Zones; P.A.Z. Pollen Assemblage Zones; Brett-Sermander Units)	L De I / L De II / L De III (Loch Lomond); IV Fa a / V Flb b / VIa Flc c / VIb Fld d / VIc Fld d / VIIa FII e / VIIb / VIII; Flandrian I / Flandrian II / Flandrian III; Preboreal / Boreal / Atlantic / Sub-Boreal / Sub-Atlantic
Major Vegetation	Herbs — Birch; Birch Pine Hazel; Hazel Pine; Pine Hazel Elm; Oak Elm Alder; Oak, Alder (heather, grasses and herbs increasing). Elm Decline 5250 B.P.
Climate	Cold — Warm (Loch Lomond Stadial: Cold); Warm and dry; Warm and wet; Cooler and drier; Cool and wet; Warmer and drier; *Cool and wet. *Sharp Deterioration
Soils	Raw, unstable soil profiles — None; Stabilised profiles / Unstable profiles; Maturing soil profiles base-rich; Stable, base-rich forest soils; Increasing acidity and podsolisation
Sea-Level Tendencies — Transgressive (+VE)	1, 2, 3, 4, 5, 6, 7, 8, 9, 10, 11, 12
Sea-Level Tendencies — Regressive (-VE)	1, 2, 3, 4, 5, 6, 7, 8, 9, 10, 11, 12
Actual Levels	Sea-Level Very Low; -20m – -15m OD; Period of Rapid Rise; 0 – +2m OD; +2 – +3.5m OD; High levels +3.5 – 5m OD. *Dune Slack Peat
Wetlands	None; Some moss peat deposition; Start of mossland formation in lake basins; Creation of coastal mosses inland raised bog growth; Extension of coastal mosses creation of meres; 'Recurrence surfaces' rapid bog growth; Truncation of bog profiles by erosion
Geology (Dated Events at SEFTON COAST; Minero/biogenic sediment'n.)	End of till deposition; Periglacial sands and gravels; Shirdley Hill Sand deposition; S.H. Sand redistribution and organic inclusions (continues through Flandrian); DH I; DH II; Marine alluvium 'Downholland Silt' / IV Sand dune building / S1 / River valley alluvium / S2 / Sand dune building. S? V? S?
Human Influence	Final Paleolithic & Mesolithic; *Poulton-le-Fylde Elk; evidence of settlement (Little Crosby); Neolithic; Early Bronze; Late Bronze + Iron Age; Rom. Occ.; Medieval. *Cereal Cultivation Locally
Clearance / Drainage / Plantation (Cl, D, Pl)	Cl; Cl; Cl; Cl; Cl; Cl D Pl

Introduction

E.F. GREENWOOD

Shortly after the National Museums & Galleries on Merseyside (NMGM) was formed in April 1986, two of the trustees with special interests in natural science, Professors A.D. Bradshaw and R.J. Berry, asked me (as Keeper of the Liverpool Museum) what particular research topic I would like to see developed.

I explained that the NMGM with its broad subject base was in an excellent position to lead a study of the interactions between humans and other living organisms on Merseyside since the last 'ice age'. The Mersey Basin embraced an area where human intervention had a long history and in the last few hundred years had caused pollution as severe as anywhere in the world. It was a profoundly altered landscape. Yet, the area also contained regions of wildscape, ranging from open moorland on the Pennine hills through woods and moors to the estuaries and Liverpool Bay, and was home to various forms of rare wildlife on at least a European scale that have found a refuge and sanctuary. I believed an integrated study would provide many lessons for the future. Furthermore, I did not believe that such a study of an urban and industrial region had taken place anywhere in the world.

During the following years interest in the wildlife of the urban environment and the ecological processes involved has developed world-wide, particularly in Europe, North America and in the UK. Often the studies reveal a lack of knowledge of what plants and animals occur or of those that do occur, how they have survived human activity or indeed to what extent their existence depends upon humans. Indeed it is possible that more is known of the remote islands of the world than of the most densely populated city regions.

Within the Mersey Basin the ten years between 1986 and 1996 saw a number of studies that made a profound difference in our understanding of the changes that took place in the prehistoric period. However, our understanding of settlement changes that took place from the 'Dark Ages' until written documents became available remains less certain with almost no information about plants and animals. Even during the last 200 years information for all but the better known groups of plants and animals is remarkably thin.

Nevertheless, there have been many recent studies in the region and, whilst it was not possible to fund an integrated research programme, it did prove possible to bring together many of those who have worked in the region during the last few years. Their work was presented at a conference attended by some 170 participants, held at the Merseyside Maritime Museum on 5 and 6 July 1996. The written versions of their papers are published in this volume.

In response to the requests of participants references are included at the end of each chapter. Also, common names are used in preference to Latin ones. Nomenclature follows the following texts: flowering plants and ferns, *New Flora of the British Isles* by C. Stace, Cambridge University Press (1991); birds, *The Birdwatcher's Yearbook and Diary 1994* edited by J.E. Pemberton, Buckingham Press (1993); mammals, *The Handbook of British Mammals*, 3rd edn, edited by G.B. Corbet and S. Harris, Blackwell Scientific Publications for Mammal Society, Oxford (1991); fish, *A list of the Common and Scientific names of fishes of the British Isles*, Academic Press, London (1992); reptiles and amphibians, A *Field Guide to the Reptiles and Amphibians of Britain and Europe* by E.N. Arnold and J.A. Burton, Collins, London (1985).

To help understanding of the major events in the region over the last 15,000 years Figure 1.1 from *The Sand Dunes of the Sefton Coast* edited by D. Atkinson and J. Houston, NMGM, Liverpool (1993) is reproduced here (Figure 0.2). However, authors have used both calibrated radiocarbon dates and calendar years. Standardised dates can be obtained by reference to the papers by M. Stuiver and P.J. Reimer (1993) Extended ^{14}C database and revised CALIB 3.0 ^{14}C age calibration program. *Radiocarbon*, **35**, 215–30. A map of the Mersey Basin is also included (Figure 0.1).

In addition the following abbreviations are used: sp for species (singular), spp. for species (plural), ssp. for sub-species (singular) and sspp. for sub-species (plural).

Figure 0.2. (left) Major events in the sub-region over the last 15,000 years in context (courtesy of J.B. Innes). Dating is in ^{14}C years BP (Years Before Present: present taken conventionally to be AD 1950). The table shows some of the inter-relationships between climate, vegetation, sea-level trends and human activity over the Holocene period. Further explanation of the table is given in chapters two, three and four of *The Sand Dunes of the Sefton Coast* (Atkinson & Houston, 1993).

Acknowledgements

Conferences and their proceedings do not just happen and I am particularly grateful to Professors A.D. Bradshaw and R.J. Berry for their help, guidance and continued support. I am also grateful to Julian Taylor, George Barker and David Atkinson who at various times with Professor Bradshaw formed an *ad hoc* steering group to organise the conference. I am also indebted to the British Ecological Society, the Mersey Basin Campaign, the University of Liverpool and NMGM for financial support and to a number of other organisations who provided support in kind. I am also very grateful to the chairs for the various sessions who provided valuable assistance by refereeing the chapters. Special thanks are due to all the contributors who made the conference so successful and who have provided a unique assemblage of information and analysis of one of the most urbanised yet fascinating parts of the world. Finally I am greatly indebted to my secretary, Barbara Rowan, for much typing and incorporating all the alterations to the scripts.

E.F. GREENWOOD

The Background Setting

F. OLDFIELD

Concern about the consequences of future climatic change and their impact on our environment and life support systems is now almost universal. Even if scepticism persists about the reality of 'greenhouse warming', the inevitability of climatic change cannot be denied – climate varies naturally on all timescales and every aspect of the world we see around us changes in response to these variations. Although they are global, the way the variations are expressed and their regional impact are highly differentiated. This makes it of exceptional importance to understand, at a regional level, the way our environment has changed in the past, and to use this knowledge realistically in any consideration of the potential effects of future climatic change. Nowhere is this more important than in a region like the Mersey Basin where the combination of a large population, a highly developed modern infrastructure and a low-lying, naturally dynamic coastal zone makes planning for a future where environmental conditions are likely to change quite dramatically a major and complex responsibility. The past does not provide simple analogues for a future world, but it is the only source of evidence we have for what has actually happened and for what may happen in response to climate changes on the timescales and of the magnitude of those predicted for the next century. We therefore value greatly the contributions to this first session of the conference and the chapters that follow, setting out the nature of past changes in the local environment and the ecological and human responses to these.

CHAPTER ONE

The influence of a changing climate

B. HUNTLEY

Introduction

In examining the ecological and landscape history of any region it is essential to consider the potential influences of climate change. Although ecologists frequently have viewed climate as an unchanging part of the environment to which organisms and ecosystems are adapted, and archaeologists frequently have taken a similarly static view of climate when examining the history of human influences upon landscapes, whenever records of environmental history are examined climate change is seen to be a ubiquitous feature. The minute-to-minute variations of the aerial environment experienced by a leaf simply are the highest frequency component of a spectrum of such changes that extends through the familiar diurnal and annual cycles and onward to much longer timescales. Ultimately, at timescales of 10^7 to 10^8 years we find records of global climate changes that relate to the rearrangement of the continents and to the major orogenic periods of geological history. In the present context, however, where our concern is with environmental history since the last glacial stage that ended about 12,000 years ago, it is those climate changes that have characteristic periodicities of 10^3 to 10^5 years that are most relevant. These are the climate changes that are associated with the alternating glacial and interglacial stages of the most recent Quaternary geological period (the last c.2.4 Ma).

These climate changes have been the subject of a considerable research effort and are now relatively well-documented, at least in the western North Atlantic region, for the last 130,000 years – the period spanning the last interglacial and glacial stages as well as the Holocene or post-glacial period. Not only have Quaternary scientists documented the changes, but they also have developed a number of hypotheses as to their underlying causes and mechanisms. These hypotheses have to some extent been evaluated by collaborative studies using atmospheric general circulation models (AGCMs) to simulate past climates (COHMAP 1988; Kutzbach & Gallimore 1988; Kutzbach & Guetter 1986; Wright et al. 1993). Two key hypotheses will be introduced below as the evolution of palaeoclimate since the last glacial maximum, some 21,000 years ago, is

described. Firstly, the Milankovitch hypothesis relates to the role of changes in the quantity and especially the seasonal distribution of solar radiation reaching the earth (Berger et al. 1984; Hays, Imbrie & Shackleton 1976), and secondly the 'ocean conveyor belt' hypothesis attempts to account for the magnitude of the resulting climate changes through a proposed oceanic circulation feedback mechanism (Broecker & Denton 1989; Imbrie et al. 1993; Imbrie et al. 1992).

In order to appreciate the approach that I have adopted it is important to see the Mersey Basin in its geographical context not only within Great Britain but also much more widely. The map presented in Figure 1.1 attempts to illustrate a number of the key features of this context; these include major features of the present-day circulation both of the North Atlantic ocean and of the atmosphere above it. The location of the Mersey Basin closer to the eastern margin of this map is an appropriate reflection of the latitudinal position in which it lies. The most characteristic feature of the atmospheric circulation in such middle latitudes is the westerly airflow, so that the climate of any location is influenced most strongly by areas to its west. Nonetheless, the map extends also to include Scandinavia because, as we shall see, 'downstream' features also can exert an influence upon climate. It is impossible to consider the influence of climatic change upon this small region of western England without seeing it in such a context, not least because only in this way can we understand which of these climate changes would be ubiquitous throughout the British Isles or even much of north-western Europe. Indeed it will become apparent that the Mersey Basin is much too small a region to have any climate changes that affect this region alone. Viewed in this context it also immediately becomes apparent that most of the climate changes to which an area such as the Mersey Basin is subjected arise entirely from causes external to the region; indeed I shall argue that some of these changes arise from causes external to Earth itself. It further should become apparent that, arising as they do from external causes, the impact of these climatic changes is inescapable, whether by ecosystems or human communities.

Thus, although this chapter is written in the context of a discussion focused upon the ecology and landscape development of the Mersey Basin, its content in general would be equally applicable to most of Great Britain. Indeed it might be sub-titled *Climate change in the western North Atlantic Region since the Last Glacial Maximum – a context for considerations of the environmental history of British landscapes*. The chapter is organised into four principal sections that deal in chronological order with the period since the last glacial maximum; it ends with a short Coda that reiterates and summarises the key points to emerge.

Climate at the last glacial maximum

The last glacial maximum was *c*.21,000 years ago;* however, conditions broadly comparable to those at the time of maximum ice extent prevailed for perhaps 6,000 years or more and the phase of most rapid melting of the major ice sheets did not start until *c*.14,500 BP. Figure 1.2 presents the palaeogeographical context for the Mersey Basin at the time of the last glacial maximum and should be compared with Figure 1.1. The most important factors that impinge directly upon the region under consideration are the local ice sheet developed within the British Isles and the lowering of sea-level that results in the present Mersey Basin becoming, at least during the early stages of deglaciation, the head-waters of a tributary to a fluvial system occupying the northern part of the Irish Sea. However, it is also important to note the changed ocean and atmospheric circulation patterns, the position of the sea-ice margin in the north-eastern North Atlantic and the fact that the British Isles are not islands but the western margin of the European mainland. In addition, we must be aware that the atmospheric concentrations of naturally occurring 'greenhouse gases', CO_2 and CH_4, were markedly lower not only than today but also than pre-industrial concentrations that prevailed during most of the Holocene (Chappellaz *et al.* 1990; Lorius *et al.* 1988). The seasonal insolation values, in contrast, were more or less the same as today (COHMAP 1988).

Although to some extent the climate at this time is rendered irrelevant by the presence of an ice sheet covering the present Mersey Basin (Boulton *et al.* 1977; Devoy 1995), it is worthwhile to note a number of features that account for the climatic conditions leading to the growth of this ice sheet, and changes in which during the subsequent deglaciation will be important causes of the climate prevailing in the area during that period. Foremost amongst these, perhaps, are the two major last glacial ice sheets developed in north-eastern North America (Laurentide ice sheet) and in Fennoscandia. The thickness of these ice sheets (1,000–2,000m) was such that they were able to influence the atmospheric circulation in much the same way as would major mountain ranges. Although more recent

palaeoclimate model results have not simulated the split in the jet stream seen in the COHMAP results (Kutzbach *et al.* 1993), the Laurentide ice sheet nonetheless had a substantial impact upon atmospheric circulation over the North Atlantic and downstream into Europe. The Fennoscandian ice sheet similarly affected the upstream circulation over the western North Atlantic and over the British Isles, as well as having an impact upon atmospheric circulation across Europe as a whole. These changes are represented on the maps (Figures 1.1 and 1.2) by shifts in the path of the winter jet stream; this in turn, however, relates to climate conditions at the surface because the predominant path taken by storm systems ('depressions') as they cross the North Atlantic and enter Europe is determined by the position of this atmospheric feature.

A second key contrast between the two maps (Figures 1.1 and 1.2) relates to the oceanic circulation within the North Atlantic. The two key classes of feature shown on the maps are the, generally more familiar, surface currents and the mid- and deep-water currents. Whereas today the North Atlantic Drift (often inaccurately referred to as the 'Gulf Stream') passes north-eastwards to the west of the British Isles and finally is dissipated in the Norwegian Sea, the equivalent surface current at the last glacial maximum crossed the Atlantic in a more or less easterly direction, reaching Europe in the latitude of southern Iberia. Although some question arises as to the extent to which a component of this warm surface flow

Figure 1.1. (right) The geographical context of the Mersey Basin at the present day
A schematic illustration of major atmospheric and ocean circulation features that affect the environment of the Mersey Basin (indicated by a black dot) today. The mean position of the January storm track is shown, along with the warm (A Gulf Stream; B North Atlantic Drift) and cold (C East Greenland Current; D Labrador Current) surface currents in the North Atlantic and the generalised path of the North Atlantic Deep Water (NADW) flow. The present extents of ice sheets and of permanent sea ice also are shown. (Redrawn and schematised from various sources.)

Figure 1.2. (right) The palaeogeographic context of the Mersey Basin at the last glacial maximum
A schematic illustration of major atmospheric and ocean circulation features that affected the Mersey Basin (indicated by a black dot) at the last glacial maximum. The mean January storm track is shown splitting around the major ice sheets as in the palaeoclimate simulation made by Kutzbach *et al.* (1993). The surface currents are schematised from various sources whilst the generalised path of a deep water flow generated by the 'boreal heat pump' (Imbrie *et al.* 1992) also is indicated. The extents of the last glacial ice sheets and of permanent sea ice are redrawn schematically from a variety of sources. The approximate position of the coastline is shown by the thin black line, the present geographical features being retained in grey for reference.

---

* Throughout this chapter ages are given in calendar years; where the ages of events have been determined by ^{14}C dating the calibration procedure of Stuiver and Reimer (1993) has been used to estimate their calendar age.

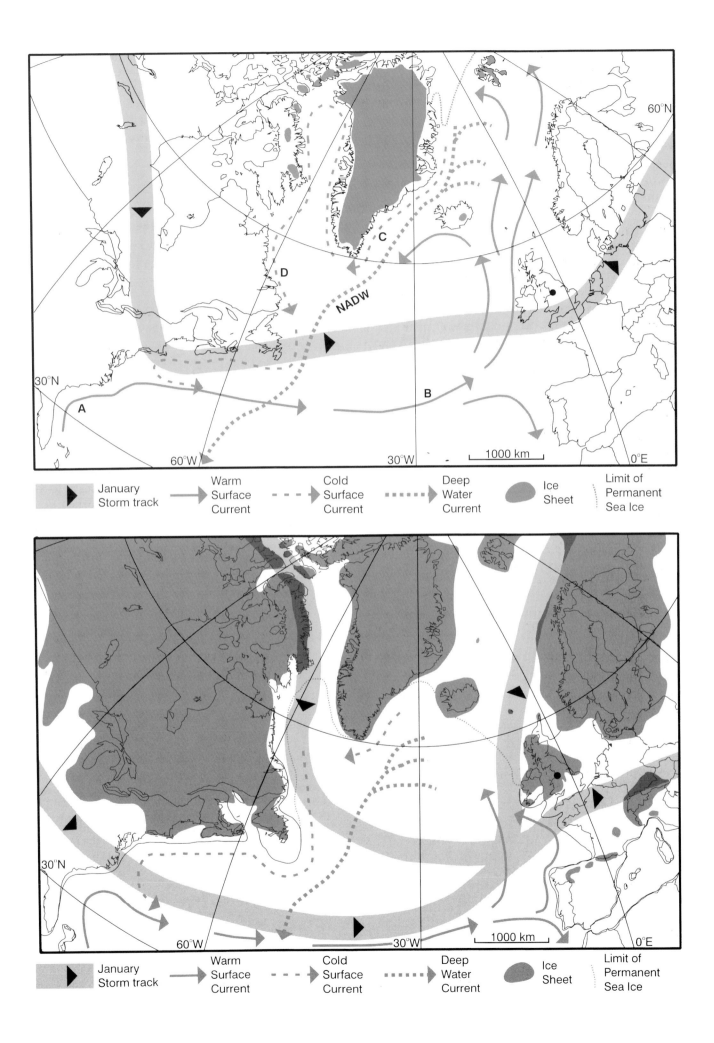

| | January Storm track | → Warm Surface Current | →‑‑‑ Cold Surface Current | ⋯▸ Deep Water Current | Ice Sheet | Limit of Permanent Sea Ice |

| | January Storm track | → Warm Surface Current | →‑‑‑ Cold Surface Current | ⋯▸ Deep Water Current | Ice Sheet | Limit of Permanent Sea Ice |

may then have travelled northwards up the western seaboard of Europe, there is no doubt as to the principal consequence of this changed surface circulation, namely that the bulk of the North Atlantic surface was much cooler during the last glacial maximum than it is today. As a result, air masses crossing the ocean arrived in Europe cooler and also drier, because less moisture was evaporated from the cooler ocean surface. The extents both of permanent and of seasonal sea ice also were much greater. This was a secondary consequence of the cooler surface, combined with a reduction in the salinity of the surface waters as compared with today, because they had not been concentrated by evaporation to the same extent as are the warmer modern surface waters derived from the tropical Atlantic and Mediterranean. The extensive seasonal formation of sea ice promoted further cooling of conditions over those areas; the winter temperatures in Europe were as a result particularly strongly affected (Atkinson, Briffa & Coope 1987; Guiot *et al.* 1993a).

In a global context North Atlantic surface temperatures exhibit greater cooling during glacial maximum times than does any other ocean region (CLIMAP Members 1976). Both as a consequence of this cooling, and of the associated reduction in surface salinity, as well as a partial cause of these phenomena, North Atlantic Deep Water (NADW) formation did not occur in the Norwegian Sea during the glacial maximum and the strength of the southward flowing deep water current in the Atlantic was reduced. However, it has been hypothesised by Imbrie *et al.* (1992) that the sinking motion that occurs in the northern North Atlantic (the 'boreal heat pump') was strengthened during the glacial maximum, leading to the formation both of an intermediate water current that flowed southward to upwell once again in the tropical Atlantic and of a deep water flow southward, albeit at lesser depths than the NADW flow during interglacial times and of much reduced volume compared to that of NADW at the present day.

The markedly lowered CO_2 concentration, $c.190$ppmv compared to a pre-industrial value of $c.280$ppmv and a present day value of $c.355$ppmv, also affected the climate as a consequence of the reduced 'greenhouse effect' and would have had direct ecophysiological effects upon terrestrial vegetation.

Together, these various factors resulted in winter temperatures at least 25°C, and perhaps as much as 35°C, cooler in north-west Europe (Atkinson *et al.* 1987; Guiot *et al.* 1993a). Summer temperatures also were markedly reduced, although not by a comparable amount. There also was a reduction in moisture supply that had its greatest impact in southern Europe and in those areas downstream of the British and especially the Fennoscandian ice sheets; orographic precipitation over the ice sheets coupled to the sinking and hence warming of the air on their downstream sides produced very dry conditions in these areas.

Climate during deglaciation and the transition to the Holocene

The onset of rapid melting of the ice sheets about 14,500 years ago heralds a period of *c.* 5,000 years of complex environmental changes and associated climate fluctuations. Amongst the relatively well-established phenomena occurring during this time we may include the following:

1. summer insolation in the northern hemisphere rose to a maximum of *c.*8% more than present at *c.*12,000 years ago, whereas winter insolation fell to a minimum of *c.*5% less than today at the same time (COHMAP 1988);

2. the principal ice sheets melted rapidly during two periods, *c.*14,000 to 12,000 years ago and *c.*11,000 to 10,000 years ago (Fairbanks 1989); between these two periods was an interval when the principal ice sheets were more or less stable (Denton & Hughes 1981). In the British Isles, however, whereas the local ice sheet had melted completely during the first major phase of melting, there was local redevelopment of an ice-cap in the western highlands of Scotland (Sissons 1981) and of corrie glaciers as far south and west as the mountains of Wales and eastern Ireland (Watts 1977). These small areas of ice rapidly melted with the onset of the second phase of melting;

3. sea-level responded to the rapid ice sheet melting, exhibiting two principal periods of particularly rapid rise coinciding with the phases of rapid melting (Fairbanks 1989). In areas of ice sheet accumulation during the glacial the effective sea-level rise was apparently higher because the isostatic rebound of the earth's crust proceeded less rapidly than the eustatic sea-level rise. Thus the maximum incursion of the sea onto areas around the Irish Sea that today are land areas occurred during the late-glacial period, i.e., during the time of deglaciation (Wingfield 1995);

4. atmospheric concentrations of CO_2 rose to their Holocene levels of *c.*280ppmv; some evidence suggests that this rise was irregular and that levels may have fluctuated quite markedly during this interval (White *et al.* 1994), perhaps even rising above their Holocene value (Beerling *et al.* 1993);

5. surface circulation patterns in the North Atlantic exhibited complex changes (Ruddiman & McIntyre 1981). Initially a north-eastward flow of warm surface water became established, although this did not penetrate as far to the north-east as does the present-day North Atlantic Drift. This trend was reversed following the first phase of rapid ice sheet melting so that during the period when ice once again accumulated in the British Isles the surface ocean circulation probably resembled most strongly that at the glacial maximum (Broecker & Denton 1989). Thereafter the surface circulation changed to something closely

resembling that which we see today. This relatively simple pattern corresponds to the simple sequence of zones recognised in many stratigraphic records, i.e. glacial – late-glacial interstadial – late-glacial stadial ('Younger *Dryas*') – post glacial (Holocene). However, many terrestrial records for this interval provide evidence for additional fluctuations within the late-glacial interstadial and/or the early post glacial (Huntley 1994; Watts 1977); the limited temporal resolution of many ocean records may mask additional changes in surface circulation patterns associated with these events.

6. both the extent of sea ice and the deep circulation of the North Atlantic also exhibit complex changes during this period that intimately link to the changes in surface circulation (Broecker & Denton 1989; Imbrie *et al.* 1992);

7. finally, atmospheric circulation and, as a result, the climate of Great Britain were undergoing substantial changes as a result of all of these other phenomena (Kutzbach & Guetter 1986; Kutzbach *et al.* 1993).

Following the retreat of the British ice sheet, the Mersey Basin, in common with much of the remainder of England, initially was occupied by tundra-like vegetation dominated by Arctic–Alpine herbaceous and dwarf-shrub taxa. During the late-glacial interstadial taller woody taxa arrived, first Juniper (*Juniperus*) and later Birch (*Betula*). Subsequently these were displaced during the late-glacial stadial; tundra-like vegetation prevailed once again at that time (Birks 1965). A wider variety of woody taxa rapidly arrived during the early Holocene, Juniper and Birch soon being displaced by more thermally-demanding tree and shrub taxa characteristic of the modern nemoral forest zone of Europe, e.g. Hazel (*Corylus*), Elm (*Ulmus*) and Oak (*Quercus*) (Huntley & Birks 1983).

In contrast to this vegetation development, fossil beetle assemblages changed in composition extremely rapidly following deglaciation; a variety of taxa that today exhibit strongly southern distributions and which in some cases no longer extend as far north as the British Isles, e.g., *Asaphidion cyanicorne* Pand., *Bembidion ibericum* Pioch. (two species found in late-glacial interstadial deposits at Glanllynnau, North Wales, see Coope & Brophy 1972), were present almost immediately. This contrast has often in the past been interpreted as indicating a lag in the response of vegetation to an early and very rapid climatic change. There even have been attempts to explain why this might occur, including the hypothesis that soil development was the limiting factor rather than dispersal and migration (Pennington 1986). Such hypotheses, however, overlook, or were formulated before we had become aware of, some of the environmental changes that characterise this period. In particular, the enhanced seasonal contrast in insolation, peaking 12,000 years ago, would have favoured the development of warm microhabitats which might

have been exploited by warmth-demanding beetles (Andersen 1993); at the same time the effective temperatures experienced by taller woody taxa may have been insufficient to meet their growth requirements, or, more likely, the extreme winter cold to which they were exposed may have prevented their survival (Huntley 1991). The relatively low atmospheric concentration of CO_2 also would have limited tree growth as well as rendering them more susceptible to moisture deficiency; this is unlikely to have had any effect upon beetles.

Thus, as the Mersey Basin became ice free it experienced a climate that was very cold in winter, with much stronger seasonal contrast in temperature and the development of microhabitats that were extremely warm in summer as a result of enhanced insolation. The amount of precipitation probably was less than today, although so too was the overall evaporative demand; nonetheless the climate probably was effectively drier than today. Sub-Arctic trees and shrubs colonised the area during the latter part of the late-glacial interstadial whilst Reindeer (*Rangifer tarandus*), Elk (*Alces alces*), Wild Horse (*Equus ferus*) and even Giant Deer (*Megaceros giganteus*) – now extinct, still roamed the landscape (Stuart 1982). Seasonally frozen ponds became very warm during the short intense summers of the interstadial, enabling them to support relatively warmth-demanding plants and invertebrates, and locally on the landscape warmth-demanding terrestrial invertebrates and even perhaps low-growing herbaceous plants occurred in suitable microhabitats at this time. The return of tundra during the late-glacial stadial, corresponding to the development of corrie glaciers in the mountains of the English Lake District and of North Wales, indicates a return of colder conditions associated with changed circulation of the North Atlantic and of the atmosphere above. This cold phase persisted for at most 1,000 years, probably not much more than 500 years, before the onset of the rapid climate change which coincides with the beginning of the Holocene.

Climate during the Holocene

The beginning of the post-glacial period is marked by one of the most rapid palaeoenvironmental changes seen during the late Quaternary. The magnitude of this event, however, has tended to obscure the fact that the environment has continued to change, albeit exhibiting changes of much smaller magnitude, throughout the Holocene. This ongoing pattern of change is what we should expect, nonetheless, given that the various components of the climate system in particular have not been static during the Holocene.

The changes in some of these components are well documented. Thus, although the seasonal contrast in insolation peaked 12,000 years ago, there was still a substantial divergence from today during the early Holocene and the difference in insolation alone 6,000 years ago remained sufficient to cause substantial differences in climate between then and now (COHMAP 1988;

Kutzbach & Guetter 1986; Kutzbach *et al.* 1993). Only during the second half of the Holocene have insolation values approached relatively closely those of today.

The major northern hemisphere ice sheets had not completely melted at the beginning of the Holocene; remnants of the Fennoscandian ice sheet persisted until less than 9,000 years ago and the Laurentide ice sheet did not completely melt until *c.*7,000 years ago. Although by the early Holocene these ice sheets no longer were thick enough to disturb atmospheric circulation directly, they remained as areas of relatively high albedo and consequently continued to impact indirectly upon atmospheric circulation in the higher northern latitudes.

Just as the ice sheets continued to melt, so the sea level continued to rise, albeit much more slowly than previously. From the view point of the Mersey Basin an important sequence of events was that associated with the evolution of the connections between the Irish Sea and the Atlantic. The connection in the north was established already during the late-glacial period *via* the deepwater channel in the Gulf of Corryvrecken (Wingfield 1995), whilst the early rapid sea-level rise also flooded the southern part of the basin, albeit temporarily. Subsequently, the isostatic rebound of the land overtook sea-level rise such that by 13,000 years ago a land connection apparently was re-established between south-west England and south-east Ireland. This land connection then migrated northward so that *c.*11,000 years ago the area of the Irish Sea into which the Mersey opens shifted from being open to the Atlantic via a relatively narrow northern channel to being open to the Celtic Sea to the south. The remaining land connection was subsequently flooded and both the northern and southern connections were open by *c.*10,500 years ago (Wingfield, 1995); thereafter sea water once again could circulate through the Irish Sea basin and, as today, it is likely that some warmer surface waters from the North Atlantic began to move through this channel. From the more general view point of Great Britain as a whole, the key events were the initial flooding of the southern North Sea *c.*11,000 years ago, apparently via the English Channel, and the subsequent isolation of Great Britain from the European mainland *c.*8,000 years ago as the waters flooding the northern North Sea met those encroaching from the south (Funnel 1995). These events had consequences both for the environment of Britain, as ocean water circulated into and through the North Sea and English Channel, and for the biota because any immigrant taxa subsequently had to cross a sea barrier in order to colonise Britain.

Changes in other components of the climate system either are unknown and/or less firmly established. Thus, for example, the Holocene record of surface conditions and overall circulation in the North Atlantic is insufficiently well resolved to be able to determine whether or not, and to what extent, there may have been a stronger north-eastward circulation during the early Holocene. Simulation experiments using a model which couples a very simple representation of the upper ocean to a rather low resolution AGCM, however, have suggested that the North Atlantic surface may have been warmer than today during the early Holocene as a result of the enhanced summer insolation and that this warming extended to the north-east leading to delayed onset of surface freezing and a reduced duration and extent of sea ice cover (Kutzbach & Gallimore 1988). Similar uncertainty surrounds the atmospheric CO_2 concentration during the Holocene. To date the published records from ice cores lack resolution in this critical interval; either they are relatively short highly-resolved records of the late Holocene or else they are long records extending back to the last interglacial or beyond. Recent work (J. Jouzel, pers. comm.), however, has shown evidence that the early Holocene levels may have been lower than previously thought, perhaps *c.*240ppmv as opposed to the immediate pre-industrial concentration of *c.*280ppmv. Although such a 40ppmv reduction may seem small, it is 45–50% of the overall change in concentration between glacial and pre-industrial times and physiologically is potentially important to many plants.

This array of ongoing changes is reflected in an ongoing pattern of change in the Holocene climate. Thus the climate of Great Britain during the first half of the Holocene was generally more strongly seasonal or 'continental' in character. Initially winter temperatures rose relatively slowly compared to an apparent rapid increase in overall growing season warmth (B. Huntley, unpublished results); this inference from the palaeovegetation record is supported independently by evidence from AGCM simulations as well as by other forms of palaeoenvironmental proxy evidence. It appears that the residual areas of the last glacial ice sheets had a persistent effect and that this reinforced the reduced winter season insolation resulting in the cooler winter conditions. High summer insolation, however, coupled to the now warmer North Atlantic surface, brought persistent warm anticyclonic conditions in summer. The same warm summer conditions also shifted the mean summer storm track northwards so that Great Britain and other areas at similar latitudes in western Europe received less summer precipitation (Guiot, Harrison & Prentice 1993b), and hence less precipitation in total. The reduced total precipitation is reflected in the records of lake levels which were relatively low across this part of Europe during the early millennia of the Holocene; that at least a significant part of the reduction was in summer precipitation is suggested both by the model results and by the evidence of an impact upon vegetation which is sensitive principally to moisture availability during the growing season (B. Huntley, unpublished results).

The subsequent overall trend during the mid- and late Holocene periods has been of reduced seasonal contrast, corresponding to the reduced seasonal insolation contrast, and of increasing moisture levels as the summer storm track has migrated towards its present mean position. Despite this overall pattern, however, we should no longer persist in using the concept of a post-glacial 'climatic optimum'. We should avoid this term for two reasons. The first is an essentially philosophical reason, namely that we cannot consider any conditions to be 'optimal' without identifying from what point of view they are optimal. Early Holocene climate in Britain may have been optimal for the development of diverse nemoral forests over most of England and Wales and even a large part of Scotland; however, they were not optimal for those warmth-demanding beetles which thrived only briefly during the early part of the late-glacial interstadial and they certainly were optimal neither for blanket mire communities in our uplands nor for the Arctic–Alpine component of our flora. More subtly, the conditions also apparently were not optimal for the expansion north-westwards of the range of nemoral trees such as Beech (*Fagus*) or Hornbeam (*Carpinus*), both of which do not appear to have colonised Great Britain until the later Holocene (Huntley & Birks 1983).

The second reason for avoiding the term relates to the observed independence amongst the principal climate variables; the time of warmest winters certainly did not correspond to the time of maximum growing season warmth, nor did the latter correspond to the time of maximum moisture availability. Thus we must define which aspect of climate is 'optimal' before we discuss a 'climatic optimum'. Because such a definition only can be made arbitrarily, it is pointless to do so.

It would be inappropriate to close any discussion of climate change during the Holocene without some consideration of the evidence for variations on centennial to millennial time scales that cannot be accounted for by the overall trends in insolation nor, to date, by any of the other known long-term changes which I have outlined above. The most recent and well-known of these variations is that which we refer to as the 'Little Ice Age' and that reached its maximum expression in the 17th and 18th centuries. This relatively cool interval lasting several centuries is but the most recent of a long series of such events which have occurred throughout the Holocene (Grove 1988). These events alternate with relatively warmer intervals of which the last was the so-called 'Medieval warm period' of the 11th to 14th centuries. Another similar phenomenon frequently discussed at least in the older palaeoecological literature is that of so-called 'recurrence surfaces' seen in peat bogs. These surfaces reflect intervals of relatively dry conditions on the bog surface, with a high degree of humification of the peat, followed apparently rather abruptly by a switch to relatively moist conditions leading to a rapid increase in the rate of accumulation of the peat and a sharp reduction in its degree of humification (Barber *et al.* 1994). Such surfaces also might reflect relatively cool periods, promoting moist conditions on the mire surface and rapid peat accumulation, alternating with relatively warmer and/or drier periods. The short-lived expansion of Scots Pine (*Pinus sylvestris*) onto blanket mire surfaces in the far north of Scotland around 4,500 years ago represents another example of the impact upon vegetation of such climate fluctuations during the Holocene (Gear & Huntley 1991).

The causes of such fluctuations remain controversial; what is clear, however, is that whatever their causes the expression we see in terms of climate variation must have been mediated through the same phenomena of atmospheric and ocean circulation as mediate the larger magnitude and longer-term changes onto which they are superimposed. Thus, whilst the most popular explanation is in terms of variations in solar output related to sunspot number, this would exercise its impact by changing the quantity of energy reaching the atmosphere as well as the land and ocean surfaces, and thus by changing the extent of heating of these different components of the climate system, which in turn would affect the atmospheric and ocean circulation. Alternative explanations that call upon injections of volcanic dust to the atmosphere or changes in the levels of naturally-occurring 'greenhouse gases' similarly rely upon the same underlying mechanisms of changes in the earth's energy balance and hence in atmospheric and ocean circulation.

In the context of the Mersey Basin it is important to remember the overall trends in Holocene climate and that these have been overlain by higher-frequency variations. The wide development of blanket peat across the southern Pennines, in the upper parts of the Mersey catchment, for example, corresponds to the development of a generally more oceanic climate during the later Holocene. In examining any records of increased or decreased human settlement, and especially activities in what today is the agriculturally marginal zone, this overall trend also must be borne in mind; so too, however, must the fluctuations around this trend which are recorded in the peat stratigraphy of western England as well as in historical documents and other such sources (Lamb 1982).

Recent, present and future climate changes

Just as climate change has provided the continuously varying backdrop to the development of the landscape and of human settlement and agricultural activities in Britain, the landscape and human activities are also vulnerable to the recent and ongoing changes in climate. The general warming since the end of the 'Little Ice Age' may well have contributed to the present-day susceptibility of Pennine blanket peat to erosion, for example. Contemporary agriculture continues to be vulnerable to the year-to-year variations in climate, as does the water-supply industry, etc. Coastal communities are affected by storms and by the continuing rise in sea-level; no part of the landscape and no human activity is completely isolated from the impact of variations in climate.

Although such interannual or even centennial climate variations may be dismissed as of limited magnitude and impact, post-industrial revolution and especially 20th-century human society in the developed world has begun to exert a major influence on the global climate (Houghton *et al.* 1996). These changes, plus associated changes in sea-level and the direct impacts of the changes in atmospheric CO_2 concentration which are a principal cause of these climate changes together will exert a profound influence upon the future development of the landscape of the Mersey Basin and upon human activities therein. Thus in the future we may see the cultivation of new crops, the establishment of new species currently with more southern distribution patterns, and the complete loss of blanket peat cover in the Pennine uplands. We also are likely to see coastal erosion and flooding. It is certain that there will be more frequent conflicts between those concerned with the conservation of the landscape and of natural ecosystems and those whose principal interest is in 'development'. Thus in the future the landscape of the Mersey Basin will face increased pressures arising both directly and indirectly from human activities.

Coda

To the extent that this account of the changing climate and its influence upon the Mersey Basin has a conclusion, it is to emphasise the extent to which factors external to the region ultimately determine its environment and thus impact upon its landscape and upon human activities in the region. It readily may be accepted that the history of the vegetation, of the landscape or of human activities in the region over the period since the last glacial maximum cannot be considered without placing them in the context of climate changes. These latter changes, however, were driven by changes in insolation, in atmospheric composition and in the circulation of both the oceans and atmosphere – changes which were external to the region. In the future the same will be true. The climate of the region will continue to change and it is likely that it will change profoundly during the next century – once again, however, the principal causes will lie outside the region itself. The fundamental conclusion is that no region may be considered to be isolated from 'Global Changes' and thus, to the extent that future changes are anthropogenic in origin, no region can afford to ignore its potential role in limiting these changes; nor, in an historical context, can the impact of climate changes be ignored by those whose focus is upon the palaeoecological, archaeological or even historical records.

Acknowledgements

I am grateful to the organisers of this conference for their invitation to present this paper and for their forbearance when the manuscript was delivered later than they had requested. The ideas presented owe much to discussions with many colleagues and students over many years; I am grateful to them all for sharing their views upon issues of global environmental change, past, present and future. A first draft of the manuscript was critically read by Jacqui Huntley.

References

Andersen, J. (1993). Beetle remains as indicators of the climate in the Quaternary. *Journal of Biogeography*, **20**, 557–62.

Atkinson, T.C., Briffa, K.R. & Coope, G.R. (1987). Seasonal temperatures in Britain during the past 22,000 years reconstructed using beetle remains. *Nature*, **325**, 587–93.

Barber, K.E., Chambers, F.M., Maddy, D., Stoneman, R. & Brew, J.S. (1994). A sensitive high-resolution record of late Holocene climatic change from a raised bog in northern England. *The Holocene*, **4**, 198–205.

Beerling, D.J., Chaloner, W.G., Huntley, B., Pearson, J.A. & Tooley, M.J. (1993). Stomatal density responds to the glacial cycle of environmental change. *Proceedings of the Royal Society of London Series B*, **251**, 133–38.

Berger, A., Imbrie, J., Hays, J., Kukla, G. & Saltzman, B. (eds) (1984). *Milankovitch and Climate*. NATO Advanced Studies Institute Series: Vol. 126. D. Reidel Publishing Company, Dordrecht/Boston, Netherlands/USA.

Birks, H.J.B. (1965). Late-Glacial deposits at Bagmere, Cheshire and Chat Moss, Lancashire. *New Phytologist*, **64**, 270–85.

Boulton, G.S., Jones, A.S., Clayton, K.M. & Kenning, M.J. (1977). A British ice-sheet model and patterns of glacial erosion and deposition in Britain. *British Quaternary Studies: Recent Advances* (ed. F.W. Shotton), pp. 231–46. Clarendon Press, Oxford.

Broecker, W.S. & Denton, G.H. (1989). The role of ocean-atmosphere reorganizations in glacial cycles. *Geochimica Cosmochimica Acta*, **53**, 2465–501.

Chappellaz, J., Barnola, J.M., Raynaud, D., Korotkevich, Y.S. & Lorius, C. (1990). Ice-core record of atmospheric methane over the past 160,000 years. *Nature*, **345**, 127–31.

CLIMAP Members (1976). The surface of the ice-age earth. *Science*, **191**, 1131–37.

COHMAP (1988). Climatic changes of the last 18,000 years: Observations and model simulations. *Science*, **241**, 1043–52.

Coope, G.R. & Brophy, J.A. (1972). Late Glacial environmental changes indicated by a coleopteran succession from North Wales. *Boreas*, **1**, 97–142.

Denton, G.H. & Hughes, T.J. (eds) (1981). *The Last Great Ice Sheets*. Wiley, New York, USA.

Devoy, R.J.N. (1995). Deglaciation, Earth crustal behaviour and sea-level changes in the determination of insularity: a perspective from Ireland. *Island Britain: a Quaternary perspective*. Geological Society Special Publication: Vol. 96 (ed. R.C. Preece), pp. 181–208. The Geological Society, London.

Fairbanks, R.G. (1989). A 17,000-year glacio-eustatic sea level record: Influence of glacial melting rates on the Younger Dryas event and deep-ocean circulation. *Nature*, **342**, 637–42.

Funnel, B.M. (1995). Global sea-level and the (pen-)insularity of late Cenozoic Britain. *Island Britain: a Quaternary perspective*. Geological Society Special Publication: Vol. 96 (ed. R.C. Preece), pp. 3–13. The Geological Society, London.

Gear, A.J. & Huntley, B. (1991). Rapid changes in the range limits of Scots Pine 4000 years ago. *Science*, **251**, 544–47.

Grove, J.M. (1988). *The Little Ice Age*. Methuen, London.

Guiot, J., de Beaulieu, J.L., Cheddadi, R., David, F., Ponel, P. & Reille, M. (1993a). The Climate in Western Europe During the Last Glacial Interglacial Cycle Derived from Pollen and Insect Remains. *Palaeogeography Palaeoclimatology Palaeoecology*, **103**, 73–93.

Guiot, J., Harrison, S.P. & Prentice, I.C. (1993b). Reconstruction of Holocene precipitation patterns in Europe using pollen and lake-level data. *Quaternary Research*, **40**, 139–49.

Hays, J.D., Imbrie, J. & Shackleton, N. (1976). Variations in the earth's orbit: pacemaker of the ice age. *Science*, **194**, 1121–32.

Houghton, J.T., Meira Filho, L.G., Callander, B.A., Harris, N., Kattenberg, A. & Maskell, K. (eds) (1996). *Climate Change 1995: The Science of Climate Change*. Cambridge University Press, Cambridge.

Huntley, B. (1991). How plants respond to climate change:

migration rates, individualism and the consequences for plant communities. *Annals of Botany*, **67**, 15–22.

Huntley, B. (1994). Late Devensian and Holocene palaeoecology and palaeoenvironments of the Morrone Birkwoods, Aberdeenshire, Scotland. *Journal of Quaternary Science*, **9**, 311–36.

Huntley, B. & Birks, H.J.B. (1983). *An atlas of past and present pollen maps for Europe: 0–13000 B.P.* Cambridge University Press, Cambridge.

Imbrie, J., Berger, A., Boyle, E.A., Clemens, S.C., Duffy, A., Howard, W.R., Kukla, G., Kutzbach, J., Martinson, D.G., McIntyre, A., Mix, A.C., Molfino, B., Morley, J.J., Peterson, L.C., Pisias, N.G., Prell, W.L., Raymo, M.E., Shackleton, N.J. & Toggweiler, J.R. (1993). On the structure and origin of major glaciation cycles. 2. The 100,000-year cycle. *Paleoceanography*, **8**, 699–735.

Imbrie, J., Boyle, E.A., Clemens, S.C., Duffy, A., Howard, W.R., Kukla, G., Kutzbach, J., Martinson, D.G., McIntyre, A., Mix, A.C., Molfino, B., Morley, J.J., Peterson, L.C., Pisias, N.G., Prell, W.L., Raymo, M.E., Shackleton, N.J. & Toggweiler, J.R. (1992). On the structure and origin of major glaciation cycles. 1. Linear responses to Milankovitch forcing. *Paleoceanography*, **7**, 701–38.

Kutzbach, J.E. & Gallimore, R.G. (1988). Sensitivity of a coupled atmosphere/mixed layer ocean model to changes in orbital forcing at 9000 years B.P. *Journal of Geophysical Research*, **93**, 803–21.

Kutzbach, J.E. & Guetter, P.J. (1986). The influence of changing orbital parameters and surface boundary conditions on climatic simulations for the past 18 000 years. *Journal of the Atmospheric Sciences*, **43**, 1726–59.

Kutzbach, J.E., Guetter, P.J., Behling, P.J. & Selin, R. (1993). Simulated climatic changes: results of the COHMAP climate-model experiments. *Global Climates since the Last Glacial Maximum* (eds H.E. Wright, Jr., J.E. Kutzbach, T. Webb, III, W.F. Ruddiman, F.A. Street-Perrott, & P.J.

Bartlein), pp. 24–93. University of Minnesota Press, Minneapolis, USA.

Lamb, H.H. (1982). *Climate, history and the modern world.* Methuen, London.

Lorius, C., Barkov, N.I., Jouzel, J., Korotkevich, Y.S., Kotylyakov, V.M. & Raynaud, D. (1988). Antarctic ice core: CO_2 and climatic change over the last climatic cycle. *Eos*, **69**, 681–84.

Pennington, W. (1986). Lags in adjustment of vegetation to climate caused by the pace of soil development: Evidence from Britain. *Vegetatio*, **67**, 105–18.

Ruddiman, W.F. & McIntyre, A. (1981). The mode and mechanism of the last deglaciation: oceanic evidence. *Quaternary Research*, **16**, 125–34.

Sissons, J.B. (1981). The last Scottish ice sheet: facts and speculative discussion. *Boreas*, **10**, 1–17.

Stuart, A.J. (1982). *Pleistocene Vertebrates in the British Isles.* Longman, London.

Stuiver, M. & Reimer, P.J. (1993). Extended ^{14}C database and revised CALIB radiocarbon calibration program. *Radiocarbon*, **35**, 215–30.

Watts, W.A. (1977). The Late Devensian vegetation of Ireland. *Philosophical Transactions of the Royal Society of London Series B*, **280**, 273–93.

White, J.C.W., Ciais, P., Figge, R.A., Kenny, R. & Markgraf, V. (1994). A high-resolution record of atmospheric CO_2 content from carbon isotopes in peat. *Nature*, **367**, 153–56.

Wingfield, R.T.R. (1995). A model of sea-levels in the Irish and Celtic seas during the end-Pleistocene to Holocene transition. *Island Britain: a Quaternary perspective.* Geological Society Special Publication: Vol. 96 (ed. R.C. Preece), pp. 209–42. The Geological Society, London.

Wright, H.E., Jr., Kutzbach, J.E., Webb, T., III, Ruddiman, W.F., Street-Perrott, F.A. & Bartlein, P.J. (eds) (1993). *Global Climates since the last glacial maximum.* University of Minnesota Press, Minneapolis, USA.

CHAPTER TWO

The land of the Mersey Basin: sea-level changes

A.J. PLATER, A.J. LONG, D. HUDDART, S. GONZALEZ
AND M.J. TOOLEY

The causes of these changes (of sea level) were then alluded to. The abrading influence of the sea, the probabilities of sudden convulsions of nature, the general rise of level of the ocean, were examined and rejected as inadequate to account for the phenomena under examination. The hypothesis which seemed to the writer the only satisfactory one was, that the land along the coast line has been for a long period in course of a gradual and slow depression in its level (Picton 1849).

Introduction

The coast of the Mersey Basin contains a wealth of evidence for past sea-level change and shoreline evolution. This evidence has attracted coastal geomorphologists, archaeologists and palaeobotanists for many years, the results of which have led to the development of a series of methodological and conceptual advances of local, regional and international importance.

In this paper we provide an introduction to the evidence for sea-level change and coastal evolution in the Mersey Basin, and explore three of the important research themes that have developed in the region. The first of these themes is the evidence for the 'Hillhouse Coast'; a shoreline variously thought to be formed during the mid-Holocene or the Late Pleistocene. The second theme considered is the pattern and cause of coastal change in the region. Here we reappraise the research in which the changes in coastal stratigraphy were attributed to oscillations in sea-level, and examine more recent (and earlier) evidence that the build-up and breakdown of coastal barriers may have been more influential in controlling coastal change. The final theme is the degree to which the past record of coastal change can be used as an analogue for the future evolution of the area.

Late-Quaternary sea-level changes

At the maximum of the last major glaciation (the Devensian cold stage) approximately 18,000 BP, global sea-level was approximately 120m lower than today (Fairbanks 1989). Consequently, vast tracts of land which are currently under water would have been sub-aerially exposed. The eastern shelf of the Irish Sea Basin, for example, would have been occupied by ice lobes which deposited glacial sediment as they retreated in response to subsequent climatic warming (Thomas 1985; Bowen et al. 1986). In the simplest model, cooler temperatures resulted in lower sea-level, through the large-scale storage of water in ice sheets and alpine glaciers during cold periods, whilst warming was characterised by sea-level rise (Fairbridge 1961).

However, in addition to the climatically-driven change in sea-level, the earth's crust has also responded to the redistribution of ice and water over its surface via isostatic adjustment. Hence, the loading of the crust by the Irish Sea ice during the Devensian brought about subsidence. This has prompted some workers (Eyles & Eyles 1984: Eyles & McCabe 1989) to suggest that much of the late-Devensian was characterised by glacio-marine deposition as crustal subsidence was of the same order of magnitude as eustatic (global) sea-level change. Although this interpretation is the subject of considerable debate (Thomas 1985; Thomas & Dackombe 1985; Austin & McCarroll 1992; McCarroll & Harris 1992; Huddart & Clark 1993), the interplay of eustatic sea-level change and glacio-isostatic crustal movements has governed sea-level trends in the Mersey Basin during the late-Quaternary (Tooley 1978a, 1982, 1985; Shennan 1989; Lambeck 1991, 1993; Wingfield 1992, 1993, 1995; Devoy 1995; Zong & Tooley 1996).

The sea-bed sediments of the eastern Irish Sea bear witness to the reworking of glacial deposits in the surf zone as sea-level rose rapidly during the early part of the present interglacial (the Holocene or Flandrian). The eastern shelf of the Irish Sea is armoured with a bed of coarse gravels and gravelly sands which remained after the fines had been winnowed out and carried eastward by the advancing surf (Wright et al. 1971). This ample supply of sediment enhanced the potential for coastal barriers to form in the nearshore zone, which, in turn, may have reduced the wave energy in the coastal zone and enabled tidal sedimentation to predominate (e.g., Eicher 1978; Leeder 1982). Consequently, the coastal lowlands of the Mersey Basin, i.e., those of the Sefton coast, northern Wirral and the Mersey Estuary, possess a sedimentary record of significant shoreline changes during the Holocene (Tooley 1978a, 1982; Kenna 1986; Innes *et al.* 1990; Bedlington 1995).

Whilst sea-level was rising as a result of ice melt and thermal expansion of the oceans, the earth's crust will also have undergone uplift as the ice sheet loading was removed. Geodynamic models of the crust, coupled with reconstructions of ice sheet thickness, enable glacio-isostatic uplift to be predicted. Lambeck (1991) suggested that the late-Pleistocene (16,000 to 12,000 BP) was characterised by rapid relative sea-level fall as the crust rose faster than the eustatic sea-level trend (Figure 2.1). This proposition, if correct, is of crucial significance in the evolution of the Mersey Basin coast.

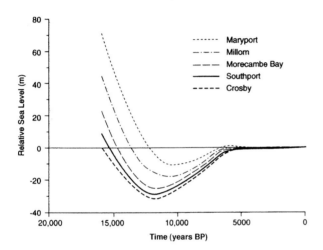

Figure 2.1. Relative sea-level curves for north-west England from geodynamic modelling of the earth's crust and eustatic sea-level trends (Lambeck, 1991).

The Hillhouse Coast

Approaches to sea-level reconstructions make use of a variety of physical evidence. In Britain, extensive use has been made of morphological and stratigraphic evidence; the former predominating (although not exclusively) in the study of Holocene sea-level change in Scotland (e.g., Sissons & Dawson 1981) and the latter in England and Wales (e.g., Tooley 1978a). Indeed, these two approaches are governed to some degree by the nature of late-Quaternary glacio-isostatic crustal movements; uplift in the north raising coastal features above the present level of the sea, and subsidence in the south resulting in the submergence of former coastal environments.

Evidence for high sea-level stands is present in the Isle of Man (Synge 1977; Thomas 1977), but raised morphological features are rare in the Mersey Basin. However, Wright (1914) postulated the presence of a post-glacial shoreline (the 25-foot raised beach) extending from the Lleyn peninsula, across the coast of the Wirral and into south-west Lancashire. Gresswell (1957) documented a morphological feature over much of the Lancashire coast which he identified as an ancient shoreline fronted by relict beach sands (the Shirdley Hill Sand) and tidal silts (the Downholland Silt). It was proposed that this

'Hillhouse Coast' marked the location of the coast during the mid-Holocene (6,000 to 5,000 BP).

Tooley (1978a) tested this hypothesis by collecting new litho-, bio- and chronostratigraphic data from the Lancashire lowlands during the 1970s, and, as a result, called the coastal origin of this morphological feature into question. First, Gresswell (1957) identified the Hillhouse Coast at altitudes above +5m O.D. (above mean sea-level), but Tooley (1976, 1978a) argued that this was too high to be a mid-Holocene shoreline since his sea-level graph for north-west England (Figure 2.2) suggested that the level of mean high water spring tides at 5,000 BP stood at approximately +2m O.D. Secondly, the Shirdley Hill Sand in south-west Lancashire extends *c.*20km inland from the present coast and is recorded at altitudes ranging from -14m to +120m O.D. Thirdly, investigation of the Shirdley Hill Sand suggested it was reworked cover sand rather than beach deposits, and that it was of late-glacial age (Godwin 1959; Tooley & Kear 1977; Wilson, Bateman & Catt 1981; Innes, Tooley & Tomlinson 1989). In contrast, evidence from the Hightown area reveals that deposits initially recorded as Shirdley Hill Sand may be coastal deposits of mid-Holocene age (Huddart 1992; Pye, Stokes & Neal 1995).

If one believes the results of recent geodynamic modelling (Lambeck 1991, 1993; Wingfield 1992, 1995), the Hillhouse Coast could be a Late Pleistocene coastal feature, with relative sea-level being close to +5m O.D. in the region of Crosby and Southport at approximately 16,000 BP (Lambeck, 1991). If this were the case, parts of the Shirdley Hill Sand could be beach sands which were reworked during the cold conditions of the late-glacial (the Younger Dryas or Loch Lomond stadial periods *c.*10,500 BP). A late-glacial coastal origin for some of the sand deposit was postulated by Tooley (1985), whilst Innes (1986) and Innes *et al.* (1989) suggested that this may have been followed by aeolian reworking. Indeed, thermoluminescence dating of Shirdley Hill Sand from Mere Sands Wood, near Rufford, revealed a minimum age for the onset of sand deposition of 11,730 ±1,510 BP, with further periods of deposition and/or reworking between 8,740 ±2,060 and 6,940 ±1,110 BP (Bateman 1995).

The Hillhouse Coast and the Shirdley Hill Sand could once again be the focus of investigation into late-Pleistocene and Holocene sea-level trends. Clearly, the geodynamic models need to be validated with reference to the available morphological and stratigraphic data. Although it might prove difficult to improve on the equivocal evidence from south-west Lancashire, research may extend northward to Cumbria where the modelled post-glacial relative sea-level raises the potential for marine inundation of the coastal lake basins of the Lake District during the late-Devensian.

Holocene coastal change

Irrespective of the status of the Hillhouse Coast, the coastal lowlands of the Mersey Basin possess a rich record of Holocene sea-level change. The buried peats of the Mersey, Wirral and Sefton were investigated more than one hundred years ago in the reconstruction of past environments (de Rance 1869, 1872, 1877; Reade 1871, 1872, 1881, 1908; Morton 1887, 1888, 1891). Attention also focused on the nature and trends of past vegetation (Travis 1908, 1922; Travis 1926, 1929; Erdtman 1928). However, the complex intercalation of biogenic and minerogenic units proved an interesting challenge, and were interpreted in the context of coastal (primarily land level) change by Binney & Talbot (1843), Picton (1849), Reade (1871), Morton (1888) and Blackburn (in Cope, 1939). Subsequently, Tooley (1970, 1974, 1976, 1978a, 1978b, 1982, 1985) undertook detailed stratigraphic and micropalaeontological work which established the West Derby region (Formby, Hightown and Downholland Moss) as one of the classic sites for Holocene sea-level reconstructions in the UK. Upward transitions from brackish-water minerogenic horizons to biogenic sediments deposited in close proximity to the shore were used as indicators of negative sea-level change, whilst the reverse provided evidence of sea-level rise. The sequence of stratigraphic intercalations in the Lancashire mosslands enabled the reconstruction of sea-level trends for north-west England from approximately 8,000 to 4,500 BP (Figure 2.2).

The resulting Holocene sea-level curve for north-west England was established using evidence from the West Derby and the Fylde in Amounderness and Morecambe Bay study areas (Tooley 1978a), and has been correlated with records from other parts of Britain and north-western Europe in the study of eustatic sea-level trends. Fairbridge (1961) had argued that minor oscillations in eustatic sea-level could be linked with fluctuations in climate, whilst Jelgersma (1966) rejected oscillating sea-level trends on the basis of the inadequacy and insufficiency of available data. Tooley (1978a) acknowledged the presence of oscillations in the north-west of England sea-level curve, but he proposed that low amplitude fluc-

tuations reflected local factors, such as sediment compaction, and high amplitude oscillations were primarily eustatic in origin. Indeed, the rapid rise in sea-level preserved in the sediments of Downholland Moss between c.8,000 and c.6,800 BP is present in other examples from north-western Europe, and has also been observed in recent work on the sedimentary record of Morecambe Bay (Zong & Tooley, 1996).

In the context of glacio-isostatic crustal movement, Shennan (1989) has shown from analysis of the sea-level curves from both southern and northern Lancashire that there was a marked decrease in the rate of uplift at c.5,000 BP. Further north, the rate of uplift in Morecambe Bay decreased at c. 6,000 BP but has continued at a low rate of rise to the present day (Shennan 1989; Zong & Tooley, 1996). Thus, it would appear that glacio-isostatic uplift is now complete in the region of the Mersey Basin. However, if uplift was related to the collapse of a glacial forebulge, subsidence may eventually follow. Wingfield (1993) believes that this forebulge-related subsidence may have passed through the region c.8000 BP, but this does not account for the changes observed in the relative sea-level trends during mid-Holocene.

There is also considerable debate concerning the reconstruction of mid- to late-Holocene sea-level trends from the stratigraphic record of the Mersey Basin. During the early-Holocene, the coast was characterised by an extensive intertidal sandflat in the west with mudflats and fringing saltmarsh to the east. There was vertical aggradation of these sedimentary environments under rising sea-level in the early- to mid-Holocene, resulting in alternating phases of marine and terrestrial sedimentation. However, Neal, Huddart & Pye (pers. comm.) suggest that more local factors could also have been important in explaining the alternating stratigraphy. Foraminiferal evidence suggests that the periodic breaching of an offshore barrier (Pye & Neal 1993a) may have provided a local control on coastal sedimentation, and enabled marked changes in the connection with the

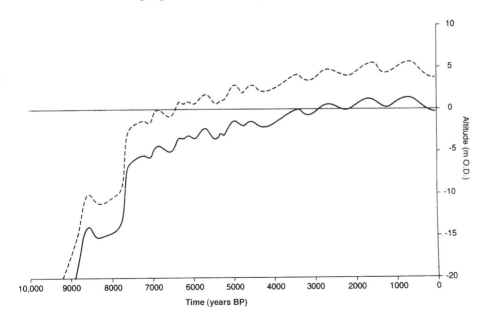

Figure 2.2. Relative sea-level curve for west Lancashire (Tooley, 1978a). The dashed line shows the trend of MHWST and the solid line represents mean sea-level.

open sea to take place. Stratigraphic sequences on the Wirral (Kenna 1986; Innes *et al.* 1990; Bedlington, 1995), have revealed a similar sequence of changes in coastal sedimentation which have been interpreted in the context of Holocene sea-level rise. Although this spatial continuity might support the interpretation of regional-scale changes brought about via sea-level trends, the Wirral and Sefton coasts may have supported similar barriers and back-barrier environments and, therefore, experienced similar periods of inundation resulting from barrier breaching. Indeed, Binney and Talbot (1843) had originally interpreted the Downholland Moss sequence in the context of coastal barrier evolution:

> After a time, a second bank of sand similar to the first one described was thrown up, which stopped the ingress of the sea and the egress of the fresh water, so as to cause a second morass which has produced the upper bed of peat.

Recent research has suggested the persistence of a barrier in the Sefton region for much of the Holocene (Pye & Neal 1993a, 1993b; Huddart 1992; Neal 1993; Neal *et al.* pers. comm.). Although radiocarbon ages from organic deposits which are overlain by dune sand range between 4,545 ±90 (Tooley 1970) and 3,380 ±60 BP (Pye & Neal 1993a), sand barrier formation was underway as early as 5,100 BP (Pye & Neal 1993a). Furthermore, the pattern of drainage channels on Altcar and Downholland Mosses associated with tidal sediments deposited between 6,000 and 5,600 BP, implies a north-south trending barrier in the Formby-Ainsdale region at this time (Huddart 1992). Indeed, Pye & Neal (1993a) propose that the earliest dunes in the region, which formed between 5,800 and 5,700 BP, may have become established on an emergent offshore sand bank which was in existence by 6,800 BP. Consequently, the main body of evidence presented by Tooley (1978a) (with the exception of the earliest period of transgression, DM-I) may have been laid down in this low energy environment which was buffered from high wave energy by the protective barrier complex.

The extent of this back-barrier environment has been the subject of recent investigation. In addition to the above evidence, the clear palaeoenvironmental and archaeological linkages between the stratigraphic records of the Wirral and Sefton coast (Gonzalez, Huddart & Roberts 1996) suggest that the coastal margins of Liverpool Bay may have been occupied by an extensive coastal plain in the lee of a discontinuous nearshore barrier complex during the mid-Holocene. In this context, the preservation of human and animal (Aurochs (*Bos primigenius*), Red Deer (*Cervus elaphus*), Roe Deer (*Capreolus capreolus*), Horse (*Equus caballus*) and crane) footprints in Holocene sediments in the foreshore at Formby Point is of particular importance in relation to the coastal evolution of the Mersey Basin. The footprints are located in the intertidal zone of the beach, with at least two stratigraphic sets of prints in sediments which exhibit a change from a nearshore, intertidal envi-ronment (lower footprints) to a terrestrial dune and dune slack (upper footprints). The stratigraphic evidence indicates a Neolithic/Bronze age for the footprints (Roberts, Gonzalez & Huddart 1996) and a ^{14}C date from an asso-ciated layer of Alder (*Alnus*) roots gives a *terminus ante quem* of 3,649 ±109 BP. The footprints show remarkable preservation (Figure 2.3), independent of sediment type, which indicates a rapid rate of sedimentation. Hence, this type of evidence is rare; Caldicot Levels in the Severn Estuary being the only other intertidal site in the UK where human and animal footprints have been reported in Holocene sediments (Aldhouse-Green *et al.* 1992).

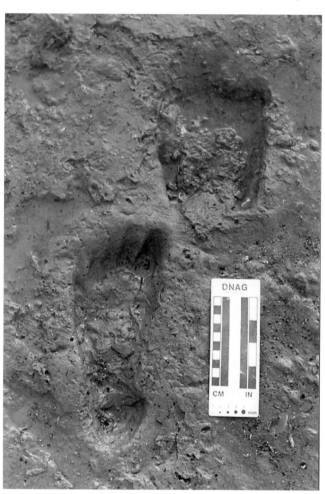

Figure 2.3. Human footprint and cattle hoofprint from Formby Point foreshore, Blundell Avenue, 17 January 1996 (*Photo: S. Gonzalez*).

The sedimentary environment in which these foot-prints were made is characteristic of a mesotidal system (Hayes 1979) in which morphology is controlled by a combination of wave energy and tidal processes in the back-barrier. If this proves to be the case, the lowland stratigraphy of the Mersey Basin will also possess a sedi-mentary record of significant palaeotidal change during the Holocene because the present spring tidal range is approximately 8.4m (Admiralty Tide Tables 1995). The required increase in tidal range (mesotidal regimes are characterised by a spring tidal range of between 2.0 and

4.0m according to Davies 1964) would necessitate a critical review of the evidence for Holocene sea-level change in north-western England. If spring tidal range has increased by up to 4m (or possibly even more) over the last 6,000 years, then eustatic mean sea-level may have been 2m higher than has been assumed for 6,000 BP from the reconstructed curves. Consequently, glacio-isostatic crustal movements could prove to be very different to those determined for the mid- to late-Holocene (Shennan 1989).

The significance of coastal barrier systems in the evolution of the Mersey Basin does not stop with the interpretation of back-barrier stratigraphy. Indeed, the youngest phase of Holocene sea-level change (Transgression V from the West Derby region) proposed by Tooley (1978a) is represented by the occurrence of a fossil dune slack peat, dated to 2,335 ±120 BP, at an altitude of +5.08m O.D. in the dunes of Formby Foreshore. Similar organic deposits have revealed ages of 2,510 ±120 (Pye 1990), 2,260 ±60 (Pye & Neal 1993a), 3,200 ±60 and 2,680 ±50 BP (Innes & Tooley 1993), and the optically-stimulated luminescence dating of podsolized dune sand yielded an age between 3,200 and 2,500 BP (Pye, Stokes & Neal 1995). These data provide evidence of dune stability which, in turn, gives equivocal evidence of sea-level change (Fairbridge 1961; van Straaten 1965; Jelgersma *et al*, 1970; Tooley 1978a, 1990; Innes & Frank 1988), i.e., there is some debate as to whether dunes are stable during periods of static or rising sea-level, and the ages give only the timing of change and not the altitude of sea-level. Further periods of dune stability are recorded in Lancashire between 1,795 ±240 and 1,370 ±85 BP, and at approximately 800 BP (Tooley 1978a, 1990; Innes & Tooley 1993). Indeed, this most recent period is also recorded in the coastal dunes of the Wirral peninsula, where dates from soils and peats overlain by aeolian sand range from 925 ±50 to 540 ±40 BP (Kenna 1986). The fact that these phases of dune stability can be correlated with similar trends found in western Europe suggests some climatic control, although this may be an indirect control via sea-level change (Tooley 1990).

The barrier control on coastal evolution is a common feature of many coasts. Consequently, the interpretation of lowland stratigraphy in the context of sea-level change is fraught with difficulty, especially when the rate of sea-level rise is low (Plater & Shennan 1992). Although the coastal lowland stratigraphy of the Mersey Basin may well be regarded as a classic record of coastal evolution brought about by Holocene sea-level rise, perhaps it should be viewed as a sedimentary record with which to investigate the relative importance of coastal processes and factors which operate over differing timescales and spatial extent.

Reverse uniformitarianism

At the end of the 18th century, James Hutton proposed that the key to interpreting the geological past was through present-day processes operating over much longer timescales than had been considered previously – the concept of uniformitarianism. The reverse of this approach has provided researchers with an essential justification for the study of late-Quaternary sea-level change. The wealth of data on sea-level change in the region of the Mersey Basin, and in particular the detail provided by the coastal dune stratigraphy for the last 3,000 years or so, provides an excellent context for assessing the consequences of present and future sea-level change. Tide gauge data from the region (Woodworth 1992) reveal rates of relative sea-level rise of the order of 2mm yr^{-1} for the latter part of the 20th century. These rates of change are likely to increase to approximately 7mm yr^{-1} by the end of the next century (Houghton, Jenkins & Ephraums 1990). The Holocene sea-level data illustrate that the predicted changes represent a marked acceleration of the most recent rates of change, but that similar rates of sea-level rise were experienced during the early part of the present interglacial.

The problem with using the early-Holocene as an analogue for what the next century holds for the coastal lowlands of the Mersey Basin is that sediment sources were less restricted at that time, the vegetation cover was very different, and the role of human societies in maintaining the coast was far less significant than today. Although humans may be able to reduce the immediate impacts of sea-level rise through the control of coastal erosion and the construction of flood defences, engineered coastlines may well be more vulnerable than previously assumed. If the response of the present-day coast to extreme events, such as storm surges, can be considered as an indication of vulnerability to sea-level rise, it would appear that the coastal dunes are more robust than anthropogenically-protected lowland environments. Indeed, coastal environments in which the sediments are able to respond to future sea-level rise may fare better than static walls and revetments. This is of some significance where the lowlands of the Mersey Estuary are concerned, and the managed retreat of society from these areas may be the only option towards the end of the next century.

A further consequence of adopting the uniformitarianism approach at the turn of the century was that catastrophic change was given less significance. However, recent research initiatives, such as IGCP 367 (late-Quaternary coastal records of rapid change: application to present and future conditions) have once again highlighted the importance of understanding low-frequency, high-magnitude events such as storm surges and tsunami. Indeed, a report by the Department of the Environment (DoE 1991) illustrated that a 0.2m rise in mean sea-level significantly increases the frequency of extreme water level events for the east and, in particular, south coasts of England. This followed the earlier findings of Rossiter (1962), who demonstrated that a 0.15m rise in mean sea-level would reduce the return interval of extreme water level events on the west coast by a factor of three.

It is unlikely that the temporal and spatial resolution

of Holocene or late-Pleistocene sea-level reconstructions will match the needs of coastal planners and managers in the immediate future. However, if research is to progress in improving the information available for sustained habitation of coastal lowlands, the sediments of the Mersey Basin offer an excellent testing ground for new techniques.

Acknowledgements

The authors should like to acknowledge all those who commented on the text, as well as the various researchers whose work is included in this review. Sandra Mather is thanked for the redrafting of the figures included in the text, Ian Qualtrough for reproducing the photograph, Frank Oldfield for his editorship, and Eric Greenwood for the opportunity to consider the research on the coastal evolution of the Mersey Basin in a broader context.

References

Admiralty Tide Tables (1995). *Volume 1 – European Waters*. The Hydrographer of the Navy.

Aldhouse-Green, S.H.R., Whittle, A.W.R., Allen, J.R.L., Caseldine, A.E., Culver, S.J., Day, M., Lundquist, J. & Upton, D. (1992). Prehistoric human footprints from the Severn Estuary at Uskmouth and Magor Pill, Gwent, Wales. *Archaeologia Cambrensis*, **141**, 14–55.

Austin, W.E.N. & McCarroll, D. (1992). Foraminifera from the Irish Sea glacigenic deposits at Aberdaron, Western Lleyn, North Wales: palaeoenvironmental implications. *Journal of Quaternary Science*, **7**, 311–17.

Bateman, M.D. (1995). Thermoluminescence dating of the British coversand deposits. *Quaternary Science Reviews*, **14**, 791–98.

Bedlington, D. (1995). *Holocene sea-level changes and crustal movements in North Wales and Wirral*. PhD thesis, University of Durham.

Binney, E.W. & Talbot, J.H. (1843). On the petroleum found in the Downholland Moss, near Ormskirk. Paper read at the Fifth Annual General Meeting of the Manchester Geological Society, 6 October 1843. *Transactions of the Manchester Geological Society*, **7**, 41–48.

Bowen, D.Q., Rose, J., McCabe, A.M. & Sutherland, D.G. (1986). Correlation of Quaternary glaciations in England, Ireland, Scotland and Wales. *Quaternary Science Reviews*, **5**, 299–340.

Cope, F.W. (1939). Oil occurrences in south-west Lancashire (with a Biological Report by K.B. Blackburn). *Bulletin of the Geological Survey of Great Britain*, **2**, 18–25.

Davies, J.L. (1964). A morphogenic approach to the World's shorelines. *Zeitschrift für Geomorphologie*, **8**, 127–42.

de Rance, C.E. (1869). *The geology of the country between Liverpool and Southport*. Explanation of Quarter Sheet 90SE of the 1 inch Geological Survey Map of England and Wales. Memoir of the Geological Survey of the United Kingdom, HMSO, London.

de Rance, C.E. (1872). *Geology of the country around Southport, Lytham Southshore*. Explanation of the Quarter Sheet 90NE. Memoir of the Geological Survey of the United Kingdom, HMSO, London.

de Rance, C.E. (1877). *The superficial geology of the country adjoining the coast of south-west Lancashire*. Memoir of the Geological Survey of the United Kingdom, HMSO, London.

Department of the Environment (1991). *The Potential Effects of Climate Change in the United Kingdom*. United Kingdom Climate Change Impacts Review Group, HMSO, London.

Devoy, R.J.N. (1995). Deglaciation, earth crustal behaviour and sea-level changes in the determination of insularity: a perspective from Ireland. *Island Britain: a Quaternary Perspective* (ed. R.C. Preece), pp. 181–208. Geological Society Special Publication No. 96.

Eicher, D.L. (1978). *Geologic Time*. Prentice-Hall, New Jersey, USA.

Erdtman, G. (1928). Studies in the post-arctic history of the forests of North-west Europe. I. Investigations in the British Isles. *Geologiska Föreningens i Stockholm Förhandlingar*, **50(2:373)**, 123–92.

Eyles, C.H. & Eyles, N. (1984). Glaciomarine sediments of the Isle of Man as a key to late Pleistocene stratigraphic investigations in the Irish Sea Basin. *Geology*, **12**, 359–64.

Eyles, N. & McCabe, A.M. (1989). The Late Devensian (<22,000 BP) Irish Sea Basin: the sedimentary record of a collapsed ice sheet margin. *Quaternary Science Reviews*, **8**, 307–51.

Fairbanks, R.G. (1989). A 17000-year glacio-eustatic sea level record. *Nature*, **342**, 637–42.

Fairbridge, R.W. (1961). Eustatic changes in sea-level. *Physics and Chemistry of the Earth*, **4**, 99–185.

Godwin, H. (1959). Studies of the postglacial history of British vegetation. XIV. Late Glacial deposits at Moss Lake, Liverpool. *Philosophical Transactions of the Royal Society of London, Series B*, **242**, 127–49.

Gonzalez, S., Huddart, D. & Roberts, G. (1996). Holocene development of the Sefton Coast: a multidisciplinary approach to understanding the archaeology. *Proceedings of the Archaeological Sciences Conference 1995* (eds. A. Sinclair, E. Slater & J. Gowlett), pp. 289–99. Oxford University Archaeological Monograph Series, Oxbow Books, Oxford.

Gresswell, R.K. (1957). Hillhouse Coastal Deposits in South Lancashire. *Liverpool and Manchester Geological Journal*, **2**, 60–78.

Hayes, M.O. (1979). Barrier island morphology as a function of tidal and wave regime. *Barrier Islands* (ed. S.P. Leatherman), pp. 1–27. Academic Press, London.

Houghton, J.T., Jenkins, G.J. & Ephraums, J.J. (eds) (1990). *Climate Change: The IPCC Scientific Assessment*. Cambridge University Press, Cambridge.

Huddart, D. (1992). Coastal environmental changes and morphostratigraphy in southwest Lancashire, England. *Proceedings of the Geologists' Association*, **103**, 217–36.

Huddart, D. & Clark, R. (1993). Conflicting interpretations from the glacial sediments and landforms in Cumbria. *Proceedings of the Cumberland Geological Society*, **5(4)**, 419–36.

Innes, J.B. (1986). The history of the Shirdley Hill Sand revealed by examination of associated organic deposits. *Proceedings of the North England Soils Discussion Group*, **21**, 31–43.

Innes, J.B. & Frank, R.M. (1988). Palynological evidence for Late Flandrian coastal changes at Druridge Bay, Northumberland. *Scottish Geographical Magazine*, **104(1)**, 14–23.

Innes, J.B. & Tooley, M.J. (1993). The age and vegetational history of the Sefton coast dunes. *The Sand Dunes of the Sefton Coast* (eds. D. Atkinson & J. Houston), pp. 3–20. National Museums & Galleries on Merseyside in association with Sefton Metropolitan Borough Council, Liverpool.

Innes, J.B., Tooley, M.J. & Tomlinson, P.R. (1989). A comparison of the age and palaeoecology of some sub-Shirdley Hill Sand peat deposits from Merseyside and south-west Lancashire. *Naturalist*, **114**, 65–69.

Innes, J.B., Bedlington, D.J., Kenna, R.J.B. & Cowell, R.W. (1990). A preliminary investigation of coastal deposits at Newton Carr, Wirral, Merseyside. *Quaternary Newsletter*, **62**, 5–12.

Jelgersma, S. (1966). Sea-level changes during the last 10,000 years. *World Climate 8000 to 0 B.C.* (ed. J.E. Sawyer), pp. 54–69. Proceedings of the International Symposium, Royal Meteorological Society, London.

Jelgersma, S., de Jong, J., Zagwijn, W.H. & van Regteren Altena, J.F. (1970). The coastal dunes of the western Netherlands; geology, vegetational history and archaeology. *Mededelingen Rijks Geologische Dienst N.S.*, **21**, 93–167.

Kenna, R.J.B. (1986). The Flandrian sequence of north Wirral (N.W. England). *Geological Journal*, **21**, 1–27.

Lambeck, K. (1991). Glacial rebound and sea-level change in the British Isles. *Terra Nova*, **3**, 379–89.

Lambeck, K. (1993). Glacial rebound of the British Isles – II. A high-resolution, high-precision model. *Geophysical Journal International*, **115**, 960–90.

Leeder, M.R. (1982). *Sedimentology: Process and Product*. George Allen and Unwin, London.

McCarroll, D. & Harris, C. (1992). The glacigenic deposits of western Lleyn, North Wales: terrestrial or marine? *Journal of Quaternary Science*, **7**, 19–29.

Morton, G.H. (1887). Stanlow, Ince and Frodsham Marshes. *Proceedings of the Liverpool Geological Society*, **5**, 349–51.

Morton, G.H. (1888). Further notes on the Stanlow, Frodsham and Ince Marshes. *Proceedings of the Liverpool Geological Society*, **6**, 50–55.

Morton, G.H. (1891). *Geology of the Country around Liverpool, including the North of Flintshire*. George Philip and Son, London.

Neal, A. (1993). *Sedimentology and morphodynamics of a Holocene coastal dune barrier complex, north-west England*. Unpublished PhD thesis, University of Reading.

Picton, J.A. (1849). The changes of sea-levels on the west coast of England during the historic period. (Abstract) *Proceedings of the Literary and Philosophical Society of Liverpool*, 36th Session, **5**, 113–15.

Plater, A.J. & Shennan, I. (1992). Evidence of Holocene sea-level change from the Northumberland coast, eastern England. *Proceedings of the Geologists' Association*, **103**, 201–16.

Pye, K. (1990). Physical and human influences on coastal dune development between the Ribble and Mersey estuaries, northwest England. *Coastal Dunes: Form and Process* (eds. K.F. Nordstrom, N.P. Psuty & R.W.G. Carter), pp. 339–59. Wiley, Chichester.

Pye, K. & Neal, A. (1993a). Late Holocene dune formation on the Sefton Coast, northwest England. *The Dynamics and Environmental context of Aeolian Sedimentary Systems* (ed. K. Pye), pp. 201–17. Geological Society Special Publication No. 72, Geological Society Publishing House, Bath.

Pye, K. & Neal, A. (1993b). Stratigraphy and age structure of the Sefton dune complex: preliminary results of field drilling investigations. *The Sand Dunes of the Sefton Coast* (eds D. Atkinson & J. Houston), pp. 41–44. National Museums & Galleries on Merseyside in association with Sefton Metropolitan Borough Council, Liverpool.

Pye, K., Stokes, S. & Neal, A. (1995). Optical dating of aeolian sediments from the Sefton coast, northwest England. *Proceedings of the Geologists' Association*, **106**, 281–92.

Reade, T.M. (1871). The geology and physics of the Post-Glacial Period, as shown in deposits and organic remains in Lancashire and Cheshire. *Proceedings of the Liverpool Geological Society*, **2**, 36–88.

Reade, T.M. (1872). The post-glacial geology and physiography of west Lancashire and the Mersey estuary. *Geological Magazine*, **9(93)**, 111–19.

Reade, T.M. (1881). On a section of the Formby and Leasowe Marine Beds, and Superior Peat Bed, disclosed by cuttings for the outlet sewer at Hightown. *Proceedings of the Liverpool Geological Society*, **4(4)**, 269–77.

Reade, T.M. (1908). Post-Glacial beds at Great Crosby as disclosed by the new outfall sewer. *Proceedings of the Liverpool Geological Society*, **10(4)**, 249–61.

Roberts, G., Gonzalez, S. & Huddart, D. (1996). Intertidal Holocene footprints and their archaeological significance. *Antiquity*, **70(269)**, 647–51.

Rossiter, J.R. (1962). Tides and storm surges. *Proceedings of the Royal Society of London, Series A*, 265, 328–30.

Shennan, I. (1989). Holocene crustal movements and sea-level changes in Great Britain. *Journal of Quaternary Science*, **4**, 77–89.

Sissons, J.B. & Dawson, A.G. (1981). Former sea-levels and ice limits in Wester Ross, northwest Scotland. *Proceedings of the Geologists' Association*, **92**, 115–24.

van Straaten, L.M.J.U. (1965). Coastal barrier deposits in South- and North-Holland, in particular in the areas around Scheveningen and Ijmuiden. *Mededelingen van de Geologische Stichting N.S.*, **17**, 41–76.

Synge, F.M. (1977). Records of sea-levels during the late devensian. *Philosophical Transactions of the Royal Society of London, Series B*, **280**, 211–28.

Thomas, G.S.P. (1977). The Quaternary of the Isle of Man. *The Quaternary History of the Irish Sea* (eds C. Kidson & M.J. Tooley), pp. 155–78. Geological Journal Special Issue No. 7, Seel House Press, Liverpool.

Thomas, G.S.P. (1985). The Quaternary of the northern Irish Sea Basin. *The Geomorphology of North-West England* (ed. R.H. Johnson), pp. 143–58. Manchester University Press, Manchester.

Thomas, G.S.P. & Dackombe, R.V. (1985). Comment on 'Glaciomarine sediments of the Isle of Man as a key to late pleistocene stratigraphic investigations in the Irish Sea Basin'. *Geology*, **13**, 445–47.

Tooley, M.J. (1970). The peat beds of the south-west Lancashire coast. *Nature in Lancashire*, **1**, 19–26.

Tooley, M.J. (1974). Sea-level changes during the last 9000 years in north-west England. *Geographical Journal*, **140**, 18–42.

Tooley, M.J. (1976). Flandrian sea-level changes in west Lancashire and their implications for the 'Hillhouse coast-line'. *Geological Journal*, **11(2)**, 37–52.

Tooley, M.J. (1978a). *Sea-level Changes: north-west England during the Flandrian stage*. Clarendon Press, Oxford.

Tooley (1978b). Interpretation of Holocene sea level changes. *Geologiska Föreningens i Stockholm Förhandlingar*, **100(2)**, 203–12.

Tooley, M.J. (1982). Sea-level changes in northern England. *Proceedings of the Geologists' Association*, **93**, 43–51.

Tooley, M.J. (1985). Sea-level changes and coastal morphology in north-west England. *The Geomorphology of North-West England* (ed. R.H. Johnson), pp. 94–121. Manchester University Press, Manchester.

Tooley, M.J. (1990). The chronology of coastal dune development in the United Kingdom. *Catena Supplement*, **18**, 81–88.

Tooley, M.J. & Kear, B. (1977). Shirdley Hill Sand Formation. *The Isle of Man, Lancashire coast and Lake District* (ed. M.J. Tooley), pp. 9–12. Guidebook for Excursion A4, X INQUA Congress, Geoabstracts, Norwich.

Travis, C.B. (1926). The peat and forest bed of the south-west Lancashire coast. *Proceedings of the Liverpool Geological Society*, **14**, 263–77.

Travis, C.B. (1929). The peat and forest beds of Leasowe, Cheshire. *Proceedings of the Liverpool Geological Society*, **15**, 157–78.

Travis, W.G. (1908). On plant remains in peat in the Shirdley Hill Sand at Aintree, south Lancashire. *Transactions of the Liverpool Botanical Society*, **1**, 47–52.

Travis, W.G. (1922). On peaty beds in the Wallasey Sand-Hills. *Proceedings of the Liverpool Geological Society*, **13(3)**, 207–14.

Wilson, P., Bateman, R.M. & Catt, J.A. (1981). Petrography, origin and environment of deposition of the Shirdley Hill Sand of southwest Lancashire, England. *Proceedings of the Geologists' Association*, **92**, 211–29.

Wingfield, R.T.R. (1992). Quaternary changes of sea level and climate. *The Irish Sea*, pp. 56–66. Irish Sea Forum Seminar Report, Global Warming and Climatic Change, Liverpool University Press, Liverpool.

Wingfield, R.T.R. (1993). Modelling Holocene sea levels in the Irish and Celtic seas. *Proceedings of the international Coastal*

Congress, ICC-Kiel '92 (eds. H. Sterr, J. Hofstede & H.-P. Plag), pp. 760–72. Peter Lang, Frankfurt am Main, Germany.

Wingfield, R.T.R. (1995). A model of sea level in the Irish and Celtic Seas during the end-Pleistocene to Holocene transition. *Island Britain: a Quaternary Perspective* (ed. R.C. Preece), pp. 209–42. Geological Society Special Publication No. 96, Geological Society, London.

Woodworth, P.L. (1992). Sea-level changes. *The Irish Sea*, pp. 21–28. Irish Sea Forum Seminar Report, Global Warming and Climatic Change, Liverpool University Press, Liverpool.

Wright, W.B. (1914). *The Quaternary Ice Age*. Macmillan, London.

Wright, J.E., Hull, J.H., McQuillin, R. & Arnold, S.E. (1971). *Irish Sea investigations 1969–1970*. Institute of Geological Sciences, Report No. 71/19, HMSO, London.

Zong, Y. & Tooley, M.J. (1996). Holocene sea-level changes and crustal movements in Morecambe Bay, northwest England. *Journal of Quaternary Science*, **11(1)**, 43–58.

CHAPTER THREE

Vegetational changes before the Norman Conquest

J.B. INNES, M.J. TOOLEY AND J.G.A. LAGEARD

Introduction

This paper summarises the history of vegetation change in the Mersey Basin from the time of the earliest available evidence until the Norman Conquest, since when humans have dominated the processes which control vegetation distributions. With the limited exception of some wetland communities, vegetation has been so greatly modified or controlled by human agricultural and industrial activities as to be virtually artificial. These recent anthropogenic plant associations will be discussed in other chapters in this volume. The balance between natural and cultural influences on the vegetation has only in the last millennium swung so decisively towards man, and for most of the past vegetation distributions were subject predominantly to external natural environmental forces and to internal processes of community succession, with human impacts localised and limited in effect. With the adoption of increasingly intensive systems of food production in the later Holocene human influence on vegetation patterns grew until the cultural factor assumed dominance over natural factors in determining the vegetation cover, laying the basis for the present artificial situation.

Over 80 pollen diagrams of varying detail are available from organic lake and mire sediments in the study area, and these are listed in Table 3.1 and shown on Figure 3.1. Despite a lack of sites in some urban areas, the distribution of these archives of palaeobotanical diversity (Barber 1993) is representative enough to allow comparative study of regional vegetation history (Behre 1981; Bradshaw 1988). Pollen nomenclature follows Moore *et al.* (1991) adjusted to follow Stace (1991). Radiocarbon dating allows broad age correlation between diagrams, but in its absence similar pollen changes can be compared ecologically, but can not be assumed to be synchronous. All dates quoted in this chapter are in uncalibrated radiocarbon years before present (BP). Vegetation changes provide a sensitive measure of the rate and scale of variation in a range of environmental factors, some like climate operating as gradual processes over longer time periods and others occurring as rapid, even catastrophic, disturbance events.

Three main natural factors in combination have deter-

mined the past vegetation in the Mersey Basin: soils, climate and drainage. The highest parts of the Pennine watershed probably escaped ice cover, but the Mersey Basin was almost entirely occupied by either lowland ice sheets or upland glaciers during the maximum Devensian cold period around 18,000 years ago, wiping clean the vegetation record. Glaciation and periglacial processes have created drift and fluvioglacial deposits which include clays, fine and coarse sands and gravels (Johnson 1985). Centres of sand and gravel sediments within the clay drift plain are the Shirdley Hill Sands of south-west Lancashire and Merseyside, gravels in the river valleys of the Irwell and other rivers, and the sands of Delamere Forest and east Cheshire. Upon these diverse parent materials formed soils of differing textures and drainage (Kear 1985) and, spatially highly variable, these form the basis for diversity in natural and cultural vegetation patterns within the region (Innes & Tomlinson 1983, 1991). Impedance of drainage and the deflection of vegetation successions into wetland pathways (Shimwell 1985; Tallis 1991) added a further important element of diversity to the floral history. Altitudinal variation and its effect on soils and climate, and hence vegetation, is another major influence on natural vegetation patterns.

The vegetation history of the Mersey Basin is considered under four major categories: a) pre-Holocene vegetation communities, b) Holocene woodland communities, c) specialised Holocene vegetation communities, and d) woodland disturbance and post-woodland vegetation communities. The first represents the long pre-Holocene period (>c.10,000 BP) with mainly open vegetation under very cold climatic conditions of varying severity. The second examines the establishment of woodland communities during the ameliorating post-glacial climate of the early and mid Holocene (c.10,000 to c.5,000 BP). The third considers natural non-woodland vegetation such as wetland, coastal or moorland and the fourth examines the late-Holocene period (<c.5,000 BP) of woodland disturbance caused by man's increasingly severe impacts compounded by natural environmental degeneration, leading to the spread of various kinds of post-woodland vegetation across almost the entire region.

Pre-Holocene communities

Pre-Late-Devensian vegetation >c.15,000 BP

The temperate Holocene period is untypical of conditions during the great majority of the past several hundred thousand years, with glacial cold phases and cold adapted vegetation lasting much longer than brief temperate interglacials. Floras from interglacials predating the Devensian ice sheet have been reported from Staffordshire on the southern fringe of the region (Worsley 1985). At Trysull deposits may be Hoxnian (c.240,000 BP), while probable Ipswichian (c.120,000 BP) peat at Four Ashes contained evidence for Oak (*Quercus*), Alder (*Alnus*), Hazel (*Corylus*), Yew (*Taxus*) and Holly (*Ilex*). From Oakwood Quarry near Chelford in Cheshire comes a pollen assemblage of terrestrial herbs reflecting a cold-climate, treeless landscape from the earlier stages of the Devensian glacial before the main ice advance. At Farm Wood Quarry, Chelford, organic sediments younger than the Oakwood deposits contained Pine

Mire sites of the Mersey Basin

1	Alt Mouth	Tooley 1978	39	Bagmere	Birks 1965a
2	Downholland Moss	Tooley 1978	40	Lindow Moss	Birks 1965b
3	Ince Blundell	Cowell & Innes 1994	41	Wybunbury Moss	Green & Pearson 1977
4	Sniggery Wood	Cowell & Innes 1994	42	Cranberry Moss	Tallis 1973
5	Flea Moss Wood	Cowell & Innes 1994	43	Peckforton Mere	Twigger 1983
6	Waterloo	Cowell & Innes 1994	44	Danes Moss	Birks 1962 unpub.
7	Rimrose Brook	Innes 1991	45	Congleton Moss	Leah *et al.* 1997
8	Moss Lake, Liverpool	Godwin 1959	46	Cock's Moss	Leah *et al.* 1997
9	Bidston Moss	Cowell & Innes 1994	47	Walker's Heath	Leah *et al.* 1997
10	Park Road, Meols	Cowell & Innes 1994	48	Mere Moss Wood	Leah *et al.* 1997
11	Newton Carr	Cowell & Innes 1994	49	Abbots Moss	Gray unpub.
12	Bull Lane, Aintree	Innes 1992	50	Whitemore, Bosley	Johnson *et al.* 1970
13	Simonswood Moss	Cowell & Innes 1994	51	Goyt Moss	Tallis 1964b
14	Knowsley Park	Cowell & Innes 1994	52	Wessenden	Tallis 1964b
15	Holiday Moss	Baxter 1983	53	Deep Clough	Tallis & McGuire 1972
16	Firswood Road	Tooley 1978	54	Snake Pass	Tallis 1964a
17	Bickerstaffe Moss	Kear 1968	55	Didsbury Intake	Tallis & Johnson 1980
18	Rainford Brook	Innes unpub.	56	Laddow Rocks	Tallis & Johnson 1980
19	Prescot	Cowell & Innes 1994	57	Bradwell Sitch	Tallis & Johnson 1980
20	Parr Moss	Cowell & Innes 1994	58	Charlesworth Landslips	Franks & Johnson 1964
21	Holland Moss	Cundill 1981	59	Seal Edge Coombes	Johnson *et al.* 1990
22	Reeds Moss	Baxter 1983	60	Lady Clough Moor	Tallis 1975
23	Mossborough Moss	Baxter 1983	61	Rishworth Moor	Bartley 1975
24	Skelmersdale	Baxter 1983	62	Tintwistle Knarr	Tallis & Switsur 1990
25	Haskayne	Baxter 1983	63	Robinson's Moss	Tallis & Switsur 1990
26	Bangor's Green	Baxter 1983	64	Alport Moor	Tallis & Switsur 1990
27	Berrington's Lane	Baxter 1983	65	Salvin Ridge	Tallis 1985
28	Red Moss	Hibbert *et al.* 1971	66	Featherbed Top	Tallis 1985
29	Hale	Baxter 1983	67	Featherbed Moss	Tallis & Switsur 1973
30	Helsby Marsh	Tooley 1978	68	Kinder	Tallis 1964a, 1964b
31	Ditton Brook	Innes unpub.	69	Bleaklow	Conway 1954
32	Risley Moss	Hibbert 1977	70	Soyland Moor	Williams 1985
33	Chat Moss	Birks 1964, 1965a	71	West Moss	Tallis & McGuire 1972
34	Nook Farm	Hall *et al.* 1995	72	Bar Mere	Schoenwetter 1982
35	Dunham Massey	Baxter 1983	73	Clieves Hills	Tooley 1978
36	Holcroft Moss	Birks 1965b	74	White Moss	Lageard 1992
37	Hatchmere	Birks unpub. 1975	75	Madeley	Yates & Moseley 1958
38	Flaxmere	Tallis 1973	76	Farm Wood, Chelford	Simpson & West 1958
			77	Oakwood Quarry, Chelford	Worsley 1985

Other mire sites mentioned in the text
78 Hoscar Moss (Cundill 1984), 79 Martin Mere (Tooley 1985a), 80 Extwistle Moor (Bartley & Chambers 1992), 81 Lismore Fields (Wiltshire & Edwards 1993)

Table 3.1. Names and major references for pollen analytical sites located on Figure 3.1.

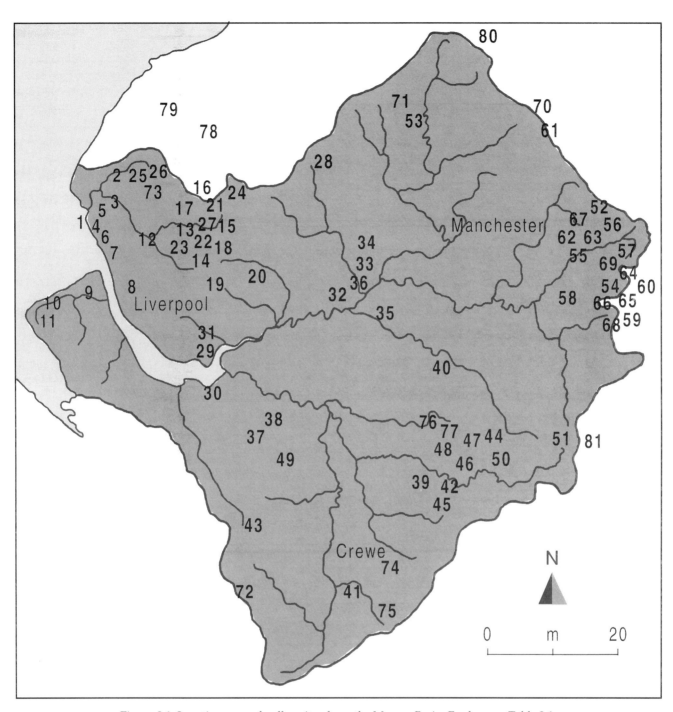

Figure 3.1. Location map of pollen sites from the Mersey Basin. For key see Table 3.1.

(*Pinus*), Spruce (*Picea*) and Birch (*Betula*) pollen and Spruce stumps, suggesting a briefly warmer climate with boreal woodland (Simpson & West 1958), radiocarbon dating suggesting a date older than 60,000 BP (Worsley 1985). Spruce is not native to Britain in the Holocene.

Late-Devensian vegetation *c.*15,000– *c.*10,000 BP

Soon after the retraction of ice cover around 15,000 BP pioneer polar desert vegetation of mosses and lichens and then turf and short herb communities with grasses (Poaceae), sedges (Cyperaceae), Clubmoss (*Lycopodium*), Dock (*Rumex*) and Mugwort (*Artemisia*) colonised the newly exposed ground surfaces (Pennington 1977),

together with a wide range of ruderal and specialised cold tolerant herbs like Rockrose (*Helianthemum*), Thrift (*Armeria*) and Goosefoot family (Chenopodiaceae). There were abundant habitats for aquatic herbs such as Alternate-flowered Water-milfoil (*Myriophyllum alterniflorum*). Rapidly changing successional vegetation characterised this initial period of environmental transition. Heath and dwarf shrub communities of Juniper (*Juniperus*), Crowberry (*Empetrum*), Dwarf Birch (*Betula nana*) and Willow (*Salix*) became established gradually, via tall herb associations, due to climatic amelioration after 13,000 BP (Walker *et al.* 1994), although herbs would still have been common as a more complete and diverse plant cover developed. In sheltered localities stands of tree birch developed and expanded until between 12,000 and 11,000 BP there was open Birch woodland in parts of the lowlands, with a rich tall herb field layer. At lowland sites like Bagmere and Chat Moss (Birks 1965a) and Moss Lake, Liverpool (Godwin 1959), mature Birch woodland did not develop in the Windermere (Allerød) interstadial between 12,000 and 11,000 BP, tree (mainly Birch) pollen reaching 25% of total pollen at most. Interstadial age pollen diagrams from beneath Shirdley Hill Sand east of Liverpool (e.g., Skelmersdale) and in north Cheshire (e.g. Dunham Massey) suggest very open herb and low shrub dominated vegetation and limited Birch woodland (Baxter 1983).

Climatic deterioration began about 12,500 BP, even as shrub and tree cover was spreading, culminating between 11,000 and 10,000 BP in the extremely cold late-glacial (Loch Lomond) stadial with unstable soils, during which tundra-type herb vegetation and Crowberry dwarf heath replaced the Juniper and Birch shrub woodland. Mugwort is characteristic of this zone, with grasses, sedges, Dock, Lesser Clubmoss (*Selaginella*) and many ruderal weed taxa. Peats beneath late-glacial coversand (Shirdley Hill Sand) reactivated at this time (Tooley 1978; Innes 1986), record pollen data of this age, as at Clieves Hills near Ormskirk at 10,455±110 BP. At higher altitude a channel at Whitemoor, Bosley (Johnson *et al.* 1970) is dominated by Dwarf Birch, grasses and sedges, with a rich herb flora including glacial taxa like Iceland-purslane (*Koenigia islandica*).

Holocene woodland communities

The development of early Holocene woodland *c*.10,000 – *c*.7,000 BP

Climatic amelioration at the end of the late-glacial swiftly replaced open ground tundra herb flora with a succession of transitional early Holocene communities. The classical pattern followed grass sward through tall herb associations with Dock, Crowberry dominated heath, Juniper shrub to Birch woodland of an increasingly closed nature. Willow and Aspen (*Populus*) were important initially until shaded out by tree Birches. Thermophilous herbs like Meadowsweet (*Filipendula*) were briefly abundant in the few centuries of warm climate before canopy closure. The immigration of forest taxa began with the arrival of Hazel, joining Birch as co-dominant, supplanting it in places, and creating a denser canopy which suppressed Juniper, Willow and the lower elements of the transitional flora. The timing of the great increase of Hazel pollen, which may include pollen of Bog-myrtle (*Myrica*), varies but the Hatchmere date of 9,580±140 BP (Switsur & West 1975) is probably representative for much of the lowlands. At White Moss in south-east Cheshire (Lageard 1992) the Hazel pollen rise occurs rather later, between 9,230±85 BP and 8,625±50 BP, with percentages of more than 25% by the latter date. Thermophilous trees Elm (*Ulmus*) and Oak, migrated into the area after 9,000 BP, out-competing Birch on the better soils, although the timing of their introduction varies from site to site. The major regional palaeobotanical event of mid-Flandrian I, however, was the spread of Pine, delaying the expansion of the broadleaf forest trees. It is difficult to assess the abundance of Pine before this spread. At White Moss Lageard (1992) showed that although large quantities of sub-fossil Pine wood were discovered during peat extraction, Pine pollen rarely exceeded 20% of the total pollen count. Low Pine pollen counts need not mean the absence of pinewoods, but it seems likely that high Pine counts must mean that the tree was locally common, and on this basis there seems to have been a diachronous spread of Pine in the Mersey Basin. At White Moss on sandy soils, Pine woodland became established at 8,625±50 BP and Pine remained common in the region until about 6,000 BP. On Shirdley Hill Sand at Knowsley Park Moss a similar early date for the Pine pollen rise of 8,880±90 BP occurred (Cowell & Innes 1994). Pine dominance was delayed on heavier clay soils, however, as at Red Moss (Hibbert *et al.* 1971) at 8196±150 BP, but other long pollen records suggest that eventual Pine abundance occurred everywhere in the region (e.g., Wybunbury Moss, Green & Pearson 1977). Pine tolerates a wide range of soil types and was not supplanted by deciduous trees until the rise of Alder and climatic and successional changes at the start of mid-Holocene Flandrian II. Although Pine persisted in some favourable areas these processes culminated with its exclusion from regional forests due to competition from broadleaved trees.

The establishment of mid-Holocene forest *c*.7,000 – *c*.5,000 BP

The optimum temperature and rainfall conditions of the mid-Holocene favoured the maximum extension of tree growth in the region. In contrast to the immigration of tree taxa in the earlier Holocene, once established the stable forest ecosystem of Flandrian II offered few opportunities for new tree species and was an almost steady state, highly stable community. It was not unchanging, for even if a long term stable 'climax' situation had been reached there was still change going on within the forest community mosaic, if only very slowly, with senescent individuals being naturally replaced by new plants and successions being set in motion. A mature, long established and undisturbed woodland may seem unchanging, but in reality the spatial distribution of woodland components must change due to individual plant mortality so that while the community structure is maintained, the distribution of different tree species and stands will gradually alter. The thermophilous Lime (*Tilia*) achieved maximum expansion during Flandrian II and was common in lowland areas, even co-dominant with Oak and Elm in places, although in the Mersey Basin it was unlikely to have been the major deciduous tree (Greig 1982). It was certainly more common than the poor representation its low pollen production affords it on pollen diagrams. The deciduous forest extended almost to the highest altitudes (Tallis & Switsur 1983, 1990), with no real altitudinal gradients in taxa abundance. Hazel, likely to be more successful if the forest canopy were more open on higher ground, shows similarly high values on the watershed (Robinson's Moss) and on lowland sandy soils (Wybunbury Moss). On heavier soils (Holcroft Moss and Lindow Moss, Birks 1965b) Hazel values tend to be lower. Site factors seem to have been generally more important than altitude.

The defining feature of Flandrian II is the rise to high pollen frequencies of Alder which occurred about 7,000 BP (7,107±120, Red Moss, Hibbert *et al.* 1971; 7403±114, Hatchmere, Switsur & West 1975), although there was considerable variation between sites (Bennett & Birks 1990), for example at White Moss where the Alder rise was delayed until 5,890±45 BP (Lageard 1992). Even allowing for its high pollen productivity, it must have been a very common tree and was favoured by the humid, oceanic conditions of the climatic optimum, seemingly replacing Pine directly in many cases. Abundant around water bodies such as the channel deposit at Whitemoor, Bosley (Johnson *et al.* 1970), and in river valleys flood plains such as the River Alt at Ince Blundell (Cowell & Innes 1994), it must also have found favourable habitats throughout the damp closed canopy forest. The more representative extra-local and regional pollen rain observed from the centres of some large raised bogs or lakes (Wybunbury Moss, Green & Pearson 1977; Holcroft Moss, Birks 1965b), or from the blanket peats near the Pennine watershed (Alport Moor, Tallis 1991) contains Oak and Alder pollen in similar proportions, as do sites from the sandy coastal plain away from centres of Alder carr growth (Sniggery Wood, Innes & Tooley 1993).

The establishment of Alder and Lime completed the assembly of the natural climax post-glacial mixed Oak forest. Site conditions, mainly edaphic factors, decided the relative importance of the components of this forested ecosystem but in most situations across the study area considerable homogeneity existed, with limited regional variability. At all sites Oak, Alder and Hazel were commonest, and most clearly so at higher altitudes, with Elm a lesser but very important member of the deciduous forest community, rivalling Oak on better soils. Ash (*Fraxinus*) was present from early times but was rare until the more open woods of later Flandrian II, being unable to supplant the other mixed Oak forest trees until the inertia of the primary forest was broken, mainly by disturbance.

In a few favourable areas Pine remained common, as on light sandy soils near the coast, on fluvioglacial outwash or shallow accumulating peats (Lageard 1992) and in some high altitude locations like the Bradwell Sitch landslip site in upper Longdendale (Tallis & Johnson 1980). In some cases Pine's replacement by deciduous trees, often Alder, was merely delayed into the first half of Flandrian II, as at White Moss where the major decline of Pine occurred at 5,890±45 BP. On a few nutrient poor sites, or where fire was a common factor, Pine remained locally important into the later Holocene.

Late-Holocene woodland (<*c*.5,000 BP)

It is difficult to evaluate any natural developments which took place in the late Holocene, Flandrian III, woodlands of the Mersey Basin, as the vegetation history of this more recent period has been so heavily influenced by the activities of man. Natural climatic and edaphic changes did occur, however, which caused woodland composition changes. Major climatic deterioration around 4,000 BP appears to have caused the final displacement of Pine from its last refuges in marginal, low nutrient locations (Bennett 1984; Gear & Huntley 1991). During Flandrian III there are pollen records from several sites (e.g., Red Moss, Hibbert *et al.* 1971), some at quite high altitudes (e.g., Franks & Johnson 1964), for the arrival of Beech (*Fagus*) and Hornbeam (*Carpinus*) in the region in low numbers, these trees being more suited to conditions well to the south. Human disturbance would have assisted their immigration and the spread of other secondary trees like Ash and Birch. At varying times around *c*.3,500 BP, Lime became much less important in the regional woodlands. This will have been due partly to climatic degeneration (Lamb 1981) during the late-Holocene, and partly as the result of human activity (Turner 1962).

Macrofossil evidence of the former forest

Although only preserved in exceptional circumstances of rapid burial by sand, alluvium or peat, sub-fossil tree stumps and trunks can be used as an indication of the former forest's extent or structure and can be used to test the conclusions reached from pollen analysis. Oak and Pine appear to have been the main taxa which colonised marginal environments throughout the region in the mid and late Holocene. Their remains are common in and beneath upland (Tallis & Switsur 1983, 1990), coastal (Kenna 1986; Pye & Neal 1993) and inland mossland (Lageard *et al.* 1995) peats. Three-dimensional recording of Pine stumps at White Moss has shown that the colonisation and demise of trees on peat sites was a protracted and complex process and not a simple single phase event. While not representative of regional dryland forest communities, sub-fossil tree remains may prove to be an important source of data on prehistoric woodland structure and ecology in the Mersey Basin.

Specialised Holocene communities

Wetland vegetation

The closed mid-Holocene forest contained few areas of open vegetation and floral diversity. Exceptions were the coastal intertidal zone, exposed locations above the tree line in the Pennines and areas where fire or other disturbance initiated successional communities. Wetlands also added a major element of long-term diversity to the vegetation mosaic. Although the distribution of surviving wetlands in the Mersey Basin suggests only a localised presence (Shimwell 1985; Howard-Davis *et al.* 1988; Hall *et al.* 1995), before systematic modern drainage wetlands of many kinds would have been widespread, controlled by climate and by local topographical and hydrological factors, each supporting different wetland plant communities (Moore *et al.* 1984). Hydroseral succession through open water aquatic, reedswamp, fen, fen-carr and fenwood vegetation (Walker 1970; Rodwell 1991) would have started at different times and proceeded at different rates in each location, creating a mosaic of wetlands at different stages of succession at any one time.

The final stage in vegetation succession for many basin wetlands would have been to mossland bog due to falling nutrient levels and increasing acidity, although recent surveys of mire stratigraphy (Wells 1992; Wells *et al.* 1993) suggest that ombrotrophic raised bog with an acid tolerant flora dominated by Bogmoss (*Sphagnum*) was not always reached. Many successions remained as intermediate mire systems, often growing together over shallow slopes from several centres, characterised by a less acid Cottongrass (*Eriophorum*) community (Hall & Folland 1970). Valley mires and river flood plain mires which were minerotrophic were maintained as mesotrophic fen and fen-carr systems, often dominated by Alder. The Gowy valley in Cheshire is a good example of long term persistence of fen-reedswamp wetland vegetation (Shimwell 1985) but before human drainage and clearance most of the lower parts of the region's watercourses would have been similar. Superabundance of Alder pollen in diagrams from such sites is common, as in the valley of the Rimrose Brook near Liverpool (Innes 1991) or at Ince Blundell in the Alt valley (Cowell & Innes 1994) where it maintains values of 80% of total pollen for thousands of years. Dense fen woodland was a long term feature of the region's coastal and mere fringe vegetation (Lageard 1992). Human clearance of the riverine Alder fen-carr and fenwood in later prehistoric and early historic times would have led to the establishment of flood plain grassland of a type determined by both hydrology and management, being maintained by grazing and adapted to inundation, both natural and due to run off effects of human activity in the catchment. Systematic drainage in the historic period has greatly impoverished the floral diversity provided by wetland habitats, probably causing the regional extinction of some wetland plants, e.g., Rannoch-rush (*Scheuchzeria palustris*) (Tallis & Birks 1965).

Coastal vegetation

The coastal zone is a transitional area of natural plant diversity, and was much more so in the past than today, when almost all the coastal wetlands have been reclaimed for agriculture and industry. Sea-level rise since deglaciation (Tooley 1978, 1985a, 1985b; Plater *et al.* 1993) established the coast near its present position by the mid-Holocene. Subsequent minor coastal changes have caused some spatial relocation of the associated plant communities.

A suite of wetland environments and vegetation types may be recognised (Tooley 1978) which have a seral relationship based upon salinity tolerance and altitude relative to tide level. Most specialised are intertidal environments in which halophyte plants are characteristic, some open ground herbs like Thrift (*Armeria*), Sea Plantain (*Plantago maritima*) and Chenopodiaceae finding favourable habitats denied to them across the landscape as a whole since the demise of the late-glacial tundra communities. The zonation of low, mid and high saltmarsh communities above the unvegetated sand and mudflat can show a complex spatial arrangement due to local intertidal topography (Shennan 1992). Halophytes of the Chenopodiaceae and Caryophyllaceae families, with Thrift and saltmarsh grasses, dominate the low saltmarsh. In the high saltmarsh less often inundated by the tide near the transition to brackish and freshwater conditions, less halophyte taxa like Lesser Bullrush (*Typha angustifolia*) dominate.

In areas of the coastal lowland above high tide the elevation of groundwater tables caused the formation of a zone of freshwater wetland environments between the higher dryland and the intertidal zone, comprising lagoons, marshes, swamps, fens and meres. This perimarine freshwater zone (Tooley 1978) supported extremely rich and diverse eutrophic successional aquatic, reedswamp, fen and carr communities (Lind 1949; Rodwell 1991), accepting nutrient rich drainage from the hillslopes inland. Many small meres came into existence, some ephemeral as a result of seasonal flooding, and although several such as Martin Mere (Tooley 1978, 1985a) became permanent features in the landscape, all were subject to rapid water level changes which caused fluctuations in their wetland vegetation type. Changes in sea level or breaching of coastal barriers often replaced this freshwater ecosystem with intertidal floras, as at Helsby Marsh and Downholland Moss (Tooley 1978), or at Ince Blundell, Crosby or Park Road, Meols (Cowell & Innes 1994). Coastal wetland vegetation has also been subject to alteration by human activity (Jones 1988; Jones *et al.* 1993) even before the modern severe drainage of most of the perimarine zone and the grazing of saltmarshes. Fire was a factor from early times and charcoal is present in many of the sediments, as at Downholland Moss (Tooley 1978) and Formby Moss (Wells 1992), particularly those of the reedswamp phase.

Blown sand dunes form a third major coastal environment, and support specialised vegetation communities (Carter 1988), originating around Liverpool Bay between 5,000 and 4,000 years ago after the stabilisation and then temporary regression of sea level (Innes & Tooley 1993; Pye & Neal 1993). The dune barriers have undergone periods of relative stability and instability since then, with wetland dune slack vegetation forming during the former and sand overblowing fringing perimarine wetlands during the latter, for example burying the westward edge of Downholland Moss at 4,090±170 BP (Tooley 1978). Distinctive dry and wet dune slack pollen floras have been recorded by O'Garra (1976) analogous to pollen data from fossil slack deposits at Lifeboat Road, Formby (Innes & Tooley 1993) and characterised by taxa like Sea-milkwort (*Glaux maritima*), Sea-buckthorn (*Hippophaë rhamnoides*), Juniper and Willow.

Heather (*Calluna vulgaris*) and Pine were common on the sandy dune fringe soils.

Heath, moorland and grassland vegetation

Heathlands and moorlands form specialised low stature vegetation communities dominated by dwarf shrubs of the heath family (Ericaceae), as well as grasses and sedges. In the Mersey Basin dry heathlands are a lowland feature (Gimingham 1972), with conditions in the cooler, wetter uplands leading to the formation of blanket mire (Chambers 1988; Moore 1988), or wet moorland with Heather and Bracken (*Pteridium aquilinum*). Crowberry heath was the dominant vegetation during colder periods of the late-glacial with less stable soils, particularly during transitional phases between open ground tundra communities and scrub woodland (Walker *et al.* 1994). The Holocene spread of closed forest over almost the entire region severely restricted the habitat of heath and moor vegetation but it is likely that soils formed on sand and gravel in Cheshire (Tallis 1973; Reynolds 1979) and north of the River Mersey (Kear 1985) maintained some heathland during the Holocene. At Knowsley Park (Cowell & Innes 1994) on the Shirdley Hill Sand of Merseyside high Crowberry pollen counts persist until about 8,650 BP, after which a continuous high Heather curve occurs throughout the rest of the Holocene. Although its composition changed, heath vegetation was able to coexist with the Holocene woodland on sandy soils as an understorey but perhaps also as climax vegetation in places. Coastal blown sand also provided refuge habitats for heath associations, as at Sniggery Wood, Sefton (Cowell & Innes 1994), and almost all of the region's lowland bogs were colonised by Heather during their drier phases from at least Flandrian II onwards. Crowberry also colonised dry bog surfaces in the later Holocene (Wybunbury Moss, Green & Pearson 1977). The survival inland of dry heath probably depended on the recurrence of fire (Wells 1992), as well as the promotion of acidification and paludification by human activity (Dimbleby 1962).

The key vegetational feature in the Mersey Basin uplands was the spread of blanket peat and Heather moor, reviewed by Tallis (1991), replacing the previously existing woodland. In places peat moor began to form in the mid-Holocene, perhaps after woodland fire disturbance (Simmons & Innes 1985), but the major expansion of Heather moor took place about 4,000 BP (Rishworth Moor, Bartley 1975; Bradwell Sitch, Tallis & Johnson 1980) probably as a response to climatic and soil deterioration (Birks 1988). Human activity (Bartley & Chambers 1992) would have accelerated the process.

Woodland disturbance and post-woodland communities

Disturbed soils

Soil instability after natural environmental disturbance was important allowing the survival of ruderal weed and early seral dryland communities within the Holocene closed forest. Landslides were an important source of such disturbance in both the late-glacial cold phase (Johnson *et al.* 1990), and throughout the Holocene (Tallis & Johnson 1980). Land subsidence due to underground salt solution (Reynolds 1979), tree windthrow after severe storms or sand dune mobilisation would have led to similar seral vegetation. Sand formations like the Shirdley Hill Sands remained naturally unstable well into the early Holocene after their main deposition in the late-glacial cold phase (Tooley 1978; Innes 1986). Innes *et al.* (1989) have described a peat lens with a tundra weed pollen flora within Shirdley Hill Sand at Holiday Moss, dated to 9,120±60 BP, when on clay soils at nearby Red Moss (Hibbert *et al.* 1971) closed Birch and Pine woodland with some Hazel was well established. In the earlier Holocene, coversands provided reservoirs of floral diversity which contrasted strongly with the forests which colonised the rest of the region. By the mid Holocene the sands supported open woodland and heathland.

Throughout the Mersey Basin in the later Holocene, cultural vegetation disturbance became by far the most common cause of soil instability and erosion and the creation of open ground habitats. Lowland mosses contain inwash stripes of eroded soil which are evidence of catchment devegetation, as at Peckforton Mere and Bar Mere (Twigger 1983) by the mid-Cheshire ridge, where much of the Iron Age and Roman period soil was eroded and redeposited, with major floral changes. Mid-Holocene woodland disturbance caused reworking of Shirdley Hill Sand (Tooley 1978; Innes 1992), promoting a scrub, heath and grassland flora. In the uplands at Deep Clough (Tallis & McGuire 1972) in Bronze Age and later times heavy erosion of catchment soils occurred. The promotion of post disturbance ruderal weed and regeneration associations is a key feature of late Holocene vegetation history.

Fire disturbance

Fire was a potent instrument of ecological change, deflecting vegetation successions, increasing plant diversity and changing vegetation patterns. Whether employed by humans or due to natural events such as lightning strike, fire was a major environmental force throughout the Holocene and was also present in earlier periods. Indeed, fire was an integral part of vegetation evolution over a very long time period and some plant taxa and associations are highly favoured by it, while occasional low intensity burning promotes most plant communities' regeneration and health.

Fire was a consistent factor in the vegetation history of the Mersey Basin, as both macro- and microscopic charcoal are preserved in sediments of all ages from all parts of the region. Abundant conifer charcoal occurred as long ago as 60,000 BP at Chelford, presumably due to natural fire and probably changing Spruce to Pine woodland (Simpson & West 1958). Microscopic charcoal particles occur in peats from the earliest Holocene onwards and often show an increased concentration at horizons where major vegetation changes occur. They may reflect longer distance transport and so measure the regularity of extra-local or regional burning. Macroscopic charcoal layers also occur, presumably due to higher intensity fires or burning very close to the sediment site (Cundill 1981). Fire probably caused major changes, particularly where pollen evidence suggests an abrupt decline of tree taxa, as at many lowland sites like Simonswood Moss (Cowell & Innes 1994) and Walkers Heath (Leah *et al.* 1997), and in the uplands as at Lady Clough Moor (Tallis 1975).

The botanical effects of fire in established woodland are to replace trees with a range of transitional habitats and successional vegetation communities. Newly open ground is rapidly colonised by pioneer weeds and ruderals, some like Cow-wheat (*Melampyrum*) and Bracken strongly responsive to post-fire conditions. Shrubs, particularly Hazel, are greatly favoured because of increased access to sunlight as members of the woodland edge community around the margins of the clearing or as part of the secondary regeneration community on the cleared area itself. Eventual regeneration of tree cover suppresses the heliophyte successional taxa. Long term alteration in woodland structure may occur, with secondary forest trees like Ash and Birch replacing primary deciduous taxa. While the effects of a single fire may appear ephemeral, the cumulative effects of many such events over millennia, as is suggested by the charcoal record from the Mersey Basin, may have been quite profound. Simmons & Innes (1985) discussed the degenerative potential of repetitive disturbance in the spread of bog and moorland, at least at altitude. A more positive consequence during the mid-Holocene may have been the replacement of steady state homogeneous deciduous forest by a mosaic structure with tracts of primary forest, areas of more open secondary woodland and patches of seral communities at various stages of regeneration, creating local centres of diversity and species richness. While individual patches would regenerate, the creation of new open areas would have maintained the mosaic structure and the overall balanced pattern of the forest.

The Elm decline

Despite intensive study the decline in Elm pollen frequencies which occurred *c.*5,000 years BP, although later in the upland (Robinson's Moss 4,875±60, Tallis & Switsur 1990) than the lowland (Knowsley Park Moss 5,290±80, Cowell & Innes 1994), remains an enigmatic feature of the Holocene pollen record. While an earlier or 'primary' Elm Decline was recognised on a few pollen diagrams (Rishworth Moor, 5,490±140, Bartley 1975) in which little but the fall in Elm pollen occurs, most Elm Declines from the Mersey Basin are accompanied by other pollen changes which indicate woodland opening. Other tree taxa also decline and weed and shrub pollen increase. This woodland clearance or 'landnam' phase implicates human activity as a likely cause. In some cases cereal pollen (Cowell & Innes 1994) or charcoal (Cundill 1984) also occur but usually indicators are restricted to

plants of grassy clearings like Poaceae, Ribwort Plantain (*Plantago lanceolata*) and Bracken. Woodland management practices, like stripping and lopping of Elm branches for foddering of stock, could have had the effect of reducing Elm pollen frequencies by reducing flowering, and once opened the woodland could have been kept open as long as beasts were herded there. Perry & Moore (1987) showed that pollen changes very like those of the Elm Decline, including a rise in Ribwort Plantain and other weeds, took place after the modern death and fall of Elms killed by disease. Disease may have caused many of the examples of Elm pollen reduction in the Mersey Basin, in places combined with human activity, and with other factors like climate change and soil deterioration of lesser importance. Whatever its cause the Elm Decline heralded real vegetation changes. The forest seems to have become less dense and secondary trees like Ash, Birch and Hazel more important.

The anthropogenic phase

Although human disturbance of woodland occurred prior to the Elm Decline, it is man's subsequent development of agricultural food production that defines the succeeding anthropogenic phase, during which the environmental results of forest clearance and the maintenance of cultural sub-climax vegetation became the dominant features of vegetation history. The history of human settlement, forest clearance and agriculture will be considered in detail by Cowell (chapter four, this volume) and so this section is restricted to a brief examination of the kind of vegetation change produced by this activity.

Forest clearance and pastoralism

Forest opening occurred in late Flandrian II and the earlier part of Flandrian III, presumably due to the agricultural activity of Neolithic and Bronze Age people. Many episodes of clearance lack cereal pollen and while negative evidence is inconclusive, cereal pollen being poorly transported and of low productivity, much of earlier prehistoric agriculture may have been pastoral, involving exploitation of a managed woodland ecosystem which included pollarding and coppicing of selected trees. The creation of grassy clearings in the woodland produces a classic 'landnam' pollen signature, with a fall in primary tree pollen like Elm and Oak, increased grass pollen and the appearance of Ribwort Plantain, encouraged by grazing and trampling by stock. Regeneration of a more open woodland with Ash, Birch and Hazel follows relaxation of grazing pressure, with Heather or Bracken colonisation in areas of poorer soils. A cycle of limited woodland opening then regeneration without cereal cultivation is familiar from the heavier clay soils of the region, as in the Bronze Age at Parr Moss, St Helens (Cowell & Innes 1994). The impact was greater in the upland, and pollen from Bronze Age soils sealed beneath barrows in the Rossendales (Tallis & McGuire 1972) shows a post clearance vegetation with some heathland on acidifying soils.

Arable cultivation

The vegetational effects of arable cultivation are very significant in that crop cultivation promotes land clearance and the spread of open ground, yet includes a tendency towards reduced vegetation diversity through the selection of a limited number of preferred food plants, so that in its later more intensive forms it aims towards areas of land maintained as almost monocultural climax vegetation. This was unachievable before modern times, and in later prehistoric periods arable land-use led to specialised diverse herbaceous associations of a kind not seen since late-glacial tundra communities. Localised cereal cultivation may have been present in the Mersey Basin since the earliest Neolithic, as cereal type pollen grains have been recorded several centuries before the dates for the Elm Decline. Although wild grasses with pollen of cereal type have been present throughout the Holocene, these early examples cluster around 5,800 BP (Cowell & Innes 1994; Williams 1985; Wiltshire & Edwards 1993) within well-defined forest clearance episodes.

The pollen data suggest that arable cultivation provided habitats for many more weeds of broken ground and ruderal conditions than did pastoralism. Even in the Neolithic, on favourable soils such as the sandstone fringing Prescot Moss about 4,520 BP (Innes & Tomlinson 1995) the weed assemblage moves beyond the Ribwort Plantain dominated grassland group and includes many which later came to be classified as arable indicators, including Dock, Mugwort, Common Knapweed (*Centaurea nigra*), Goosefoot family, Thistle (*Cirsium*), Greater and Hoary Plantains (*Plantago major* and *P. media*), Mayweed (*Matricaria*)-type and Chickweeds (*Stellaria* spp.). This arable community is well illustrated in Cheshire in the Iron Age pollen spectra from Lindow Moss (Oldfield *et al.* 1986; Branch & Scaife 1995), and on the sandstone ridges around Bar Mere and Peckforton Mere (Twigger 1983) where major replacement of oakwoods by cultivation occurred. Later crop plants diversified from Wheat and Barley (*Triticum/ Hordeum*) types into other cereals such as Oat (*Avena*) and Rye (*Secale*), and other crops such as Hemp (*Cannabis sativa*), Flax (*Linum usitatissimum*) and Broad Bean (*Vicia faba*), and crop weeds which had earlier occurred sporadically, such as Cornflower (*Centaurea cyanus*), Corn Spurrey (*Spergula arvensis*), Knotgrass (*Polygonum aviculare*) and spurges (*Euphorbia*), become more common. Arable communities declined between Roman times and the Norman Conquest. One aspect of arable agriculture which promoted vegetation diversity was the creation of fields and their boundaries to keep stock from the valuable crop. Shrubs such as Holly (*Ilex*), Hawthorn (*Crataegus*), Blackthorn (*Prunus*) and others, which could be hedge taxa, often increase during arable pollen phases.

Conclusions

Significant elements of floristic diversity were present in the Mersey Basin before major human disturbance of ecosystems occurred in the later Holocene. Altitude, geology and soil variations encouraged differing patterns of natural vegetation across the region even in pre-Holocene times, and the increasingly severe human effects were superimposed upon a landscape already a complex mosaic of vegetation communities. Some areas like the mere and mire landscape of Cheshire were floristically very rich and are important on a more than regional basis, requiring more detailed research. Concentration is needed in parts of the region which have few pollen records (Figure 3.1), and the systematic work of the North West Wetlands Survey should identify sites of high potential here and elsewhere. The present record relies on a few researchers to whom a great vote of thanks is owed, but who naturally focused upon their own research areas. A more even distribution of data is required from which to plan future work. Where sediment distribution allows, past time periods for which little botanical information exists should receive closer attention in future.

Acknowledgements

We thank Colin Wells and Elizabeth Huckerby (North West Wetlands Survey) for making unpublished data available. JGAL acknowledges the support of a NERC research studentship, grants from Cheshire County Council and the British Ecological Society (Small Ecological Project Grant No. 1145). Vicki Innes produced the manuscript. Figure 3.1 was drawn in the Department of Geography, University of Durham.

Dedication

This paper is dedicated to the memory of Elizabeth Emma Innes.

References

Barber, K.E. (1993). Peatlands as scientific archives of past biodiversity. *Biodiversity and Conservation*, **2**, 474–89.

Bartley, D.D. & Chambers, C. (1992). A pollen diagram, radio-carbon ages and evidence of agriculture on Extwistle Moor, Lancashire. *New Phytologist*, **121**, 311–20.

Bartley, D.D. (1975). Pollen analytical evidence for prehistoric forest clearance in the upland area west of Rishworth, West Yorkshire. *New Phytologist*, **74**, 375–81.

Baxter, J. (1983). *Vegetation History of the Shirdley Hill Sand in Southwest Lancashire*. PhD thesis. U.C.W. Aberystwyth.

Behre, K-E. (1981). The interpretation of anthropogenic indicators in pollen diagrams. *Pollen et Spores*, **23**, 225–45.

Bennett, K.D. & Birks, H.J.B. (1990). Postglacial history of Alder (*Alnus glutinosa* (L.) Gaertn.) in the British Isles. *Journal of Quaternary Science*, **5**, 123–33.

Bennett, K.D. (1984). The Postglacial history of *Pinus sylvestris* in the British Isles. *Quaternary Science Reviews*, **3**, 133–55.

Birks, H.J.B. (1964). Chat Moss, Lancashire. *Memoirs and Proceedings of the Manchester Literary and Philosophical Society*, **106**, 1–24.

Birks, H.J.B. (1965a). Late-Glacial deposits at Bagmere, Cheshire and Chat Moss, Lancashire. *New Phytologist*, **64**, 270–75.

Birks, H.J.B. (1965b). Pollen analytical investigations at Holcroft Moss, Lancashire and Lindow Moss, Cheshire. *Journal of Ecology*, **53**, 299–314.

Birks, H.J.B. (1988). Long term ecological change in the British uplands. *Ecological Change in the Uplands* (eds M.B. Usher & D.B.A. Thompson), pp. 37–56, Blackwell, Oxford.

Bradshaw, R.H.W. (1988). Spatially-precise studies of forest dynamics. *Vegetation History* (eds B. Huntley & T. Webb III), pp. 725–51, Kluwer Academic Publishers, Dordrecht Netherlands.

Branch, N.P. & Scaife, R.G. (1995). The stratigraphy and pollen analysis of peat sequences associated with the Lindow III bog body. *Bog Bodies: New Discoveries and New Perspectives* (eds R.C. Turner & R.G. Scaife), pp. 19–30. British Museum Press, London.

Carter, R.W.G. (1988). *Coastal Environments*. Academic Press, London.

Chambers, F.M. (1988). Archaeology and the flora of the British Isles: the moorland experience. *Archaeology and the Flora of the British Isles* (ed. M. Jones) pp. 107–15, Oxford University Committee for Archaeology Monograph 14, Oxford.

Conway, V.M. (1954). Stratigraphy and pollen analysis of southern Pennine blanket peats. *Journal of Ecology*, **42**, 117–47.

Cowell, R. & Innes, J.B. (1994). *The Wetlands of Merseyside*. North West Wetlands Survey 1. Lancaster Imprints 2, Lancaster.

Cundill, P.R. (1981). The history of vegetation and land use of two peat mosses in south-west Lancashire. *The Manchester Geographer*, **2**, 35–44.

Cundill, P.R. (1984). Palaeobotany and archaeology on Merseyside: additional evidence. *Circaea*, **2**, 129–31.

Dimbleby, G.W. (1962). *The Development of British Heathlands and their Soils*. Oxford Forestry Memoirs 23.

Franks, J.W. & Johnson, R.H. (1964). Pollen analytical dating of a Derbyshire landslip: the Cown Edge Landslides, Charlesworth. *New Phytologist*, **63**, 209–16.

Gear, A.J. & Huntley, B. (1991). Rapid changes in the range limits of Scots pine 4000 years ago. *Science*, **251**, 544–47.

Gimingham, C.H. (1972). *Ecology of Heathlands*. Chapman & Hall, London.

Godwin, H. (1959). Studies in the Postglacial history of British vegetation. XIV. Late Glacial deposits at Moss Lake, Liverpool. *Philosophical Transactions of the Royal Society of London B*, **242**, 127–49.

Green, B.H. & Pearson, M.C. (1977). The ecology of Wybunbury Moss, Cheshire II. Post-Glacial history and the formation of the Cheshire mere and mire landscape. *Journal of Ecology*, **65**, 793–814.

Greig, J. (1982). Past and present limewoods of Europe. *Archaeological Aspects of Woodland Ecology* (eds M. Bell & S. Limbrey), pp. 23–55. British Archaeological Reports International Series, 146, Oxford.

Hall, B.R. & Folland, C.J. (1970). *Soils of Lancashire*. Memoirs of the Soil Survey of Great Britain 5, Harpenden.

Hall, D., Wells, C.E. & Huckerby, E. (1995). *The Wetlands of Greater Manchester*. North West Wetlands Survey 2. Lancaster Imprints 3, Lancaster.

Hibbert, F.A., Switsur, V.R. & West, R.G. (1971). Radiocarbon dating of Flandrian pollen zones at Red Moss, Lancashire. *Proceedings of the Royal Society of London B*, **177**, 161–76.

Howard-Davis C., Stocks C. & Innes, J.B. (1988). *Peat and the Past. A Survey and Assessment of the Prehistory of the Lowland Wetlands of North-West England*. English Heritage and Lancaster University, Lancaster.

Innes, J.B. (1986). The history of the Shirdley Hill Sand revealed by examination of associated organic deposits. *Proceedings of the North of England Soils Discussion Group*, **21**, 31–43.

Innes, J.B. (1991). A preliminary report on pollen analyses from Rimrose Brook. *An Archaeological Assessment of the Rimrose Valley, Sefton* (eds R.W. Cowell & S.M. Nicholson), pp. 3–4. Liverpool Museum, Liverpool.

Innes, J.B. (1992). Pollen analysis of a radiocarbon dated peat bed within Shirdley Hill sand at Aintree, Merseyside. *The Manchester Geographer*, **11**, 44–51.

Innes, J B. & Tooley, M.J. (1993). The Age and Vegetational History of the Sefton Coast Dunes. *The Sand Dunes of the Sefton Coast* (eds D. Atkinson & J. Houston), pp. 35–40. National Museums & Galleries on Merseyside in association with Sefton Metropolitan Borough Council, Liverpool.

Innes, J.B. & Tomlinson, P.R. (1983). An approach to palaeobotany and survey archaeology in Merseyside. *Circaea*, **1**, 83–93.

Innes, J.B. & Tomlinson, P.R. (1991). Environmental Archaeology in Merseyside. *Journal of the Merseyside Archaeological Society*, **7**, 1–20.

Innes, J.B. & Tomlinson, P.R. (1995). Radiocarbon dates from Warrington Road, Prescot. *Journal of the Merseyside Archaeological Society*, **9**, 61–4.

Innes, J.B., Tooley, M.J. & Tomlinson, P.R. (1989). A comparison of the age and palaeoecology of some sub-Shirdley Hill Sand peat deposits from Merseyside and south-west Lancashire. *Naturalist*, **114**, 65–9.

Johnson, R.H. (1985). The imprint of glaciation on the west Pennine uplands. *The Geomorphology of North-west England* (ed. R.H. Johnson), pp. 237–62. Manchester University Press, Manchester.

Johnson, R.H., Franks, J.W. & Pollard, J.E. (1970). Some Holocene faunal and floral remains in the Whitemoor meltwater channel at Bosley, east Cheshire. *North Staffordshire Journal of Field Studies*, **10**, 65–74.

Johnson, R.H., Tallis, J.H. & Wilson, P. (1990). The Seal Edge Coombes, north Derbyshire – a study of their erosional and depositional history. *Journal of Quaternary Science*, **5**, 83–94.

Jones, C.R., Houston, J.A. & Bateman, D. (1993). A history of human influence on the coastal landscape. *The Sand Dunes of the Sefton Coast* (eds D. Atkinson & J. Houston), pp. 3–20. National Museums & Galleries on Merseyside in association with Sefton Metropolitan Borough Council, Liverpool.

Jones, R.L. (1988). The impact of early man on coastal plant communities in the British Isles. *Archaeology and the Flora of the British Isles* (ed. M. Jones) pp. 96–106. Oxford University Committee for Archaeology Monograph 14, Oxford.

Kear, B.S. (1968). *An investigation into soils developed on the Shirdley Hill Sand in south-west Lancashire*. MSc thesis, University of Manchester.

Kear, B.S. (1985). Soil development and soil patterns in north-west England. *The Geomorphology of North-west England* (ed. R.H. Johnson), pp. 80–93, Manchester University Press, Manchester.

Kenna, R.J.B. (1986). The Flandrian sequence of North Wirral (N.W. England). *Geological Journal*, **21**, 1–27.

Lageard, J.G.A. (1992). *Vegetational history and palaeoforest reconstruction at White Moss, south Cheshire, UK*. PhD Thesis, Keele University.

Lageard, J.G.A., Chambers, F.M. & Thomas P.A. (1995). Recording and reconstruction of wood macrofossils in three-dimensions. *Journal of Archaeological Science*, **22**, 561–67.

Lamb, H.H. (1981). Climate from 1000BC to 1000AD. *The Environment of Man: the Iron Age to the Anglo Saxon Period* (eds M. Jones and G. Dimbleby), pp. 53–65. British Archaeological Reports (British Series), **87**, Oxford.

Leah, M., Wells, C.E., Huckerby, E. & Appleby, C. (1997). *The Wetlands of Cheshire*. North West Wetlands, Survey 4. Lancaster Imprints 5, Lancaster.

Lind, E.M. (1949). The history and vegetation of some Cheshire meres. *Memoirs and Proceedings of the Manchester Literary and Philosophical Society*, **90**, 17–36.

Moore, P.D. (1988). The development of moorlands and upland mires. *Archaeology and the Flora of the British Isles* (ed. M. Jones), pp. 116–22. Oxford University Committee for Archaeology Monograph 14, Oxford.

Moore, P.D., Merryfield, D.L. & Price, M.D.R. (1984). The vegetation and development of British mires. *European Mires* (ed. P.D. Moore) pp. 203–35, Academic Press, London.

Moore, P.D., Webb, J.A. & Collinson, M.E. (1991). *Pollen Analysis*. Blackwell, Oxford.

O'Garra, A. (1976). *Dune slack systems – vegetation and morphological development at Ainsdale*. BSc thesis, Department of Geography, University of Liverpool.

Oldfield, F., Higgit, S.R., Richardson, N. & Yates, G. (1986). Pollen, charcoal, rhizopod and radiometric analysis. *Lindow Man, the Body in the Bog* (eds I.M. Stead, J.B. Bourke and D. Brothwell), pp. 82–5. Guild Publishing, London.

Pennington, W. (1977). The late-glacial flora and vegetation of Britain. *Philosophical Transactions of the Royal Society of London B*, **280**, 247–71.

Perry, I. & Moore, P.D. (1987). Dutch elm disease as an analogue of Neolithic Elm Decline. *Nature*, **326**, 72–73.

Plater, A.J., Huddart, D., Innes, J.B., Pye, K., Smith, A.J. & Tooley, M.J. (1993). Coastal and sea-level changes. *The Sand Dunes of the Sefton Coast* (eds D. Atkinson & J. Houston), pp. 23–34. National Museums & Galleries on Merseyside in association with Sefton Metropolitan Borough Council, Liverpool.

Pye, K. & Neal, A. (1993). Stratigraphy and Age Structure of the Sefton Dune Complex: Preliminary Results of Field Drilling Investigations. *The Sand Dunes of the Sefton Coast* (eds D. Atkinson & J. Houston), pp.41–4. National Museums & Galleries on Merseyside in association with Sefton Metropolitan Borough Council, Liverpool.

Reynolds, C.S. (1979). The limnology of the eutrophic meres of the Shropshire–Cheshire plain. *Field Studies*, **5**, 93–173.

Rodwell, J.S. (ed.) (1991). *British Plant Communities*. Volume 2 *Mires and Heaths*. Cambridge University Press, Cambridge.

Schoenwetter, J. (1982). Environmental archaeology of the Peckforton Hills. *Cheshire Archaeological Bulletin*, **8**, 10–11.

Shennan, I. (1992). Late Quaternary sea-level changes and crustal movements in eastern England and eastern Scotland: an assessment of models of coastal evolution. *Quaternary International*, **15/16**, 161–73.

Shimwell, D.W. (1985). The distribution and origins of the lowland mosslands. *The Geomorphology of North-west England* (ed. R.H. Johnson), pp. 299–312, Manchester University Press, Manchester.

Simmons, I.G. & Innes, J.B. (1985). Late Mesolithic land-use and its environmental impacts in the English uplands. *Biogeographical Monographs*, **2**, 7–17.

Simpson, I.M. & West, R.G. (1958). On the stratigraphy and palaeobotany of a Late-Pleistocene organic deposit at Chelford, Cheshire. *New Phytologist*, **57**, 239–50.

Stace, C. (1991). *New Flora of the British Isles*. Cambridge University Press, Cambridge.

Switsur, V.R. & West, R.G. (1975). University of Cambridge Natural Radiocarbon Measurements XIII. *Radiocarbon*, **17**, 35–51.

Tallis, J.H. (1964a). The pre-peat vegetation of the southern Pennines. *New Phytologist*, **63**, 363–73.

Tallis, J.H. (1964b). Studies on southern Pennine peats. I. The general pollen record. *Journal of Ecology*, **52**, 323–31.

Tallis, J.H. (1973). The terrestrialisation of lake basins in North Cheshire, with special reference to the development of a 'Schwingmoor' structure. *Journal of Ecology*, **61**, 537–67.

Tallis, J.H. (1975). Tree remains in southern Pennine blanket peats. *Nature*, **256**, 482–4.

Tallis, J.H. (1985). Mass movement and erosion of a southern Pennine blanket peat. *Journal of Ecology*, **73**, 282–315.

Tallis, J.H. (1991). Forest and moorland in the south Pennine upland in the mid-Flandrian period. III. The spread of moorland – local, regional and national. *Journal of Ecology*, **79**, 401–15.

Tallis, J.H. & Birks, H.J.B. (1965). The past and present distribution of *Scheuchzeria palustris* (L.) in Europe. *Journal of Ecology*, **53**, 287–98.

Tallis, J.H. & Johnson R.H. (1980). The dating of landslides in Longdendale, north Derbyshire, using pollen-analytical techniques. *Timescales in Geomorphology* (eds R.A. Cullingford, D.A. Davidson & J. Lewin), pp. 189–205, Wiley, Chichester.

Tallis, J.H. & McGuire, J. (1972). Central Rossendale: the evolution of an upland vegetation. I. The clearance of woodland. *Journal of Ecology*, **60**, 721–37.

Tallis, J.H. & Switsur, V.R. (1973). Studies on the southern Pennine peats. VI. A radiocarbon dated pollen diagram from Featherbed Moss, Derbyshire. *Journal of Ecology*, **61**, 743–51.

Tallis, J.H. & Switsur, V.R. (1983). Forest and moorland in the south Pennine uplands in the mid-Flandrian period. I. Macrofossil evidence of the former forest cover. *Journal of Ecology*, **71**, 585–600.

Tallis, J.H. & Switsur, V.R. (1990). Forest and moorland in the south Pennine uplands in the mid-Flandrian period. II. The hillslope forests. *Journal of Ecology*, **78**, 857–83.

Tooley, M.J. (1978). *Sea-Level Changes in North West England During the Flandrian Stage*. Clarendon Press, Oxford.

Tooley, M.J. (1985a). Sea-level changes and coastal morphology in north-west England. *The Geomorphology of North-west England* (ed. R.H. Johnson), pp. 94–121, Manchester University Press, Manchester.

Tooley, M.J. (1985b). Climate, sea level and coastal changes. *The Climatic Scene. Essays in Honour of Gordon Manley* (eds M.J. Tooley & G.M. Sheail), pp. 206–34, George Allen and Unwin, London.

Turner, J. (1962). The *Tilia* decline: an anthropogenic interpretation. *New Phytologist*, **61**, 328–41.

Twigger, S.N. (1983). *Environmental Change in Lowland Cheshire*. Unpublished report. Department of Prehistoric Archaeology, University of Liverpool.

Walker, D. (1970). Direction and rate of some Post-glacial hydroseres. *Studies in the Vegetational History of the British Isles* (eds D.Walker and R.G. West), pp. 117–39. Cambridge University Press, Cambridge.

Walker, M.J.C., Bohncke, S.J.P., Coope, G.R., O'Connell, M., Usinger, H. & Verbruggen, C. (1994). The Devensian/ Weichselian Late-glacial in northwest Europe (Ireland, Britain, north Belgium, The Netherlands, north-west Germany). *Journal of Quaternary Science*, **9**, 109–18.

Wells, C.E. (1992). Stratigraphic survey in Merseyside and West Lancashire borderlands. *North West Wetlands Survey Annual Report 1992* (ed. R. Middleton), pp. 43–7. Lancaster University Archaeological Unit, Lancaster.

Wells, C.E., Huckerby, E. & Hall D. (1993). Archaeological and palaeoecological survey in Greater Manchester. *North West Wetland Survey Annual Report 1993* (ed. R. Middleton), pp. 29–35, Lancaster University Archaeological Unit, Lancaster.

Williams, C.T. (1985). *Mesolithic Exploitation Patterns in the Central Pennines. A Palynological Study of Soyland Moor*. British Archaeological Reports (British Series), 139, Oxford.

Wiltshire, P.E.J. & Edwards, K.J. (1993). Mesolithic, early Neolithic, and later prehistoric impacts on vegetation at a riverine site in Derbyshire, England. *Climate Change and Human Impact on the Landscape* (ed. F.M. Chambers), pp. 157–68, Chapman & Hall, London.

Worsley, P. (1985). Pleistocene history of the Cheshire–Shropshire plain. *The Geomorphology of North-west England* (ed. R.H. Johnson), pp. 201–21, Manchester University Press, Manchester.

Yates, E.M. & Moseley, F. (1958). Glacial lakes and spillways in the vicinity of Madeley, North Staffordshire. *Quarterly Journal of the Geological Society of London*, **113**, 409–28.

CHAPTER FOUR

The human influence to the Norman Conquest

R.W. COWELL

Introduction

This paper covers 9,000 years since the last Ice Age and incorporates three distinct phases of human exploitation of the Mersey Basin environment. It is a time, however, which has left little direct impact on today's landscape, other than that seen in the bare peat covered uplands of the Basin (chapter eleven, this volume) and in some shadowy links between the nature of the agricultural lowlands and landscape changes taking place in the final part of the period covered here. The prehistoric and early historic land-use of the area therefore largely provides a contrasting picture to the nature of the landscape in the succeeding period, which was the one in which many features of today's landscape originated (chapter five, this volume).

The earliest forest c.9,500–c.3,200 BP

The occupation of this part of England during the late stages of the last Ice Age has left no trace within the Mersey Basin, although sites of mobile hunter-gatherers of this period are known in the uplands of Derbyshire and North Wales, and parts of northern Lancashire and Cumbria (Jacobi 1980). The earliest evidence for inhabitants in the Mersey region comes from shortly after the final retreat of the ice c.10,000 years ago, with the development of the wooded landscape of the Mesolithic period.

There were probably two main ways in which the mobile hunter-gatherer communities of this period exploited the landscape. One was in residential areas, places where a family, or group of families, might stay for a period of time; the second was in various types of smaller, specialist sites associated with hunting and gathering expeditions away from the residential areas (Smith 1992).

The coastal areas of Sefton and Wirral (Figure 4.1) are the main residential locations in the Basin (Cowell & Innes 1994). Mesolithic settlement of the Basin coincided with loss of land due to rising sea-level, so that by c.6,000 BP the approximate line of the present coast would have been reached for the first time (chapter two, this volume). The coastline provided a rich variety of resources that would have been available almost all year round. Estuarine muds would have been important for wildfowl breeding and nesting, while inshore fishing would have been possible in the estuaries. Backing onto these habitats were freshwater reedswamps and fens where wildfowl, fish, and aquatic animals would have been plentiful.

Apart from rich coastal environments, after c.7,500 BP, the local hunter-gatherers lived in an environment where large areas of swamp and subsequently fen, now covered by peatland, occupied inland hollows and waterlogged gradients across large areas of the lowlands north of the Mersey (Birks 1964, 1965; Cowell & Innes 1994; Hall et al. 1995; Hibbert, Switsur & West 1971; Howard-Davis, Stocks & Innes 1988; chapter three, this volume) and within smaller basins south of the river in Cheshire. By c.7,000 BP, Alder fenwood was common around wetlands, rivers and on the lower slopes of the uplands.

Elsewhere during this period, mixed deciduous woodland had come to dominate the landscape, resulting in a dense forest that must have hindered easy movement within it. Deciduous woodland also clothed the Pennines, in general to altitudes above c.500m, above which Hazel (*Corylus*) and Birch (*Betula*) scrub probably predominated, although in some poorly drained areas, particularly at higher altitudes, the increased rainfall of this period also led to the initiation of peat (chapter eleven, this volume).

This expanse of dryland woods, interspersed with carr wetlands and the mixed vegetation along the upper tree limit in the Pennines, would have provided rich habitats with plentiful grazing for Aurochs (*Bos primigenius*), Wild Boar (*Sus scrofa*), and Deer as well as edible plants, berries, and nuts for humans. The exploitation of this landscape may have been different in the lowlands and the adjoining uplands from the coasts. In the interior lowlands of the Basin, the nature of Mesolithic activity is generally dispersed and largely concentrated in the main river valleys, such as those of the Rivers Mersey and Weaver and tributaries like the Ditton (Barnes 1982; Cowell 1991, 1992). Most of these sites may represent small camps for specialist task groups.

The Pennines have produced the greatest concentra-

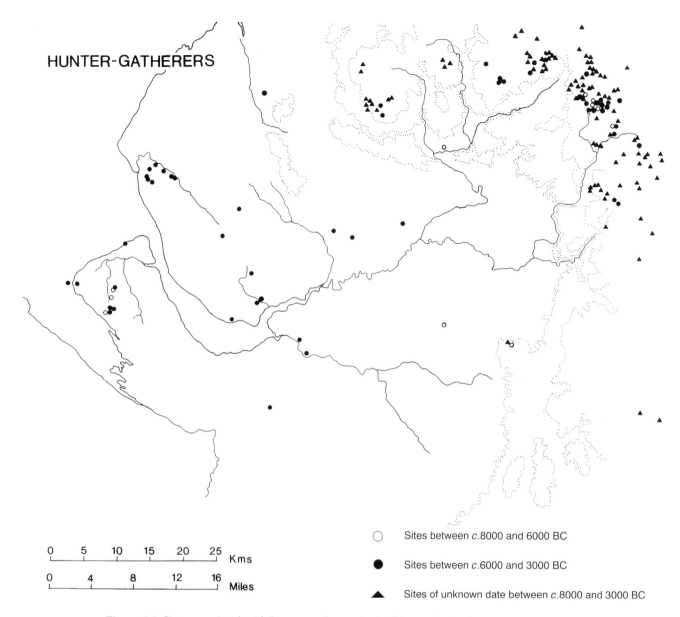

HUNTER-GATHERERS

○ Sites between *c*.8000 and 6000 BC

● Sites between *c*.6000 and 3000 BC

▲ Sites of unknown date between *c*.8000 and 3000 BC

Figure 4.1. Sites associated with hunter-gatherers in the Mersey Basin, from *c*.9,500–5,500 BP.

tion of Mesolithic sites in the country. They are found mainly between the 366m and 488m contours (Jacobi, Tallis & Mellars 1976), with the greatest concentration in a fairly restricted area between Saddleworth and Marsden, where the Pennines are at their narrowest (Barnes 1982; Stonehouse 1989, 1994; Wymer & Bonsall 1977). Most of the upland sites are interpreted as summer hunting camps, occupied by a limited number of people (Mellars 1976), although Williams (1985) argues for an intensive form of managed grazing over many centuries in parts of the Pennines.

Pollen diagrams suggest that in the early part of the period the woodland cover was not being altered to any degree (Cowell & Innes 1994; Hibbert *et al*. 1971; chapter three, this volume). Disturbance to the woodland cover in the lowlands of the Mersey Basin before *c*.6,000 BP is only found in coastal areas. This suggests that hunter-gatherers were undertaking activities at the natural clearings around the edges of swamps or fens without

affecting the woodland or disturbing the ground to any great degree. It is only during the 6th millennium BP that evidence from pollen diagrams shows broken ground and associated small dips in the woodland cover becoming more common. This occurs both in the coastal and central areas of the Basin.

There is evidence that burning of the woodlands may have been taking place in the uplands (chapter eleven, this volume) as well as in the lowlands of Merseyside (Cowell & Innes 1994) and Greater Manchester (Hall *et al*. 1995), probably after *c*.7,000 BP. In the uplands, hunting sites are often associated with an increase of Heather (*Calluna*) and charcoal flecks in the artefact layers at altitudes which probably mark the upper limits of tree cover. Manipulation of the edges of the forest, where browse is more plentiful, would have been accomplished more easily and there may have been a deliberate policy of woodland management rather than the use of accidental natural events (Mellars 1976).

In the more thinly exploited interior lowlands, if the extensive evidence for burning represents a deliberate policy of firing woodland, then it would have to have been undertaken by small groups of hunters, operating away from the main residential areas, part of whose function might have been to clear scattered areas for future hunting expeditions.

Whether or not the woodland burning during this period was associated with human actions, it appears to have had an effect on the ecological succession of these habitats (chapter eleven, this volume; Hall *et al.* 1995, p. 116). In the uplands, mire development may have resulted naturally at the higher altitudes, but may have been encouraged by human activity at the forest margins where clearance activity was more concentrated, leading to impoverishment of the soil and the onset of peat formation during the deteriorating climate of this period. At lower altitudes and across the lowlands generally woodland regeneration was the norm.

The impact of agriculture: after *c*.5,400 BP

The introduction of agriculture, traditionally associated with the Neolithic period, is a potentially important development in the vegetational history of the Basin. A number of sites, mainly coastal, in north-western England show woodland disturbances accompanied by cereal type pollen at the beginning of the 6th millennium BP. This is interpreted as the adoption of one part of the agricultural process by Mesolithic hunters (Cowell & Innes 1994; Williams 1985), before the adoption nationally of the full Neolithic cultural and economic repertoire many centuries later. Thus, small areas might have been planted as a slightly more predictable adjunct to the normal round of seasonal hunting and gathering.

The adoption of cultivation in the Mersey Basin, however, seems to have had little effect on the vegetation cover of the lowlands, which remained much as it had been in the Mesolithic for several millennia (Cowell & Innes 1994), although there is some quickening of the pace of clearance and a probable expansion in land-use in the 4th millennium BP (early Bronze Age).

Although the pattern of small woodland disturbances becomes more widespread in the Merseyside part of the Basin in the Neolithic, the nature of each disturbance episode is still limited in extent. Each episode is also succeeded by reasonably long periods of forest regeneration. This is also the pattern in the uplands with only small, occasional breaks in woodland cover (Barnes 1982).

An 'Elm Decline' is present in several dated pollen diagrams from the Basin towards the end of the 6th millennium BP (Cowell & Innes 1994; Howard-Davis *et al.* 1988). In other areas of the country, this often marks the beginning of new trends of woodland clearance associated with agriculture. In much of the Basin this is not the case and the trend may be natural, although at Holcroft Moss and Lindow Moss in the Mersey valley and Hatchmere, Cheshire potentially significant clearance activity is present after the 'Elm Decline'

(Birks 1965; Howard-Davis *et al.* 1988). Other natural changes were also taking place. Woodland cover was still prevalent up to *c*.350m in the Pennines but blanket peat continued to spread, particularly in areas with poor drainage (Barnes 1982). In some localities in the lowlands, wetter conditions may also have become more prevalent from the incidence of pool peats in some of the Greater Manchester peat bogs (Hall *et al.* 1995, p. 117).

The archaeological evidence for the late-6th to late-4th millennium BP largely mirrors the pollen evidence. Neolithic settlements are found in the same localities as in the Mesolithic and of a similar form, suggesting exploitation of the landscape had changed little. The recent discovery by National Museums & Galleries on Merseyside of a wooden trackway of *c*.4,900 BP on the beach at Hightown near Formby, Merseyside highlights the continuing importance of the coast. The incidence of Neolithic stone axes, some of which may have a link with woodland clearance, tends to reinforce this pattern (Figure 4.2). The axes from the uplands are, however, found mainly along the lower fringes, particularly in the river valleys, in areas where little Mesolithic material is known. A number of Mesolithic sites at higher altitudes, particularly in the Saddleworth/Marsden area do though include Neolithic flintwork, especially arrowheads.

No major changes took place during the next thousand years, covering the earlier Bronze Age, as the inland forest remained relatively dense and damp. In the lowlands, small, infrequent woodland disturbances are recorded between *c*.4,000–3,400 BP in Merseyside, similar to those in the Neolithic (Cowell & Innes 1994). These are quite low intensity occurrences, mainly recognised by the appearance of weeds, plantains (*Plantago*), Bracken (*Pteridium*) and grasses, with only very limited reductions in woodland cover. Settlement evidence for this period is limited, with few core areas recognised although scattered chance finds are more numerous than in the Neolithic, suggesting that Bronze Age communities were still mobile to some degree. Upland areas provide a similar pattern, although the clearings become more widespread, which with the increased number of finds, suggests they and their fringes were being used more extensively than in the Neolithic (Barnes 1982). Flint arrowheads are common from the higher parts of the uplands suggesting that hunting was still important in these landscapes.

Elements of a more visible social cohesion are found in the Basin from the distribution of burial sites during this time (Figure 4.2). Neolithic examples are found along the upland fringes in Cheshire (Longley 1987) and Lancashire (Bu'lock 1959), and in the lowlands in Merseyside (Cowell & Warhurst 1984). In the early Bronze Age, earthen burial mounds (barrows) are common in the uplands and major river valleys. These remained as fixed points over many centuries, as witnessed at Winwick in the Mersey valley (Freke & Holgate 1990). Their existence presupposes some signif-

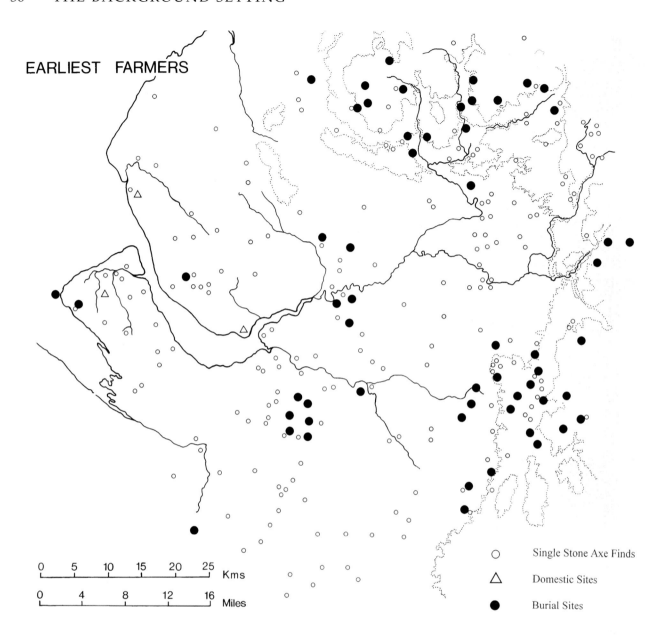

EARLIEST FARMERS

Single Stone Axe Finds

Domestic Sites

Burial Sites

Figure 4.2. Sites and findspots belonging to the first farmers of the Neolithic and Bronze Age periods
in the Mersey Basin, from c.3,500–550 BP.

icant reduction in woodland by c.3,800 BP, as they must have acted as visible markers in the landscape. In the uplands, soil deterioration was taking place, as barrows were erected over soils with a developing podsol structure with local heathland development within Birch and Hazel woodland (Tallis & McGuire 1972). Some areas of the lowlands may also have become heathland during this period with Heather and Birch becoming more common at the margins of the wetland (chapter three, this volume).

The first settled farming communities c.3,200 BP–c.AD 500

At the beginning of the 3rd millennium BP increased rainfall and a fall in temperature of c.2°C, to a summer average c.0.5°C lower than that of today (Lamb 1981) may have contributed to changes in the archaeological

pattern. Raised bog became prevalent in the major lowland valleys of the Basin, particularly the middle Mersey and in the uplands (Tallis & Switsur 1973), and in inland mires such as at Simonswood and Parr Moss, Burtonwood, during the middle to later part of the millennium (Cowell & Innes 1994), leading to areas of low agricultural productivity. Recurrence surfaces in local bogs imply two horizons of particular deterioration, typically dated at Chat Moss, Greater Manchester to the end of the 4th millennium BP and to c.2,600 BP (Nevell 1992). Nationally, this climatic deterioration is seen as being a possible cause for the archaeological changes seen at the end of the earlier Bronze Age at c.3,200 BP (Burgess 1974).

The main artefactual evidence in the region for the period after c.3,000 BP comes from metalwork of late Bronze Age date. Nationally, this is commonly found around wetland areas and rivers, particularly in the form

Figure 4.3. Early Neolithic (*c.*4,900 BP) wooden trackway under excavation on the beach at Hightown, near Formby, Merseyside.

of hoards (Bradley 1984). In the north-west, hoards are rare, the pattern being dominated by single finds with a trend for them to be located in the middle to upper reaches of the main rivers in the region, contrasting with the pattern in the early Bronze Age. This dislocation of the pattern of metalwork finds from the lower Mersey to the upper reaches of the rivers in the later Bronze Age is interpreted as a response to the worsening of conditions on the lower ground (Davey 1976).

It is during this period of the late Bronze Age, however, that the pattern of small, fairly widespread, but infrequent woodland disturbances alters in the western part of the Basin. The areas around the Merseyside central mosslands were for the first time the focus of more concerted farming activity during the early to mid-3rd millennium BP. This represents the first relatively substantial clearance of woodland in these areas and the best indications of the adoption of mixed agriculture (Cowell & Innes 1994). In the east of the Mersey Basin, however, such activity is not noticed until nearer the end of the millennium (Howard-Davis *et al.* 1988), although small scale, low impact, short lived activity is still seen

in some areas of the Pennine fringes preceding this (Tallis & Switsur 1973).

The first substantial farm settlements became established during the 3rd millennium BP, although close dating for them is poor at the moment. They are in the form of oval or sub-rectangular ditched enclosures, mainly confined to the Triassic sandstone belt in the Wirral and southern half of Merseyside (Philpott 1994), extending to the middle reaches of the Mersey valley (Nevell 1989a), and across central Cheshire (R. Philpott pers comm., Figure 4.4). They may be several centuries later (later Iron Age) than the early to mid-3rd millennium BP (late-Bronze Age/early-Iron Age) clearance horizon in the pollen diagrams of the western part of the Basin. The nature of the environmental changes seen at this horizon, the presence of contemporary metalwork and of sites such as the late-Bronze Age hill settlement at Beeston, Cheshire (Keen & Hough 1993) in the western part of the Basin, however, provide a context that hints that some ostensibly late Iron Age sites could be earlier. This would not accord with the argument for settlement dislocation towards the upper reaches of the main valleys in the Basin for climatic reasons.

Until, and if, part of the lowland settlement pattern of the local Iron Age can be shown to have its general origins in the period of worsening climate in the late-Bronze Age/early-Iron Age, the first identifiable archaeology of settled farmsteads in the lowlands of the Basin dates to the period towards the end of the 3rd millennium BP (late-Iron Age). Three main farmstead enclosures have been excavated in the Basin, at Great Woolden, Halewood, and Irby, all of which suggest that late-Iron Age settlement was, to differing degrees, succeeded by Romano-British occupation on the same site (Philpott 1993; Cowell & Philpott 1994; Nevell 1989b). This suggests continuity of settlement into the Romano-British period may have been a common feature of the area. It is at this time, beginning in the second half of the millennium, that the first concerted clearances associated with cereals are found in the eastern part of the Basin, as at Chat Moss and Lindow Moss (Nevell 1992) and on the fringes of the uplands (Tallis & Switsur 1973). This coincided with an improvement in climate, becoming warmer and drier (Lamb 1981). The pollen sites from the western part of the Basin also show clearance activity with some cereal evidence subsequent to the mid-3rd millennium BP, but none are dated absolutely.

The impact of the Roman conquest of Britain

The impact of the Roman legions on the rural landscape in the 1st century AD was directly limited to the network of roads through the region (Margary 1967) linking the major forts, towns and industrial settlements such as Wilderspool, Warrington, and other settlements such as at Wigan. A network of farmsteads existed across southern parts of the Basin, probably representing an increase in the density of the settlement pattern from the

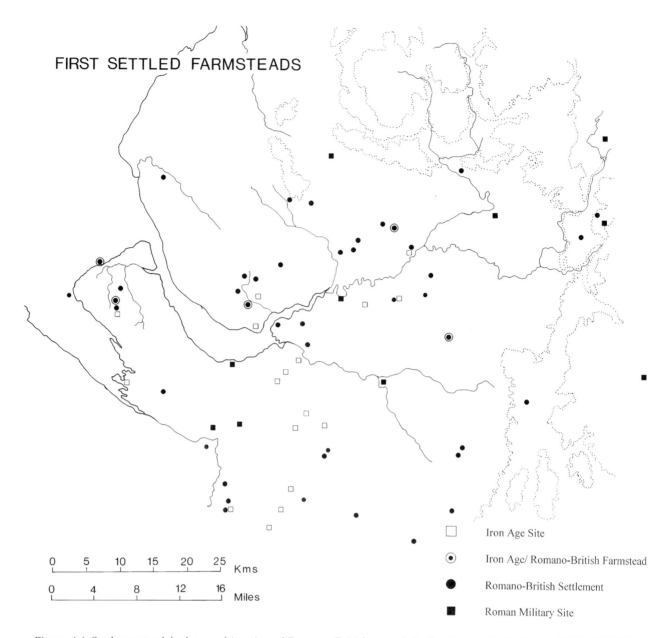

FIRST SETTLED FARMSTEADS

0   5   10   15   20   25   Kms

0   4   8   12   16   Miles

☐   Iron Age Site

◉   Iron Age/ Romano-British Farmstead

⬤   Romano-British Settlement

◼   Roman Military Site

Figure 4.4. Settlements of the late-prehistoric and Romano-British periods in the Mersey Basin, from c.3,000–1,500 BP.

Iron Age. The important coastal settlement of Meols, at the mouth of the Dee, may have acted as a focus for the distribution of a series of scattered farmsteads around the sandstone ridges in central Wirral (Philpott 1993).

The pollen evidence for this period is very limited but the Domesday Survey suggests that by the late-10th century AD woodland still covered large areas of the Basin. Cropmark enclosures are now recognised, however, on the heavier soils in the south-east of the Basin through programmes of aerial photography (Collens 1994; Philpott 1994; Nevell 1989b) and as earthworks in some foothill locations (Nevell 1992) which imply scattered clearings in the woodland were relatively common. Recent excavations have shown that even this distribution may be a great underestimate of the extent of Romano-British rural settlement in the area (Cowell & Philpott 1994).

The scale of clearance around each farmstead is not clear for many sites. At Featherbed Moss in the eastern foothills (Nevell 1992; Tallis & Switsur 1973) relatively substantial clearance associated with cereal farming, sustained over several centuries, is attested.

In the coastal, and sandy areas in the north and north-west of the Basin such sites are proving more difficult to locate, suggesting that areas formerly attractive to settlement had now become more marginal. The fact that some of the mosslands were acid raised bog at this time may have made some areas of the north central part of the Basin less attractive for mixed farming. The pollen diagram, however, from Knowsley Park Moss suggests clearance was taking place on an increased scale to that of the late-Bronze Age/early-Iron Age, with mixed farming being carried out somewhere in the vicinity at c.AD 300. Although the activity appears to be fairly intense, the extent of the clearance was probably limited (Cowell & Innes 1994, p. 129). Other sites may also have been placed adjacent to mosslands (Cowell & Innes 1994; Nevell 1989b).This suggests that mossland resources, perhaps unconnected with agriculture, were still being exploited but the reason for settlement location would seem to be the conjunction of a number of complementary resources, with sandstone and clay areas being more favoured for settlement than the lighter, previously attractive sandy soils. With the trend in the region for late prehistoric sites to be on the same locations as the Roman ones, this pattern may have much earlier origins.

The emergence of new settlement patterns *c.*AD 500–1,000

After the withdrawal of the Roman army from Britain in the early 5th century AD the period up to the Norman Conquest is often known as the Dark Ages. Direct evidence for settlement and landuse becomes very scanty, although it is a period when many important developments, visible in the archaeology of the post-Conquest period became established (chapter five, this volume). The details of these developments are, however, not well understood in the region. The best evidence is the Domesday Survey from which it is possible to extrapolate backwards to identify trends and developments.

By the late medieval period (11th–15th century AD) the landscape witnessed several major changes. Some areas, particularly where sandstone dominates the local geology, had developed what is regarded as the typical medieval landscape of townships (territories made up of several estates), large communal arable fields, and smaller, privately owned estates, nucleated hamlets or villages alongside isolated settlements, increasing clearance of woodland, and the growth of scattered towns (chapter five, this volume). One of the main problems of the pre-Conquest period is to identify to what extent these major developments had originated during this period.

The earliest Saxon (pre-Conquest) settlers may have reached the area during the 7th century from Cheshire (Thacker 1987) with many of the place names belonging to the 'tun' form e.g. Huyton, Denton, etc., which is regarded as a possible later phase of settlement than the earliest post-Roman phase. The 7th and early 8th centuries also marked a fall in tree cover in parts of the south-east of the Basin (Nevell 1992) which may be seen as an expression of this expansion of settlement.

The densest pattern of Saxon place names, which probably include many pre-Conquest settlements, is found in the same areas where most of the Romano-British farmsteads are known. A pattern of land ownership and landuse based on large agricultural estates is generally associated with the Saxon period (Sawyer & Thacker 1987). Large areas were still heavily wooded, particularly in the north and the east of the Basin, and clearance and settlement of many of these areas only properly started a century or so after the Conquest (Cowell 1982; Lewis 1982), so that the landscape of many

Saxon estates would have included much woodland. Some areas, e.g., the large township of West Derby, probably retained their wooded status to function as royal late-Saxon estates which may have been retained for hunting (Shaw 1956). There are grounds for thinking that some Saxon estates may be related to Romano-British ones and that a degree of conservatism may have held in landuse and settlement pattern (Cowell & Philpott 1994), suggesting a degree of continuity in site location and perhaps landuse and land tenure in some locations. The only pollen diagram which covers this period, however, at Featherbed Moss, shows that regeneration of woodland took place during the 6th century AD after a period of Romano-British farming (Tallis & Switsur 1973). This was also a period when the climate deteriorated, with increasing rainfall and a decline in average temperatures between the 5th and 7th centuries AD.

For those townships in the medieval period which had developed a mixed landscape including settlement nucleation and communal resources, the major problem is to define what social, economic or even political conditions lay behind these changes and at what rate the process took place. Centres of population with parish churches such as Huyton, Childwall or Prescot, in Merseyside may be examples of a relatively widespread trend, which is likely to be pre-Conquest in date associated with the introduction of Christianity into the area after the 6th or 7th centuries AD (Cowell 1982). Small townships on the Wirral suggest that by the time township boundaries came to be defined in the landscape, certainly by the late medieval period (11th–15th century AD), there was greater competition for land between communities here than to the east of the River Mersey, but there is no evidence that this had led to earlier general nucleation in the Wirral. Townships with present-day villages with Scandinavian names, such as Irby, Greasby, or Frankby, are likely to have received an influx of settlers by way of Ireland in the 10th century (Chitty 1978). Again there is no evidence to show whether this influx was a contributory factor to the growth of nucleation or whether it was assimilated into a pre-existing dispersed or nucleated pattern.

Although a number of townships in the Mersey Basin retained a pattern of dispersed settlement during the late-medieval period, which might reflect land organisation from a much earlier period in some cases, many more developed a form of common field system which would have transformed the appearance of the landscape. In its classic form this was a late-Saxon feature of midland and eastern England, but to what extent it developed in a pre-Conquest context in north-western England is unclear. There is evidence, though, that some common systems may have developed quite late in the late-medieval period (chapter five, this volume).

Away from the belt of potential early settlement, in the uplands and in large areas of the lowlands, particularly around the mossland in the north and east of the Mersey Basin, land remained largely undeveloped and marginal into the medieval period. This contrast conditioned the appearance of parts of the landscape well into the post-medieval period, in a way retaining some small echo of its prehistoric past.

Thus, the landscape pattern for approximately 9,000 years before the Domesday Survey was one rooted in woodland, fen, swamp, peat bog, and long sweeps of coast. Early human intervention in this environment, due to low population pressure and the sustainable nature of the human activity, meant that the effects were limited both in scale and distribution. The vegetation in some areas, such as the coasts or the uplands, may have been more seriously affected during this time, but this is often likely to have been the result of the broad natural changes taking place with human interference contributing either to speed up or extend the process. Even when the nature and density of settlement changed, c.3,000 years ago, for the first half of this period the natural landscape would not have been unfamiliar to a hunter of the Mesolithic. It is not until the post-Roman period that the pattern of settlement and landuse recognisable today probably started to be defined, but those links are still very obscure and many areas of the Mersey Basin would still have been dominated by woodland and moss. It is only with the population rise of the late medieval period that many areas of the Mersey Basin were developed in a way that still has an impact on the nature of today's landscape.

References

Birks, H.J.B. (1964). Chat Moss, Lancashire. *Memoirs and Proceedings of the Manchester Literary and Philosophical Society*, **106**, 22–45.

Birks, H.J.B. (1965). Pollen Analytical Investigations at Holcroft Moss, Lancashire and Lindow Moss, Cheshire. *Journal of Ecology*, **53**, 299–314.

Barnes, B. (1982*). Man and the Changing Landscape*. Merseyside County Museums,University of Liverpool Department of Prehistoric Archaeology, Liverpool. Work Notes, **3**. Merseyside County Council.

Bradley, R. (1984). *The Social Foundations of Prehistoric Britain*. Longman, London.

Bu'lock, J.D. (1959). The Pikestones: A Chambered Long Cairn of Neolithic Type on Anglezarke Moor, Lancashire. *Transactions of the Lancashire & Cheshire Antiquarian Society*, **68**, 143–45.

Burgess, C. (1974). The Bronze Age. *British Prehistory* (ed. C. Renfrew), pp. 165–233. Duckworth, London.

Chitty, G.S. (1978). Wirral Rural Fringes Survey Report. *Journal of the Merseyside Archaeological Society*, **2**, 1–25.

Collens, J. (1994). Recent Discoveries from the Air in Cheshire. *From Flints to Flower Pots, Current Research in the Dee–Mersey Region* (ed. P. Carrington), pp. 19–25. Chester Archaeological Service Occasional Paper **2**, Chester.

Cowell, R.W. (1982). *Liverpool Urban Fringe Survey*. Unpublished report at Liverpool Museum, National Museums & Galleries on Merseyside.

Cowell, R.W. (1991). The Prehistory of Merseyside. *Journal of the Merseyside Archaeological Society*, **7**, 21–61.

Cowell, R.W. (1992). Prehistoric Survey in North Cheshire. *Cheshire Past*, **1**, 6–7.

Cowell, R.W. & Innes, J.B. (1994). *The Wetlands of Merseyside*. North West Wetlands Survey **1**, Lancaster Imprints 2, Lancaster.

Cowell, R.W. & Philpott, R.A. (1994). Excavations along the route of the M57–A562 Link Road, Merseyside, in 1993. Unpublished report at Liverpool Museum, National Museums & Galleries on Merseyside.

Cowell, R.W. & Warhurst, M.H. (1984). *The Calderstones: a Prehistoric Tomb on Merseyside*. Merseyside Archaeological Society, Liverpool.

Davey, P.J. (1976). The Distribution of Bronze Age Metalwork from Lancashire and Cheshire. *Journal of the Chester Archaeological Society*, **59**, 1–13.

Freke, D.J. & Holgate, R. (1990). Excavations at Winwick, Cheshire in 1980: 1. Excavation of two 2nd millennium BC mounds. *Journal of the Chester Archaeological Society*, **70** (for 1987–8), 9–30.

Hall, D., Wells, C., Huckerby, E., Meyer, A. & Cox, C. (1995). *The Wetlands of Greater Manchester*. North West Wetlands Survey **2**, Lancaster Imprints 3, Lancaster.

Hibbert, F.A., Switsur, V.R. & West, R.G. (1971). Radiocarbon dating of Flandrian zones at Red Moss, Lancashire. *Proceedings of the Royal Society of London B*, **177**, 161–76.

Howard-Davis, C., Stocks, C. & Innes, J. (1988). *Peat and the Past*. Lancaster University, Lancaster.

Jacobi, R.M. (1980). The Early Holocene Settlement of Wales. *Culture and Environment in Prehistoric Wales* (ed. A.J. Taylor), pp. 131–206. British Archaeological Reports (British Series) 76, Oxford.

Jacobi, R.M., Tallis, J.H. & Mellars, P. (1976). The Southern Pennine Mesolithic and the Ecological Record. *Journal of Archaeological Science*, **3**, 307–320.

Keen, L. & Hough, P. (1993). *Beeston Castle, Cheshire, a report on the excavations 1968–85*. Archaeological Report no. **23**. Historic Buildings and Monuments Commission, London.

Lamb, H.H. (1981). Climate from 1000 BC to 1000 AD. *The Environment of Man: the Iron Age to the Anglo Saxon Period* (eds M. Jones & G. Dimbleby), pp. 53–65. British Archaeological Reports (British Series) **87**, Oxford.

Lewis, J.M. (1982). *Sefton Rural Fringes Survey Report*. Unpublished report at Liverpool Museum, National Museums & Galleries on Merseyside.

Longley, D.M.T. (1987). Prehistory. *A History of the County of Chester*, Vol. I (eds B.E. Harris & A.T. Thacker), pp. 36–114. The Victoria History of the Counties of England, published for the Institute of Historical Reseach by Oxford University Press, Oxford.

Margary, I.D. (1967). *Roman Roads in Britain* (revised edn). John Baker, London.

Mellars, P. (1976). Fire Ecology, Animal Populations and Man: a Study of Some Ecological Relationships in Prehistory. *Proceedings of the Prehistoric Society*, **42**, 15–45.

Nevell, M.D. (1989a). An Aerial Survey of Southern Trafford and Northern Cheshire. *Greater Manchester Archaeological Journal*, **3** (for 1987–8), 27–34.

Nevell, M.D. (1989b). Great Woolden Hall Farm: Excavations on a Late Prehistoric/Romano-British Native Site. *Greater Manchester Archaeological Journal*, **3** (for 1987–8), 35–44.

Nevell, M. D. (1992). *Tameside Before 1066*. Tameside Metropolitan Borough Council, Manchester.

Philpott, R.A. (1993). A Romano-British Farmstead at Irby, Wirral, and its Place in the Landscape: An Interim Statement. *Archaeology North West: Bulletin of Council for British Archaeology North West*, **5**, 18–24.

Philpott, R. A. (1994). New Light on Roman Settlement: Recent Aerial photography in Cheshire. *Cheshire Past*, **3**, 6–7.

Sawyer, P. H. & Thacker, A.T. (1987). The Cheshire Domesday. *A History of the County of Chester*, Vol. I (eds B.E. Harris & A.T. Thacker), pp. 293–370. The Victoria History of the Counties of England, published for the Institute of Historical Reseach by Oxford University Press, Oxford.

Shaw, R. C. (1956). *The Royal Forest of Lancaster*. The Guardian Press, Preston.

Smith, C. (1992). *Late Stone Age Hunters of the British Isles*. Routledge, London.

Stonehouse, P. (1989). Mesolithic Sites on the Pennine Watershed. *Greater Manchester Archaeological Journal*, **3** (for 1987–8), 5–17.

Stonehouse, P. (1994). Mesolithic Sites on the Pennine Watershed Part II. *Archaeology North West: Bulletin Council for British Archaeology North West*, **8** (Vol. 2, Part II), 38–47.

Thacker, A. T. (1987). Anglo Saxon Cheshire. *A History of the County of Chester*, Vol. I (eds B.E. Harris & A.T. Thacker), pp. 237–292. The Victoria History of the Counties of England, published for the Institute of Historical Reseach by Oxford University Press, Oxford.

Tallis, J. H. & McGuire, J. (1972). Central Rossendale: the Evolution of an Upland Vegetation. *Journal of Ecology*, **60**, 721–51.

Tallis, J. H. & Switsur, V.R. (1973). Studies on southern Pennine Peats VI. A radiocarbon dated pollen diagram from Featherbed Moss, Derbyshire. *Journal of Ecology*, **61**, 743–51.

Williams, C. T. (1985) *Mesolithic Exploitation Patterns in the Central Pennines, A Palynological Study of Soyland Moor.* British Archaeological Reports (British Series), 139, Oxford.

Wymer, J.J. & Bonsall, C. J. (1977). *Gazetteer of Mesolithic Sites in England and Wales.* Council for British Archaeology: Research Report, **20**, London.

SECTION TWO

The Tide of Change

J.F. HANDLEY

The first section of this volume concluded that, for about 9,000 years before Domesday, the landscape pattern of the Mersey Basin was 'one rooted in woodland, fen, swamp, peat bog and long sweeps of coast'. Human influence is there to see but these landscapes were essentially shaped by natural processes. This section shows that the tide of change was to quicken dramatically during the next millennium when landscape change was increasingly driven by cultural influences.

Chapter five documents the influence of agriculture on the landscape beginning with widespread forest clearance to feed a steadily growing population. Climate change and disease in the 14th century caused social and economic dislocation and with it a switch in land-use from arable to pastoral. As population growth resumed urban centres began to take shape within an increasingly 'enclosed' landscape of hedgerows and dikes.

The 18th century witnessed further profound change with the coming of the industrial revolution and the emergence of a modern economy powered by fossil fuel rather than wind, water and biomass. The new industrial prosperity brought with it major environmental impacts. The consequences of this quickening 'tide of change' are documented in contrasting accounts of what remains today of the original natural habitats and what new opportunities for wildlife were created by industrialisation. The balance sheet shows significant losses but also some surprising gains.

The final contribution analyses the most recent evidence for land-use change and suggests that the period since the 1970s has been one of relative stability. However, what is striking is the paucity of systematically gathered information, which is geographically specific and sufficiently fine-grained to illustrate landscape dynamics in the Mersey Basin today. At the end of a millennium which began with the Domesday survey there is an urgent need to remedy this deficiency.

CHAPTER FIVE

The human influence: the Norman Conquest – Industrial Revolution

R.A.PHILPOTT AND J.M.LEWIS

Introduction

In the mid-17th century Lancashire was described as a 'close county full of ditches and hedges' (Walker 1939, p.72), a description which might equally well have been applied to Cheshire. The landscape by that time was an intensely partitioned and virtually entirely man-made one. The process by which such a landscape developed across these two counties by 1700 involved a combination of factors, social, demographic, climatic, and economic, through the preceding 600 years.

The landscape at Domesday (AD 1086)

The pattern of human settlement in the Mersey Basin by the late Anglo-Saxon period consisted largely of single farms and small hamlets. At Domesday the region had only one town, Chester, although the Cheshire salt 'wiches' should perhaps qualify as urban, and there appear to have been few villages in the classic Midland sense of a nucleated settlement with parish church and manor house. Many places which are mentioned in the Domesday Book and which had become villages by the 16th or 17th century when the earliest maps become available were probably no more than single farms or small clusters of dwellings in 1086: Domesday surveyed resources and estates not villages (Sawyer & Thacker 1987). Here, as elsewhere in northern England, the majority of villages may well have developed in the post-Conquest period but nucleated settlements certainly did become established at some places which were settled early, often because they were locally important as parish or administrative centres. They are often dry-point sites set on sandstone rises, with good potential for arable land in the vicinity, and their place-names often contain classic Saxon or Norse habitative elements, such as -tūn (Figure 5.1), -ham, -bury, or -by (Gelling 1978, p. 143).

The distribution of settlement in the early period is broadly indicated by the location of estates and their inhabitants in the Domesday Book. Domesday refers to, without necessarily naming, nearly 450 places in southern Lancashire and Cheshire (Figure 5.2), although some places contained more than one estate. The Domesday population of both Cheshire and Lancashire was low with perhaps under 10,000 in each county. Overall, Domesday records a low population of 1,524 individuals for Cheshire in 1086 (Terrett 1962a, p. 350) – to be multiplied by a factor of four or five for actual population (i.e., approximately 6,000–7,500) and it has been claimed that the county was the poorest of the Marcher Shires of England competing only with Derbyshire for the lowest population density in the region (Higham 1993, p. 203). The most densely occupied areas of Cheshire were Wirral and the Dee valley with perhaps three to four persons per square mile (1 square mile = 2.59 square kilometres); by contrast, the Pennine slopes of east Cheshire were very sparsely populated with only one person per square mile and some uninhabited vills. There are also significant gaps such as the Kingsley area in what was later to become the Royal Forest of Delamere and Carrington Moss in northern Cheshire where woodland, poor soil or mossland were unfavourable to agriculturally-based settlement. The information for Lancashire is much less detailed but the estimated population of 1,780 individuals for the region between the River Ribble and the River Mersey suggests a density of no more than two people per square mile and a total of approximately 7,000–8,700 once multiplied (Terrett 1962b, p. 407). Here the exceptional size of the ecclesiastical parishes and of their component townships demonstrates both the low population and the poor quality of the land, particularly where they embraced tracts of uninhabited woodland or mossland.

Domesday allows us to observe the effect of one major historical event which had a serious, if temporary, impact on the landscape. The 'harrying of the North', the campaigns of 1069–70 when William the Conqueror's army brought the Midlands and the North under the full control of the king, seems to have been deliberate devastation as reprisal for resistance to Norman rule (Sawyer & Thacker 1987, pp. 336–37; Higham 1982, p. 20). The result can be seen in the large number of vills which are still recorded in 1086, some seventeen years later, as 'waste', i.e., land which had gone out of cultivation. Eastern Cheshire appears to have been particularly hard-hit, with numerous underpopulated or deserted settlements, the west of the county recovered quickly, while

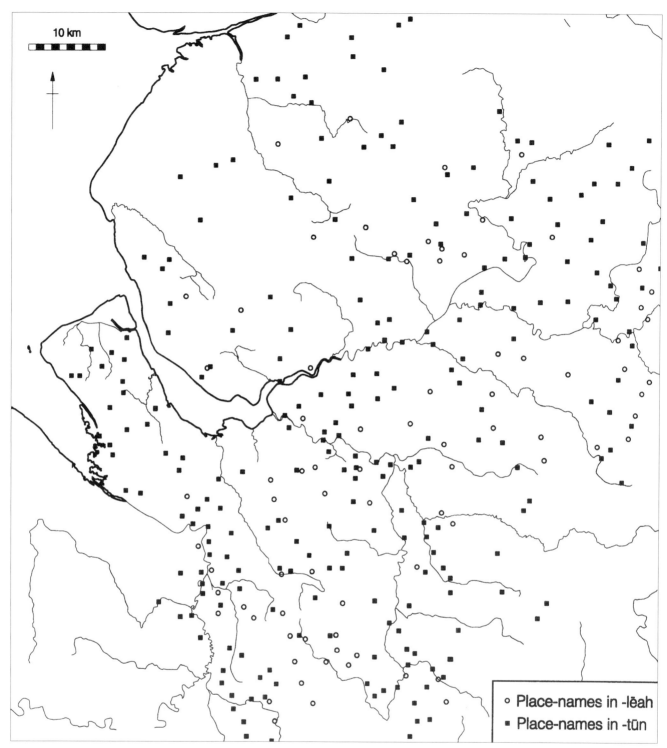

Figure 5.1. Place-names containing Anglo-Saxon elements -*tūn*, indicating settlement, and -*lēah*, indicating woodland clearance (after Kenyon 1989).

southern Lancashire seems to have escaped the worst effects of the destruction of 1070.

Already by Domesday the region had developed a complex subdivision of estates and landholdings and it is the township which provides the physical and administrative framework for the organisation of the landscape. Townships had developed in the late Anglo-Saxon period and are often, although by no means always, identical with the manor. The township was the 'economic area' capable of supporting a settlement or group of hamlets (Roberts 1979, p. 77). Each township had a variety of resources, managed for the benefit of the lord or lords and his tenants, usually consisting of variable proportions of waste, woodland, meadow and arable. Regulation by the manorial court ensured the equitable distribution of different types of land or access to it and defined the rights and obligations of the inhabitants. Neighbouring townships could share resources on their mutual margins such as mossland and woodland.

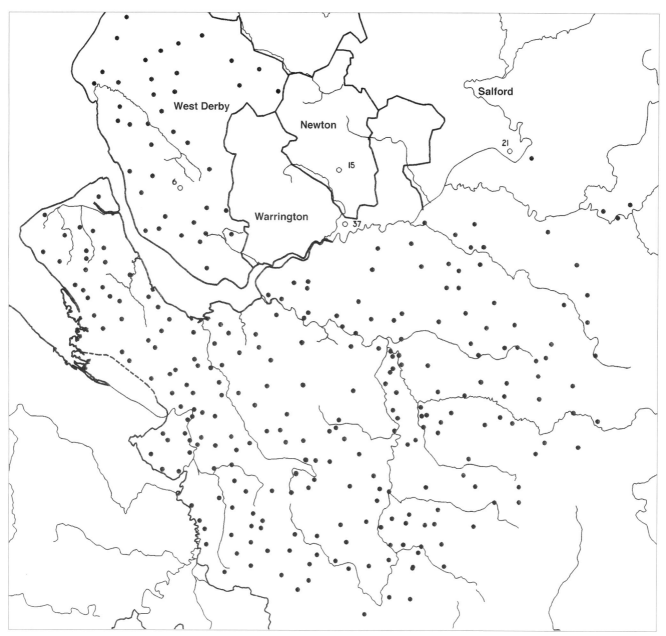

Figure 5.2. The location of settlements named in Domesday Book 1086; hundredal centres north of the Mersey have the total of unnamed places (after Terrett 1962).

The agricultural systems like the settlements themselves have a distinctive regional character. Discrete estates farming their own land in severalty from a single farmstead were common, and in this fairly conservative area with low population pressure they probably represent the descendants of an ancient form of land organisation and settlement pattern which may trace its ancestry from Romano-British enclosed farmsteads. In many places at Domesday it is likely that the population was too small and draught animals too scarce to allow the development or introduction of two- or three-field common arable rotation on a strip basis (Higham 1993, p. 203). Such a system may have come in with increasing population from the later 11th century onwards, but it was by no means universal, and even neighbouring townships might develop very different agricultural systems through circumstances of ownership or early

land-use (Figure 5.3). In form the Mersey Basin open field developed elements quite distinct from the well-known Midland-type system (Sylvester 1950, 1957, 1959; Chapman 1953; White 1983, 1995; Youd 1962; Chitty 1978; Cowell 1982; Lewis 1982, 1991). In Cheshire and Lancashire common arable, for example, did not necessarily have equal shares for all tenants in all fields, but fields could be intermixed between small numbers of tenants. Labour services too were light by comparison with the Midlands and were frequently and early commuted to cash payments. In aspect, a township might contain a diverse range of open fields, enclosed crofts farmed in strips, and crofts held in severalty in proportions that varied from township to township.

The location of townships with an open field system is closely matched by the distribution of early settlement. The settlement pattern was itself determined in part by

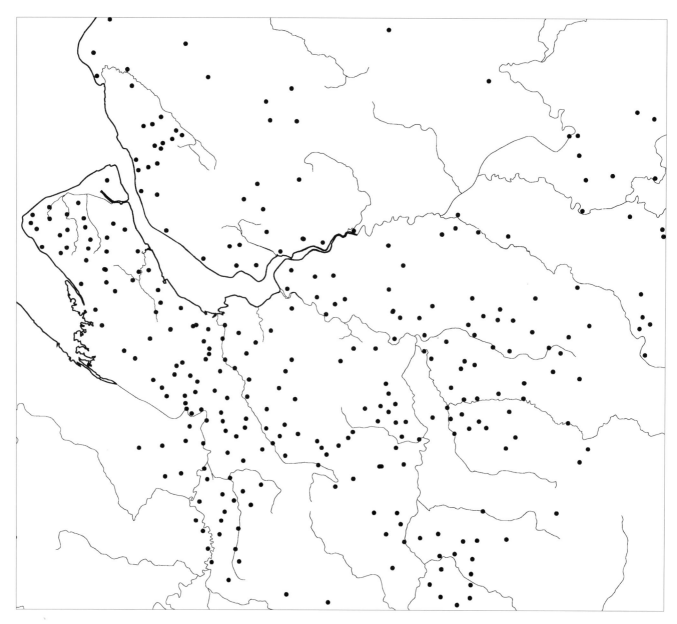

Figure 5.3. The location of open field systems, indicating organised common agriculture
(after Sylvester 1963; Morris 1983, and unpublished sources).

the presence of favourable soils, some derived from Shirdley Hill Sands, but also over sandstone and boulder clay (Shaw 1956, p. 296; Youd 1962, p. 5). In Cheshire, at Domesday, the land was understocked with ploughteams, with on average only half the teams that the land could have supported being used, but there are marked differences between the west and east. The density of teams in 1086 is ten times higher in Wirral and the Dee valley than the east and north-east of the county, with a belt of intermediate density in central and northern Cheshire. There is a striking correlation between those areas where Domesday has a high density of ploughteams and those which subsequently show some evidence of open arable cultivation. West Cheshire, the anciently-settled Dee valley and Wirral, were largely dominated by open field agriculture while in central Cheshire a belt of common arable field townships was bordered to east and west by the forests of Macclesfield

and Delamere 'in which pastoralism predominated and individual assarts were more numerous than common field strips' (Elliott 1973, p. 50).

North of the River Mersey, the western and central part of West Derby hundred consisted of an extensive arable belt in which nucleated settlements farmed open fields. Usually on the margins of the townships were areas of carr or marshes for meadow and grazing (Shaw 1956, p. 345). Inland from the coastal arable belt was an extensive woodland region extending south from Burscough to West Derby, Upholland, to the Prescot area. In the east of Lancashire open fields occur mostly in river valleys (Youd 1962) while few of the mossland townships of Halsall, Kirkby and Simonswood, had common arable fields (Tables 5.1 and 5.2).

In the early post-conquest period many communities practised a mixed agricultural regime but the precise balance depended on local circumstances and condi-

Township	Arable	Oxen/cows	Sheep	Goats	Horses	Pasture	Pigs	Wood	Meadow	Marsh	Fishery
Great Crosby	•									•	
Formby	•	•									•
Hale/Halewood	•	•			•		•	•	•		•
West Derby	•						•		•		
Toxteth					•	•	•				•
Downlitherland	•										
Widnes		•			•				•		
Little Crosby	•										
Poulton with Fearnhead	•	•				•	•		•		•
Woolston with Martinscroft		•				•	•	•	•		•
Altcar	•	•	•		•		•		•		
Aigburth	•						•		•		
Kirkdale	•								•		•
Speke	•		•			•	•				
Ravenmeols	•	•	•		•	•				•	•
Ainsdale	•		•				•		•	•	•
Garston	•	•	•								•
Bootle	•						•		•		
Ince Blundell	•	•	•							•	
Rixton with Glazebrook	•										•

Table 5.1. Summary of medieval land-use (coastal and riverine settlements north of the River Mersey).

tions. Livestock were needed for various purposes, as draught or riding animals, for meat or wool. Some areas through their particular geographical or climatic situation were ideally suited to animal husbandry. The south Lancashire coastal belt has a strong emphasis on sheep farming in the monastic grants of the 13th century, while the woodland belt including Melling in Merseyside and the forests of Delamere and Macclesfield in Cheshire have frequent mention of pannage for pigs and a wider range of stock.

Clearance and the expansion of agriculture

Any expansion of arable land took place at the expense of the waste and woodland. Woodland is considered in more detail elsewhere in this volume (chapter ten) but some points with a direct bearing on the landscape development are discussed briefly. At Domesday, woodland was extensive in the Mersey Basin although it is not recorded consistently. Well-wooded areas included that east of St Helens (Newton hundred with its measure of woodland larger than the hundred itself and probably including Warrington's assessment), the Melling/ Lydiate area, Delamere and in eastern Cheshire, in Macclesfield Forest. Place-names, often dating to the later Saxon period, confirm their generally wooded nature (Figure 5.1). However, much woodland had probably been cleared already. In some areas, notably the west of Cheshire, very little is mentioned and early clearance names are scarce. By the Norman Conquest, forest is already recorded in West Derby, and in the later 11th century extensive areas of land in both Lancashire and Cheshire (forests of Lancaster, Wirral, Macclesfield and Delamere) were designated as Forest and reserved for royal hunting. They were subject initially to draconian forest law, which imposed severe restrictions on clearances, but once these were relaxed by the 13th-century arable expansion and settlement do not seem to have been unduly impeded.

Progressive clearance eroded the extent of woodland, and by the 16th century commentators lament the scarcity of timber. Replanting took place under the initiative of individual landowners in the 17th and 18th centuries, but much of the woodland which can be seen on the 18th- and 19th-century maps of the region is post-medieval plantation.

The clearance that made such inroads into the woodland and waste of the Mersey Basin continued a process which was under way in the late Saxon period. Through the late 11th to mid-14th century the landscape saw a series of major developments. The main factor was the rapid growth in the population. England's population is thought to have increased from 1.1 million in 1086 to 3.3 million just before the Black Death in the mid-14th century (Beresford & Hurst 1989, p. 7), but there are no means of calculating accurately population growth either in Lancashire or in Cheshire. The rise of 360% at Burton in Wirral between 1086 and the early 14th century suggests the national pattern was mirrored here (Booth 1981, pp. 2–3). The population growth led to increased demand for agricultural land. This was met in two main ways, expanding the open arable fields at the expense of the waste or marsh lying immediately around them, or

Township	Arable	Oxen/cows	Sheep	Goats	Horses	Pasture	Pigs	Wood	Meadow
Burtonwood	•	•	•		•		•	•	•
Rainford							•		
Fazakerley	•								
Aughton/Litherland								•	
Walton-on-the-Hill	•	•	•			•	•	•	•
Knowsley	•					•			
Huyton	•					•		•	
Kirkby	•							•	
Little Woolton	•	•	•	•	•		•	•	
Roby	•								
Sutton	•						•		
Maghull							•	•	
Tarbock							•		
Eccleston	•	•				•	•	•	
Cronton	•								
Newton	•								•
Dalton	•						•	•	
Allerton	•	•	•						
Sefton, Netherton, Lunt	•								•
Thornton	•								•
Whiston	•								
Billinge	•						•	•	
Bickerstaffe	•							•	
Upholland							•	•	
Haydock	•						•	•	•
Lathom	•						•	•	
Melling	•						•	•	
Bold	•					•	•	•	•
Hurlston	•					•			•
Great Sankey	•					•		•	•
Lydiate			•	•	•		•	•	
Windle							•		

Table 5.2. Summary of medieval land-use (inland settlements north of the River Mersey).

clearance of more extensive pockets of woodland or waste around the margins of the townships. Large areas were still being cleared into the early 14th century, with monastic houses who were amongst the largest landowners in the counties active in the forefront of the movement. The cleared land might be added to the open fields or farmed in severalty as private estates. Numerous separate estates were created during the 12th–14th centuries, often on the edge of townships, as a reward for services to overlords, or for the younger sons of manorial families. Common arable field systems probably developed for the first time in many areas during this period, although they are found by no means everywhere (Figure 5.3). In addition, areas of the region, which were uninhabited or sparsely populated were colonised; eastern Cheshire may have been settled partly through an act of policy by Norman earls of Chester.

The impact on settlement too was marked. Existing rural settlements increased in size and population but new subsidiary settlements were established on the margin of existing townships. New chapels of ease were created to serve distant areas of large parishes. During the 13th and earlier 14th centuries the increasing need for cash by peasants to pay rents and rise in cash- rather than service-based transactions stimulated trade and exchange, which led to the development of towns (Philpott 1988). Market and borough charters were granted to a network of places, often parish or administrative centres, where a scattered population took the opportunity to trade, but in a few cases, such as Tarbock, centrally within a dispersed non-nucleated settlement. During the late medieval period economic downturn and decline in trade ended the burghal status of a number of small market towns, some failing altogether like Hale or Roby, but population increase in the more favourably located and larger towns ensured their survival through the post-medieval period. The influx of rural poor in the 16th and 17th centuries who suffered hardship from static wages and rising prices swelled the local populations of specialist craftsmen and artisans, whose dependence on agriculture was diminished if not completely broken.

Throughout most of the medieval and early post-medieval period, towns played a minor role in the development of the landscape. They were small in extent, Liverpool by the 1660s having no more than seven streets, and in many cases retained a strong agricultural component until well into the post-medieval period. Early importance was no guarantee of continued success. A hundredal centre at Domesday such as West Derby might be overshadowed as early as the 13th century by a more prosperous or better situated neighbour (Liverpool). Conversely, in the industrial and commercial expansion of the 18th or 19th centuries minor agricultural settlements could grow into major modern towns such as Birkenhead or St Helens.

During the 13th–14th centuries there was a rise in the conspicuous marks of status within the landscape such as construction of moated houses and the creation of hunting parks for manorial use. The park was the preserve of the local landowners, and the region follows the national pattern in seeing an upsurge in their creation in the period 1200–1350. In the later medieval period, the expense of upkeep proved too great for many landowners and parks were often turned over to pasture (Cantor 1983, p. 3). During the 16th to mid-17th centuries many parks were enclosed and farms created within their bounds (Cantor & Hatherly 1979, p. 79). However, in some cases existing medieval parks were greatly expanded in the late medieval period and hunting parks became more ornamental and recreational by the 17th century. Large landowners often landscaped parks with follies and pools. Many parks were declining in the later medieval period and were often converted to grazing or even arable and sold off. Others such as Knowsley survived into the post-medieval period to become amenity parks, landscaped as the graceful setting for large country houses.

The limits of arable expansion

Settlement expansion and population growth probably reached its height in the late 13th century and continued into the early 14th century. However, the population growth began to slow down early in the 14th century as a result of a combination of sudden climatic downturn, a series of disastrous harvests, outbreaks of disease amongst livestock, and social unrest. A more severe blow was to follow with the devastation of the bubonic plague in 1348/9, and further outbreaks later in the same century. The impact of the Black Death is hard to gauge accurately but nationally the population fell by an estimated 25–40% (Smith 1992, p. 209). Labour became scarce and land plentiful. Prices declined, arable cultivation became less profitable, and demesnes were leased out by lords of the manor. Pastoral farming which was less labour-intensive gained ground rapidly, although it

had always been strong in certain areas, and a new emphasis in documents can be traced on the values for pasture land and herbage where previously arable commanded high prices.

Climatic change had an impact on low-lying coastal regions. Locally-experienced storms and periods of sea-level change had some effect on settlement patterns and farming practices in the coastal and riverine areas. Periods of dune instability at a time of low sea-level contrast with those of high sea-level and a consequent increase in the water table; in general terms each would affect the local community, the former by impoverishing if not obscuring cultivable soils and the latter by reducing hard-won reclaimed mosslands and marshlands to wet, summer pasture. In the early 13th century land was lost to the sea on Wirral and at Ince on the Mersey Estuary (Hewitt 1929, p. 5), and the ancient coastal settlement at Meols may also have been a consequence of inundation rather than long-term depopulation (Chitty 1978, p. 21). On the south Lancashire coast Argarmeols (Birkdale), had disappeared due to inundation by 1346 and the amount of arable land in Ravenmeols, a little further south, had been reduced by 1289 (Lewis 1982, pp. 61–62). The Cistercian abbey at Stanlow, founded on the Mersey marshes in 1178, was eventually moved to Whalley, in Lancashire, in 1296 after several inundations. Further episodes of severe flooding and encroachment were recorded along the Mersey marshes and Formby coast during the 14th and 15th centuries (Greene 1989, p. 31; Hewitt 1929, p. 5 note 11; Lewis 1982, p. 62).

Depopulation resulting from the Black Death hastened rather than initiated a process which was under way by the early 14th century towards enclosing the landscape by hedges and ditches. Enclosures were versatile; they could be used for both arable cultivation and control of livestock. Assarts farmed in severalty were often enclosed from the outset, but increasingly too there was a move towards the exchange of strips in outlying fields in a township to create consolidated blocks. The limited numbers of owners or tenants in the Lancashire and Cheshire fields and the low degree of intermixing within individual fields meant that agreement was easily reached. The process in the region was largely achieved by the 16th century, without the need for Parliamentary Enclosure Acts found in the Midlands. Nor was there the misery of clearance from the land of the poor which caused such hardship and desertion of settlements elsewhere.

The post-medieval landscape

By the 16th century, much of the region was under grass, and arable cultivation was practised only as required to feed the family and animals. In the north of Cheshire, cattle rearing and fattening dominated whilst the south and west of the county, with the exception of Wirral, was given over to dairy farming. The hinterland of the post-medieval towns were increasingly dominated by pasture set aside to supply animals within them.

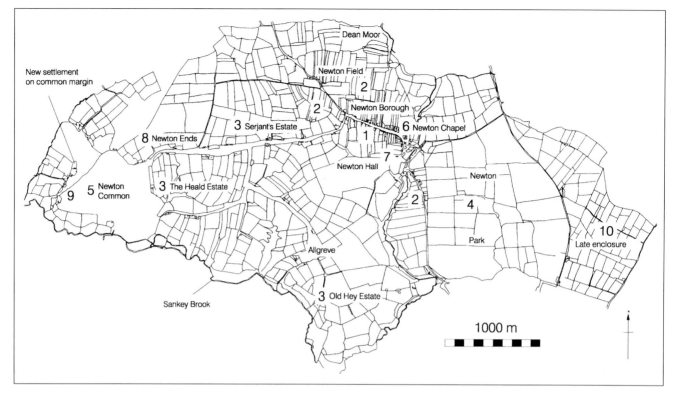

Figure 5.4. Newton-in-Makerfield (now Newton-le-Willows) in 1745.

In 1745 the township retained fossilised elements of the medieval landscape.
1. Decayed nucleated settlement (the borough, with market charter of 1257).
2. Relict open fields, including a townfield.
3. Three medieval estates recorded in the 14th century (two discrete, one scattered).

4. Former hunting park, in existence by 1322.
5. Unenclosed common.
6. Chapel, recorded by 1284.
7. Manor house, recorded by 1465.
8. Subsidiary hamlet, Newton Ends.
9. New settlement on the margin of the common.
10. Late rectilinear enclosure of mossland.

Throughout the late medieval period the exchange and consolidation continued so that by the 16th century the landscape of the Mersey Basin had acquired its 'closed' aspect (Figure 5.4). New farmsteads were constructed in the blocked holdings and a new tendency developed towards dispersal of farms around the margins of townships. This was strengthened by the population rises of the 16th and 17th centuries: with the first reliable figures in the 1560s Cheshire and southern Lancashire together numbered 112,000, but over the next century grew rapidly at near the national average rate to 182,000 (Phillips & Smith 1994, p. 5). By 1500 the landscape still contained much unused or under-used land, with one estimate of extent of waste at the end the medieval period at nearly 50% (Rodgers 1955, p. 88). This came under renewed pressure from the rising population, leading to the taking into cultivation of increasingly marginal land – the moor, waste, moss and forest – and the creation of cottages on the edges of the much-eroded common land.

Alongside the change in the physical disposition of the landscape came agricultural improvements in the drainage and reclamation of land, in its fertilisation and in the increasing specialisation of production. The post-medieval period saw the continued reclamation and cultivation of the mosslands, which had their beginnings in initiatives of the monastic houses in the Sefton area as early as the 14th century. The main efforts in the low-lying marshy western portion of the south Lancashire plain required large-scale engineering works of the 18th century, when Martin Mere was finally drained and the seasonal flooding of the Alt was improved by the Alt Drainage Act of 1779 (19 Geo III, Cap. 33). It was only in the 19th century that arable management of these poorly-drained lands behind the coastal sand dunes became economically viable. Other small wetland areas in Cheshire and Lancashire such as Newton Carr in Wirral, were drained, surveyed and partitioned with character-istic straight boundaries in the 18th and 19th centuries. The creation of long narrow strips called 'moss rooms' for turbary demonstrates the increasingly tight manage-ment of the resource, while drainage provided valuable additional arable land to help sustain the continuing population increase of the later 17th-18th centuries. However, some mosses, notably those of northern Merseyside, remained unreclaimed by the mid-19th century (Figure 5.5).

The use of marling to improve soil fertility on

Figure 5.5. Yates' map of Lancashire 1786, showing the developed post-medieval landscape, with a network of isolated farms, small nucleated villages, parks and extensive mosses surviving in the Simonswood-Kirkby area. Liverpool is rapidly expanding but remains distinct from Everton.

cultivable lands and to bring less tractable lands into production is recorded at least as early as the 13th century, and it became important for improvement of reclaimed mosslands soils in the 18th century (Hewitt 1920, pp. 22–23; Elliott 1973, p. 60). The extraction has left its mark in the form of pits, often grouped together in boulder clay regions.

The industries which proved so important in the region in the 18th and 19th centuries had their origins in the post-medieval or medieval periods, but their impact on the landscape was minor and localised (Morris 1983, pp. 19–21). Extractive industries in the countryside such as coal-mining or clay digging for pottery production had begun in the medieval period, but are documented increasingly commonly from the 16th century. Like textile production, these industries initially appear to be a part-time activity for smallholders with a few livestock, but began to attract capital investment during the 17th century. Salt, which had been exploited from the Iron Age in Cheshire, during the period discussed here was always an urban industry with far-reaching trade networks. During the 17th century production was stimulated by demands of dairying and provision of salt food for shipping. The textile industry, too, is recorded from the medieval period but by the 17th century wool, cotton and linen trades were strongly represented in south-east Lancashire. The development of burgeoning industries, notably coal and cloth in southern Lancashire and Cheshire salt, was hampered by poor communications, although turnpike trusts and canals early in the 18th century (Phillips & Smith 1994, p. 85) began to make improvements to major routes.

By the beginning of the 18th century the landscape was heavily enclosed, pasture was dominant and Cheshire's dairying reputation was already firmly established. Many new farms were created on blocks of land amalgamated out of scattered medieval holdings. Part of the rapid population increase was accommodated in cottages encroaching on the reduced common land, the others in enlarged villages and towns, but the majority of the population continued to depend on agriculture for their livelihood.

Conclusions

At the outset of the period the Domesday Book demonstrates that the Mersey Basin had a low dispersed population living in individual farms or small hamlets, with few truly nucleated settlements, and a great under-exploitation of the agricultural potential. Three centuries of population increase resulted in colonisation of woodland and waste, and an expansion of the arable land by clearance. Disease exacerbated by climatic deterioration proved decisive in halting the population increase. The resulting social and economic disruption led to the demise of the feudal system and hastened incipient landscape changes. The distinctive regional variant of the open field system, which however was never universally adopted, began to decay and a shift in balance can be

detected in the late medieval period from arable towards pastoral farming, which was less labour-intensive and favoured enclosed landscapes. Enclosure and consolidation by exchange was common throughout the later medieval period so that much of the landscape by 1700 was held in separate holdings, although some towns and villages retained shrunken open fields. In some places, notably the towns where ownership was most complex, the fossilised relic of the medieval open fields, the town-field, survived long enough to be recorded in a growing number of estate maps. Major landowners at all stages could retard or instigate change but the region was characterised by a high proportion of freemen, often farming discrete estates, and in the post-medieval period by a rising yeoman class. Despite technical advances in both agriculture and industry, the impact on the landscape remained relatively localised and superficial throughout the period. The towns failed to make a serious effect on the agricultural landscape until the post-medieval period, when they stimulated demand for fodder and other produce, whilst preserving by extreme subdivision the common arable fields in their immediate environs. The region continued to depend heavily on agriculture for its livelihood but by 1700 the stage was set for the major changes of the Industrial Revolution which were to see an intensification of agricultural production to supply the markets of the expanding towns.

References

Beresford, M.W. & Hurst, J.G. (1989). *Deserted Medieval Villages*. Lutterworth, Guildford/London.

Booth, P.H.W. (1981). *The financial administration of the lordship and county of Chester 1272–1377*. Chetham Society, Manchester.

Cantor, L.M. (1983). *The Medieval Parks of England: A Gazetteer*. Loughborough University of Technology, Loughborough.

Cantor, L.M. & Hatherly, J. (1979). The Medieval Parks of England. *Geography*, **64**, 71–85.

Chapman, V. (1953). Open Fields in West Cheshire. *Transactions of the Historic Society of Lancashire and Cheshire*, **104**, 35–59.

Chitty, G. (1978). Wirral Rural Fringes Survey. *Journal of the Merseyside Archaeological Society*, **2**, 1–25.

Cowell, R.W. (1982). *Knowsley Rural Fringes Report*. Merseyside County Museums, Liverpool (unpublished report deposited in the Department of Archaeology and Ethnology, Liverpool Museum, National Museums & Galleries on Merseyside).

Elliott, G. (1973). Field Systems of North West England. *Studies of Field Systems in the British Isles* (eds A. Baker & R. Butlin), pp. 41–81. Cambridge University Press, Cambridge.

Gelling, M. (1978). *Signposts to the Past: Place-names and the history of England*. J.M. Dent and Sons, London.

Greene, P. (1989). *Norton Priory*. Cambridge University Press, Cambridge.

Hewitt, W. (1920). Marl and Marling in Cheshire. *Proceedings of the Liverpool Geological Society* Part 1, **13** (for 1919–20), 24–8.

Hewitt, H.J. (1929). *Medieval Cheshire – An Economic and Social History*. The Chetham Society, New Series, 88. Manchester.

Higham, N.J. (1982). Bucklow Hundred: the Domesday Survey and the Rural Community. *Cheshire Archaeological Bulletin*, **8**, 15–21.

Higham, N.J. (1993). *The Origins of Cheshire*. Manchester University Press, Manchester.

Kenyon, D. (1989). Notes on Lancashire Place-Names 2; the Later Names. *English Place-Names Society Journal*, **21**, 23–53.

Lewis, J. (1982). *Sefton Rural Fringes Survey*. Merseyside County Museums, Liverpool, (unpublished report deposited in the Department of Archaeology and Ethnology, Liverpool Museum, National Museums & Galleries on Merseyside).

Lewis, J.M. (1991). Medieval Landscapes and Estates. *Journal of the Merseyside Archaeological Society*, **7**, (for 1986–7), 87–104.

Morris, M.G. (1983). *The Archaeology of Greater Manchester. Volume I: Medieval Manchester: A Regional Study*. Greater Manchester Archaeological Unit, Manchester.

Phillips, C.B. & Smith, J.H. (1994). *Lancashire and Cheshire from AD 1540*. Longman, London and New York.

Philpott, R.A. (1988). *Historic Towns of the Merseyside Area: a study of urban settlement to c.1800*. Liverpool Museum Occasional Paper 3. National Museums & Galleries on Merseyside, Liverpool.

Roberts, B.K. (1979). *Rural Settlement in Britain*. Hutchinson, London.

Rodgers, H.B. (1955). Land Use in Tudor Lancashire: The Evidence of the Final Concords, 1450–1558. *Transactions of the Institute of British Geographers*, **21** (1955), 79–97.

Sawyer, P.H. & Thacker, A.T. (1987). The Cheshire Domesday. *A History of the County of Chester* Vol. I (eds B.E. Harris & A.T. Thacker), pp. 293–341. The Victoria History of the Counties of England. Published for the Institute of Historical Research by Oxford University Press, Oxford.

Shaw, R.C. (1956). *The Royal Forest of Lancaster*. Guardian Press, Preston.

Smith, R. (1992). Human Resources. *The Countryside of Medieval England* (eds G. Astill A. Grant), pp. 188–212. Blackwell, Oxford.

Sylvester, D. (1950). Rural Settlement in Cheshire: Some Problems of Origin and Classification. *Transactions of the Historic Society of Lancashire and Cheshire*, **101**, 1–37.

Sylvester, D. (1957). The Open Fields of Cheshire. *Transactions of the Historic Society of Lancashire and Cheshire*, **108** (for 1956), 1–33.

Sylvester, D. (1959). A Note on Medieval Three-Course Arable Systems in Cheshire. *Transactions of the Historic Society of Lancashire and Cheshire*, **110**, 183–86.

Terrett, I.B. (1962a). Cheshire. *The Domesday Geography of Northern England* (eds H.C. Darby & I.S. Maxwell), pp. 330–91. Cambridge University Press, Cambridge.

Terrett, I.B. (1962b). Lancashire. *The Domesday Geography of Northern England* (eds H.C. Darby & I.S. Maxwell) pp. 392–418. Cambridge University Press, Cambridge.

Walker, F. (1939). *Historical Geography of South West Lancashire before the Industrial Revolution*. The Chetham Society, New Series, 103, Manchester.

White, G. (1983). On Dating of Ridge-and-Furrow in Cheshire. *Cheshire History*, **12**, (Autumn 1983), 20–23.

White, G. (1995). Open fields and rural settlement in medieval west Cheshire. *The Middle Ages in the North-West* (eds T. Scott & P. Starkey), pp. 15–35. Leopard's Head Press, Oxford.

Youd, G. (1962). The Common Fields of Lancashire *Transactions of the Historic Society of Lancashire and Cheshire*, **113**, 1–41.

CHAPTER SIX

Where there's brass there's muck: the impact of industry in the Mersey Basin c.1700–1900.

A.E. JARVIS AND P.N. REED

Introduction

The presentation of even a narrative, *a fortiori* an explanation, of the industrialisation of the Mersey Basin would be far too lengthy to include in this chapter. What is offered instead is a sketch of the connections between different developments to form the background to an explanation of some of the effects of 'The Industrial Revolution'.

That expression is enclosed in quotes because there is now a fair measure of debate whether the events to which it has been applied were actually revolutionary, or whether the term arose only from a perceived discontinuity of development which in turn arose only from our former ignorance of the connections. That is a debate for another place: for present purposes we may leave it aside and simply note that developments in such fields as the application of water power to processes other than flour milling increased both in rate of change and extent of application. The availability of water-power on the River Derwent was a significant factor in the location of the Lombe brothers' silk mill in Derby, an enterprise which can be argued to constitute a 'factory' and thus to oust Arkwright from his place as the supposed inventor of the factory system. The River Goyt provided the power for the important silk mills of Macclesfield, and water continued to be the main prime mover for factory driving well into the 19th century. For further details and extensive bibliographies see Hudson (1992) and Reynolds (1981).

Cheshire silk and Lancashire cotton are the prototype factory industries in which the production process is broken down into stages, each carried out by specialised machinery. Because the machinery is expensive, the unit of production must grow in size, which both demands and enables the de-skilling of the individual steps of the process. That simplifies the recruitment and training of workers, which enables a further increase in size of operation. There is, however, only a finite number of foot-pounds of work in an entire river system, which means that further growth may be constrained by either the size or the number of mills on any particular part of it. The remedy already existed, namely the rotative steam engine: more expensive than water-power, it was gradually adopted as and when continued expansion made it profitable. The mere availability of technology did not guarantee its adoption – the Newcomen engine continued with remarkable tenacity long after Watt had quadrupled its thermal efficiency (Hills 1989).

The nature of industrialisation

The steam engine changed the rules. It was now possible to employ almost any amount of power in any location where coal was obtainable at an acceptable price. Again, the necessary technology (for obtaining coal) was already in existence. The raw materials – cotton and silk – which dominated this phase of industrial development could not be produced in any meaningful quantity in this country. At least as important was the fact that the large amounts of homogeneous products of the cotton mills very soon exceeded local consumption and became dependent on export markets before the installed horse-power of steam engines came to exceed that of water-wheels. Whatever the parentage of invention, necessity is undoubtedly the mother of investment, and the second half of the 18th century saw widespread road improvements whose importance is often under-rated (Bird 1969) and the building of most of the principal canals of the country. Needless to say, several of these were geared to meeting the industrial needs of southern Lancashire and northern Cheshire, and the first industrial canal, the first canal to cross a river valley and the first to cross a watershed (the two latter 'firsts' extending only to England), connected with the River Mersey (Hadfield & Biddle 1971). The Trent & Mersey canal may well have been the first to establish a 'linear habitat' along which characteristic canalside plants and animals proliferated, though it must be remembered that both the water and the margins soon became heavily polluted for reasons given below.

The greatest transport need was for a port. This was in part to serve coastal shipping, whose importance tends to be underestimated, but principally to provide overseas raw materials and serve export markets. During the 18th century, the Port of Liverpool grew at an increasing rate: between 1757 and 1857 the number of vessels using the port rose sixteenfold and the revenue rose one hundred and sixtyfold, despite cuts in the rates of dues. In 1757 the physical extent of the docks was under 8 acres (3.24ha), while by 1857 it was 290 acres (117.45ha) – with another 160 acres (64.8ha) in Birkenhead. Round it there grew up a series of satellite ports – Runcorn, Widnes, Ellesmere Port, Tarleton to name but a few, which linked the canal hinterland with Liverpool through the medium of the ubiquitous Mersey Flats. All these developments were interdependent, and the best general guide to them is still Hyde (1971), with more specialised contributions from Hadfield & Biddle (1971), Porteous (1977), Stammers (1993) and Jarvis (1996).

There were further links. Textile industries have many ancillary processes which, traditionally, were not integrated with spinning or weaving. These include bleaching, dyeing and printing, and the demand for these services naturally increased in proportion to the growth of output of yarns and textiles. Dyeing did not change as early as the others, but the adoption of chlorine-based 'bleaching powder' and the rapidly increasing demand for alum as a mordant probably affected the chemical industry as much as the textile industries (Clow 1952). The adoption of steam printing of *The Times* in 1814 opened the way for bulk printing of textiles (and other products like wallpaper).

These changes required increasingly large and complex items of plant. Down to perhaps 1830, we find that top-flight craftsmen, especially millwrights, retained a good deal of control in the design and construction of plant. Where they were subordinated to people who might already be termed professional engineers, those engineers had often risen from the ranks of millwrighting or allied occupations. A rapidly-increasing application of scientific theory and predictive methods, especially in thermo-dynamics and strength of materials, brought about a growing separation of design and construction and more particularly a growing specialisation. Lancashire became a major centre of the manufacture of factory machinery, the boilers and engines to drive it and the machine tools to make it. The enhanced design and precision of execution made machinery faster and more reliable, but more expensive too. These issues have been the subject of long-running historical debate, to which the latest contribution is made by Jarvis (1997).

There used to be a romantic view that gas lighting was introduced as a paternalist benefit to enable thrifty housewives to spend their evenings sewing and self-helping intelligent artisans to read Plato. Those were side-effects: the real purpose of gas lighting was to enable more efficient mills to become more efficient still by means of shift working, and it should come as no surprise that the most rapid spread of the gas industry follows shortly after the 'ten hours act' of 1847, encouraging the employment of two shorter shifts in place of one longer one. Gas, with its large and constant demand for coal and its enabling of greatly enhanced return on invested capital, can be seen as a significant catalyst in what some

1715	Liverpool's first dock.
1718	The Lombe brothers' silk mill, Derby.
1733	The flying shuttle.
1737	Sulphuric acid.
1757	Sankey navigation.
1761	Bridgewater Canal.
1764	The spinning jenny.
1771	Arkwright's mill at Cromford.
1777	Trent & Mersey Canal.
1779	Crompton's mule.
1781	Boulton & Watt's rotative steam engine.
1787	Cartwright's power loom.
1789	Arkwright's water frame.
1799	Bleaching powder.
1802	Gas lighting introduced at Watt's Soho Works.
1814	Steam printing.
1816	Leeds & Liverpool Canal completed. (at last!)
1817	The Jaquard loom.
1818	Liverpool Gas Light Company.
1828	Muspratt's 70.43m. chimney (for HCl).
1830	Liverpool & Manchester Railway.
1832	First cholera pandemic.
1847	Liverpool appoints James Newlands Borough Engineer.

Table 6.1. Chronological summary of the major events in the industrialisation of the Mersey Basin.

have termed the 'Second Industrial Revolution'. Williams (1981) provides a general history of the gas industry, while Griffiths (1992) considers the origins in greater detail.

All of these changes fed back into each other, and into others as yet unmentioned. Warrington was a traditional centre of hand wire-drawing, but its position in relation to the most productive parts of the Lancashire coalfield assured it a major role in the large-scale manufacture of iron (later steel) winding cables for the new deep mines which were enabled by better pumping, ventilating and winding machinery.

The Mersey Basin also played a key role in the development of railways, especially through the building of the first real main-line railway, the Liverpool & Manchester (L&MR). Railways eventually interacted with the supplies and the produce of practically every local industry, but their initial impact was mainly in passenger carriage (Lardner 1850). The L&MR made it possible to do business in person on a daily basis and to dispatch trade samples in the knowledge that they would arrive within a day. Before the introduction of the penny post and the electric telegraph, this was an innovation of the utmost importance to all the wheelers and dealers on whom every industry, then as now, depended for both purchases and sales. Railways, through their consumption of coal, were a major pollutant but they were also a major provider of linear habitats.

Virtually all the changes mentioned implied a growth of towns. The steam engine enabled, probably eventually required, the appearance of places like Oldham which were entirely dominated by mills and factories. The engineering industry is necessarily gregarious, as machinists need suppliers of castings, materials and tools nearby. It is all too easy to forget the merchants and brokers: they needed market places in Liverpool and Manchester, and they needed extensive wholesaling infrastructures. Ellison (1886) gives an idea of the intricacies of the 'networking' system. The results of these needs, coupled with rapid population increase, are well known: rapid urbanisation leading to over-populated squalor and assisting in the spread of epidemic diseases on a scale unknown since the last visitation of the Black Death – see chapter five, this volume.

There is one consequence which was inevitable. Every industrial development mentioned together with many others affected the natural environment to a greater or lesser extent. Even comparatively clean industries like machine tool manufacture produced smoke, and the obstruction of watercourses by ash and cinders was a long-term problem in the Manchester area (Report of the Royal Commission 1870). Mining caused subsidence, polluted underground watercourses and shifted water-tables, as did the use of wells for municipal water supplies and industrial process or cooling water. Perhaps the greatest disruption of this kind was caused by brine pumping in the Cheshire saltfield, damaging buildings and bringing about significant landscape and habitat changes through the formation of flashes (Report of the Select Committee 1890–91). Brine evaporation was particularly messy because it required a low sustained heat, which was most economically achieved by the fairly slow combustion of the worst grades of Lancashire coal. These coals could produce appalling smoke even when burned in relatively 'high-tech' boilers – and in *Hard Times* Dickens remarks on the way this was viewed as a symbol and an indicator of prosperity in Coketown.

Because early process industries were relatively inefficient, large amounts of waste were simply piled up or quietly released to atmosphere or into rivers. These were the direct results of industrialisation, but the indirect effects were possibly worse. Millions of households burned local soft coal in highly inefficient grates, but the largest pollutant of all was sewage. There is a perceived image of the Victorians as being 'hung up' about sex: perhaps they were, but they still bred like 'bunny-rabbits' (see Table 6.2). Concentrate millions of people in a relatively small river system and the millions of tons of sewage they produce each year substantiates the title of this paper. Before the construction of the Manchester Ship Canal, the natural channel of the River Irwell and the artificial one of the Mersey & Irwell Navigation were so full of 'solids' that Salford regularly got what it deserved by being flooded feet deep in sewage. (Though Salford naturally, and possibly defensibly, claimed that the Mancunians were to blame, owing to the fact that the river flowed more swiftly on the Manchester bank,

resulting in the deposition of solids – of whatever provenance – on the Salford side.)

The Ship Canal had many and various other effects. It effectively severed the area known as Moss Side (between Warrington and Runcorn) from the rest of the world, thereby creating an entirely new habitat for present day bird watchers. The downside of its improvements to the drainage of Salford was that it needed constant dredging, which resulted in the creation of a small desert of sludge lagoons at Weston Marsh. Dredging of the Mersey approaches to allow access across the Bar for larger vessels and over a longer period of the tide began in 1890, and just before the Great War involved the constant employment of all of the four largest dredgers in the world. The impact of their work has yet to be understood.

Population: (thousands)

	British Isles	Lancashire
1751	6,467 (est)	
1801	8,893	673
1851	17,928	2,301
1901	32,528	4,373
	Liverpool	Manchester
1801	82	75
1851	376	303
1901	685	544
	Oldham	Bolton
1801	12	18
1851	53	61
1901	137	168

Estimated Coal Production: Lancashire and Cheshire (thousand tons)

1700	80	1750	350
1800	1,400	1850	9,600
1900	28,700 (Lancashire only)		

Coal Consumption: UK manufacturing (million tons)

1816	4.9	1855	18
1887	46	1903	62

Soap Duties: amount of soap charged on (million lbs)

1713	24.4	1750	28.4
1800	46.9	1850	164.2

Paper Duties: England and Wales (tons)

1713	2,583	1750	4,115
1800	12,394	1850	44,159

Exports of cotton goods: UK totals (£ thousands)

1700	28	1750	20
1800	5,851	1850	28,300
1900	69,800		

Table 6.2. The industrialisation of the
Mersey Basin: a few key statistics taken from Mitchell (1988).

Pollutants and the environment

Later in the 19th century the beginnings of what is now the environmental lobby could be identified. It was mostly ineffectual despite the movement towards pollution control which is described below. What really made the difference was the realisation that pollution was waste: black smoke was unburned calories; sulphur dioxide from copper smelting could be reduced and sold as flowers of sulphur; sewage could be de-watered and sold as fertiliser.

Nevertheless, the effects of industrialisation on the natural environment changed over the period 1700–1900 and were mitigated by the ever increasing sophistication of the chemicals and materials being produced and the terms of controlling legislation.

Table 6.3 shows some of the major industrial sectors of the Mersey Basin together with some illustrative examples of the pollutants associated with each sector. The range of industrial activity was high as would be expected from an area so closely involved with the Industrial Revolution but also, the 'cocktail' of chemicals/materials impacting on the natural environment was very considerable. At certain times over this period, e.g., the early years of the 19th century, the mixture of pollutants was remarkably varied. Through the 19th century as a whole the impact of industrial activity on the environment was probably at its greatest and industry, local government and central government struggled to provide a legislative framework to reduce the effects of these pollutants.

The chemical industry

In this context it is useful to consider the chemical industry in more detail. Today the economic viability of a chemical process is paramount, and this will be determined by a number of factors including the choice of process, the costs of raw materials, the cost of energy and the recycling of chemicals from one process to another to provide value for money for the plant overall. However, before 1850 little or no attempt was made to recycle waste products from one process as raw materials for another process. Any products surplus to needs were dumped on the surrounding land if they were solids, fed into rivers or streams if they were liquids or released into the atmosphere if they were gases. It does not take a very creative imagination to picture the kind of natural environment that resulted. Without some kind of intervention the situation was only going to get steadily worse.

A good example to illustrate these issues is the Leblanc process introduced into the UK in the early years of the 19th century for the production of alkali (in the form of sodium carbonate) from salt. Before this process alkali was produced from natural sources – usually from the ashes of kelp or from barilla derived from a Mediterranean plant. When the Peninsular War prevented supplies of barilla reaching the UK, other

Source/Industry	Pollutants
Households	Sewage
	Coal Smoke
Chemical	Hydrogen chloride
	Hydrochloric acid
	Sulphur waste
	Ammonia
	Oxides of nitrogen
	Nitric acid
	Sulphur dioxide
	Sulphuric acid
	Organic residues
Bleaching	Bleaching powder
	Hydrochloric acid
	Sulphuric acid
Paper	Alkali liquors
	Rags
Tanneries	Spent tan liquor
	Lime liquor
Cotton	Bleaching powder
	Lime
Dyeworks	Organic residues
	Cyanides
Gas works	Tar residues
Agriculture	Fertilisers
	Pesticides
Engineering	Acids
	Coal smoke
	Oils
Metal refining	Arsenic
	Cadmium
Petrochemicals	Various organic compounds
Woollen works	Vat liquors for printing
	Nitrogenous organic waste
	Soap suds
Nuclear related	Radioactive radiation
	Radioactive waste

Table 6.3. Major industries of the Mersey Basin and their pollutants

sources had to be found urgently because so many parts of industry at that time depended on the availability of cheap alkali, e.g., textiles, soap and glass.

The Leblanc process required coal, salt and limestone. All these were readily available in the Mersey Basin or in the case of limestone, brought in from the surrounding areas. From about 1820 until the 1880s the Mersey Basin was a centre for the alkali industry using the Leblanc process with the main production being centred at various times in Liverpool, St Helens, Newton le Willows, Runcorn and Widnes. When James Muspratt, a Dublin chemical manufacturer, moved to Liverpool in 1822 to take advantage of the opportunities for the Leblanc process, it marked a new phase of the development of the chemical industry with the changeover from a small scale trade to the large tonnage alkali production. The dramatic increase in production brought increased threats to the natural environment.

An outline of the process is given in Figure 6.1. For every tonne of salt decomposed half a tonne of hydrogen chloride gas (a very pungent gas known as muriatic acid gas in the alkali trade) was produced with about two tonnes of alkali waste. In the Merseyside area in the 1840s it is estimated that over 100,000 tonnes of salt were converted into alkali yielding about 60,000 tonnes of hydrogen chloride gas. There were limited attempts to disperse the gas with most manufacturers releasing the gas from tall chimneys (the chimney of Muspratt's works in Vauxhall Road, Liverpool was reputed to be over 70m high), and the number of such chimneys (Figure 6.2) in an area became a barometer of industrial activity. Probably they should have been seen as 'ominous landmarks' for the damage their gases were causing.

In 1836 William Gossage developed a method of condensing the muriatic acid gas using a derelict windmill packed with bracken and twigs through which he ran water. The water was brought into maximum contact with the gas (which is very soluble in water) producing hydrochloric acid. Unfortunately, until the 1860s this supply of acid far exceeded the demand, and the acid was run off into surrounding rivers and streams with debilitating effect. In the 1860s the hydrochloric acid was used as a source of chlorine in the bleaching of esparto grass during the manufacture of paper. This is an early example of the by-products of one process being used as the raw materials for another, and as chemical understanding of the processes increased and the competitive nature of the industry grew, so the routing of materials for as many purposes as possible was more widely adopted.

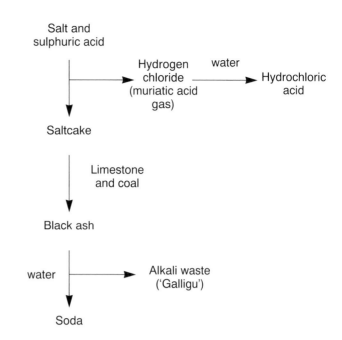

Figure 6.1. Leblanc process for the production of soda.

Even after Gossage's invention few manufacturers adopted his 'acid tower' with which to condense their muriatic acid gas. Many did not understand the principles involved as very few chemical works had qualified chemists before the 1880s. The manufacturers were content to release the gas into the atmosphere and were not concerned about the social or environmental aspects.

Figure 6.2. The smoke environment in Widnes during Leblanc period.

Few saw the need to invest in the towers. After all, they were not required by law to do so, and the extra cost might put them at an economic disadvantage against other manufacturers.

While the ever-increasing amounts of gas released into the atmosphere in the Mersey Basin exacerbated the already poor environmental conditions, the law did try to intervene. Muspratt found himself in court on a number of occasions to face charges of causing a nuisance. In one case in 1838 he escaped with a fine of one shilling having 'proved' that muriatic acid gas possessed beneficial properties. He also prevented most prosecution witnesses from giving evidence. More worrying was the difficulty both the judge and jury had in admitting scientific evidence into the proceedings when the evidence from defence and prosecution was directly conflicting. In many court cases of the period this evidence was eliminated from the proceedings. Another difficulty was attributing responsibility for damage to particular factories or works, especially where a number of such works were grouped together as was the case in most towns.

By the early 1860s the weight of opinion from landowners, farmers, doctors and the public (very much centred on the Mersey Basin area where the damage from the Leblanc process was at its greatest) forced the government to intervene and move away from their previous *laissez-faire* approach. The lobby achieved such a momentum that in 1862 Lord Derby moved for an inquiry in the House of Lords. A Committee was set up and moved quickly, hearing evidence from over 45 witnesses including eminent scientists, farmers, local authorities, doctors and manufacturers. The *Report from the Select Committee of the House of Lords on the Injury from Noxious Vapours* was published in August 1862, and in March 1863 a private bill was brought by Lord Stanley of Alderley which was passed in July 1863 as the Alkali Works Act for an initial five-year period. The provision of the legislation included the requirement for all alkali works to condense at least 95% of their muriatic acid gas, the appointment of an Inspector and a team of sub-inspectors and the recovery of damages by civil action brought by the Inspector in the County Court.

Evidence before the Select Committee confirmed that while manufacturers were aware of the Gossage tower and the inherent benefits, few had made use of it. Yet Robert Angus Smith (the first Inspector under the Alkali Works Act) was able to show in his First Annual Report in 1865 that the provision of Gossage towers was a sound investment when compared with the cost of fines for causing a nuisance. By this stage (and even before enforcement of the legislation) the manufacturers had changed their stance, most had built towers and most were achieving levels of condensation of the gas in excess of the legal requirements of 95%.

The relationship between Smith and his inspectors on the one hand and the manufacturers on the other, was key to the successful enforcement of the Act. The Alkali Inspectorate under Smith's leadership and direction showed sensitivity in dealing with the manufacturers and developing a mutually beneficial partnership. The inspectors were providing a peripatetic service for the manufacturers and assisting them in being more efficient in the operation of their works without providing necessarily an economic advantage to one of them – a very difficult tightrope to walk. At an early stage Smith dismissed as unscientific and unsystematic the suggestion that some people were sufficiently skilled as observers to enable them to differentiate between a 5% escape of gas and a 6% escape; instead he saw the need for rigorous chemical analysis of the gases released from the chimneys of chemical works. Sophisticated self-acting aspirators were designed and built so that collecting and analysing the gases could continue at regular intervals throughout the day and night under 'sealed' conditions, where the manufacturers and their staff were unable to interfere and falsify results. This greatly aided the inspection process where there were only four inspectors plus Smith to cover the whole country.

In 1868 the Alkali Works Act legislation was reviewed and an assessment made of its effective enforcement. The levels of muriatic acid gas had been greatly reduced with the accompanying reduction in the environmental effects. In July 1868 a new Bill was introduced to extend the 1863 legislation. This gave Smith's work new momentum and even greater commitment to applying effective legislation to other sectors of the chemical industry. There were two immediate concerns: a percentage measure was outdated (5% of several 100,000 tonnes is a considerable quantity) and a volumetric measure was required; as the Leblanc process was being replaced by the ammonia-soda process, the nuisance from alkali works was from sulphurous and nitric acids and ammonia, and the terms of any legislation needed to reflect these changes. Because of the success Smith had in enforcing the earlier legislation, Parliament felt able to adopt measures to take account of the developing situation. As the chemical industry changed (as with other industry sectors), so it was necessary for the terms of any legislation to change to provide the inspectors with the necessary powers to reduce damage to the natural environment, while enabling these important industries to function within an increasingly competitive and international market.

Table 6.4 illustrates how legislation changed over the period to provide better protection of the natural environment and improve the conditions for day-to-day living. It is interesting to note that running parallel to these chemical pollutants in this period was the damage from black coal smoke which predates most of these chemical pollutants and was not really satisfactorily controlled until 1956 when the Clean Air legislation was put into effect. Fortunately, smogs are now a phenomenon of the past in the UK.

Pollutants were not confined to the air. As with the case of muriatic gas there was frequently a link between air and water pollution. While people were generally

1770	Charles Roe and Company prosecuted by Liverpool Corporation for causing nuisance, and copper works in Liverpool closed down.
1831	James Muspratt prosecuted by Liverpool Corporation for causing nuisance, and 1838 alkali works effectively closed down, but moved works to St Helens, Newton-le-Willows and Widnes.
1853	Liverpool Corporation formed Smoke Prevention Committee to control smoke from coal burning.
1863	Alkali Works Act to control emission of hydrogen chloride gas from alkali works; setting up of Alkali Inspectorate.
1874	Alkali Works Amendment Act: additional constraint on amount of hydrogen chloride emissions.
1881	Alkali, etc., Works Regulation Act: extended to manufacture of sulphuric acid, chemical manures, gas liquors, nitric acid, sulphate of ammonia and chlorine and bleaching powder.
1892	Alkali, etc., Works Regulation Act: further extension to include extraction of zinc ore, tar distilling and recovery of alkali waste.
1956	Clean Air Act: prohibited black smoke, regulated grit and dust, prescribed the design of new furnaces, introduced smokeless zones and smoke-control areas.
1996	The Environment Agency established.

Table 6.4. Major events in the control of air pollution.

more aware of air pollution because of breathing the obnoxious fumes and the associated smell or taste, water pollution was not so immediately evident, except if you were fishing in a local river or stream (Jarvis 1995). Nevertheless, as with the air so there was an increasingly complex 'cocktail' of pollutants being dispersed through the water system. It is, however, important to put the impact of industrialisation into perspective. Household sewage had been the main source of water pollution since the early part of the 19th century, coinciding with the greatly increased population of towns. The effective disposal of sewage waste then began to run parallel to the need to provide safe drinking water. Although improvements in sewage disposal have and continue to be made it is still a major source of pollution.

Table 6.5 shows some of the key pieces of legislation in the control of water pollution and improved drinking water supplies. Although the control of air pollution led the way as it was demonstrated that air quality could be improved by legislation and the appointment of inspectors, so the emphasis began to focus on the control of water quality. As with air pollution, there was a gradual tightening of the terms of the legislation to meet ever-more stringent requirements of water quality. With the Rivers Pollution Act of 1876 it was almost certain, even before the legislation was passed, that Smith would be asked to take responsibility for enforcing the legislation in parallel with his work on air pollution. Smith brought the same rigor to the enforcement of this legislation that he had brought to the Alkali Works Act, and achieved similar improvements.

Conclusions

Pollution from industrial activity is often naïvely thought to be a post-Second World War phenomenon, and while the sophistication of the chemical and materials waste has grown, the origins are very much part of the activities associated with the Industrial Revolution. The waste products from industrialisation are varied in their potential danger and in their quantity and control of them is even more important now than at the start of the Industrial Revolution. The legislation adopted in the middle of the 19th century and the subsequent extensions, together with the methodology set in place by Robert Angus Smith, have carried through during the 20th century to the Alkali Inspectorate and more recently to the Health and Safety Executive, the Environment Agency and other government agencies with responsibility for the natural environment. Nevertheless, the effects of the pollution (together with the use of land for a variety of industrial activities) from earlier phases of industrial development have left a legacy which is evident in the diverse habitats within the Mersey Basin area. This legacy has left its mark in the effect of air and water pollution, the development of large urban and industrial areas, derelict land often involving heaps of waste material, creation of subsidence lakes (flashes) and linear transport routes of canals, railways and roads, some of which, e.g., motorways, are still being created.

Today the de-industrialisation of the Mersey Basin provides the opportunity to take stock. Industrialisation has produced some hideous results. But man is not unique in altering the environment, e.g., Beavers (*Castor* spp.), can make major changes to medium sized river systems. However, the speed with which 'nature'

1786	*From Improvement Acts –*	
	1786	Liverpool
	1813	Warrington
	1833	Birkenhead
	1845	St Helens
	1852	Runcorn
1846	Liverpool Sanitary Act	
1847	Liverpool appoints first full-time Borough Engineer (James Newlands) at same time as first Medical Officer of Health.	
1848	Public Health Act.	
1848	Liverpool Corporation buy out two local water companies and extend supply.	
1876	Rivers Pollution Act.	
1876	Local Government Board was allowed to set up Mersey and Irwell Joint Committee.	
1885	Rivers Pollution Prevention Act	
1887	Rivers Pollution Prevention Act Amendment: allowed for an inspector to be appointed by the Local Government Board.	
1892	Mersey and Irwell Prevention of Pollution Act.	
1996	The Environment Agency.	

Table 6.5. Major events in the control of water pollution

recovers from all but the most toxic remains of industrialisation suggests that in the long term there may not be too much to worry about.

References

Bird, A. (1969). *Roads and Vehicles.* David & Charles, Newton Abbot.

Clow, A. & N. (1952). *The Chemical Revolution.* Batchworth, London.

Ellison, T. (1886). *The Cotton Trade of Great Britain.* Reprinted by Cass, London, 1968.

Griffiths, J. (1992). *The Third Man.* Deutsch, London.

Hadfield, C. & Biddle, G. (1971). *Canals of the North West.* David & Charles, Newton Abbot.

Hills, R.L. (1989). *Power from Steam.* Cambridge University Press, Cambridge.

Hudson, P. (1992). *The Industrial Revolution.* Arnold, London.

Hyde, F.E. (1971). *Liverpool and the Mersey.* David & Charles, Newton Abbot.

Jarvis, A. (1995). An historical backwater : the fishing and fish trading of Liverpool. *Northern Seas Yearbook,* **1995**, 51–76.

Jarvis, A. (1996). *The Liverpool Dock Engineers.* Sutton Publishing, Stroud.

Jarvis, A. (1997). 'Theory versus Practice in Dock Engineering'. *Transactions of the Newcomen Society,* **69**, 57–68.

Lardner, D. (1850). *Railway Economy.* Taylor, London.

Mitchell, B.R. (1988). *British Historical Statistics.* Cambridge University Press, Cambridge.

Porteous, J.D. (1977). *Canal Ports,* Academic Press, London.

Report of the Royal Commission on The Best Means of Preventing Pollution of Rivers, British Parliamentary Papers 1870 (37) XLI.

Report of the Select Committee on Brine Pumping, British Parliamentary Papers 1890–91 (206) XI. 219.

Reynolds, T.S. (1981). *Stronger than a Hundred Men: the History of the Vertical Water Wheel.* Johns Hopkins University Press, Baltimore.

Stammers, M.K. (1993). *Mersey Flats and Flatmen.* Dalton, Lavenham.

Williams, T.I. (1981). *History of the Gas Industry.* Clarendon Press, Oxford.

CHAPTER SEVEN

Natural habitats of the Mersey Basin: what is left?

L. WEEKES, T. MITCHAM, G. MORRIES AND G. BUTTERILL

Introduction

This chapter reviews the status of the remaining semi-natural habitats of the Mersey Basin. By the middle of the last century, the wildlife and habitats of the Mersey Basin were as thoroughly explored as any area of comparable size in the world. By contrast, they are relatively poorly known and documented today. This must be due, in part at least, to the great loss of semi-natural habitats in the area over the last 150 years. Interest started to revive with the general growth of environmental awareness in the 1960s and 70s, but it was not until the late 1980s that any systematic surveys of habitat distribution were undertaken. The information presented here regarding what semi-natural habitats remain is largely based on this survey work.

Mossland

Intact lowland raised bogs are one of Europe's rarest and most threatened habitats. The importance of the lowland mosslands of the Mersey Basin can be appreciated as, after the great expanse of fenland peat surrounding the Wash, the central Lancashire and Greater Manchester area is home to the most extensive lowland peat deposits in Britain. These once extensive areas of mossland have been considerably affected by man, and the majority lost. What remains should be a high priority for conservation effort.

The Pennine uplands once had extensive areas of blanket bog, which have also been considerably altered by human activities. This topic has been dealt with else-where in this volume (chapter eleven) and so this chapter concentrates on lowland bogs.

Serious exploitation of the mosslands began with the rapid population rise of Manchester and Liverpool. Large-scale drainage of mosslands for agriculture in the area was first attempted in 1692 by Thomas Fleetwood at Martin Mere. An intensive phase of drainage, exploitation and development ensued. Landfill is still a major issue on the mosses; urbanisation and infrastructure development are a constant threat and the industrial extraction of peat for the horticultural industry is a major pressure on this delicate ecosystem.

Today, the lowland raised bogs of the Mersey Basin have all but disappeared. The mosslands now largely comprise a complex patchwork of agricultural land with drainage networks and a relatively small number of remnant bogs (Table 7.1). All of these bog remnants are highly disturbed and occur in a range of shapes and sizes; few are capable of supporting stable populations of characteristic mossland plants and animals.

The degree of loss of the Mersey Basin's mossland can be illustrated by the fact that in Lancashire, Merseyside and Greater Manchester there were 10,728ha of untouched wet mossland remaining in the mid-19th century. By the mid-20th century just 2,804ha of greatly modified mossland remained (Greater Manchester Countryside Unit 1989).

Soil survey records indicate that Chat Moss originally occupied 2,650ha (English Nature 1992) (Figure 7.1). It has rapidly declined over recent years with 800ha of raw mossland surviving in 1958, which had reduced to 440ha in 1984 and to 233ha in 1989. Only 68ha of remnant moss-

	Altcar deposit	Chat Moss deposit	Risley Peat deposit	Simonswood complex
Area of peat deposit	1,470	2,650	810	2,140
Area of SSSI	0	230	100	0
Extract peat for horticulture	0	310	0	135
Area under agriculture	1,440	1,900	660	1,700

After English Nature (1992).

Table 7.1. Estimates of area (ha) for the dominant land-uses on the four major peat deposits of the Mersey Basin.

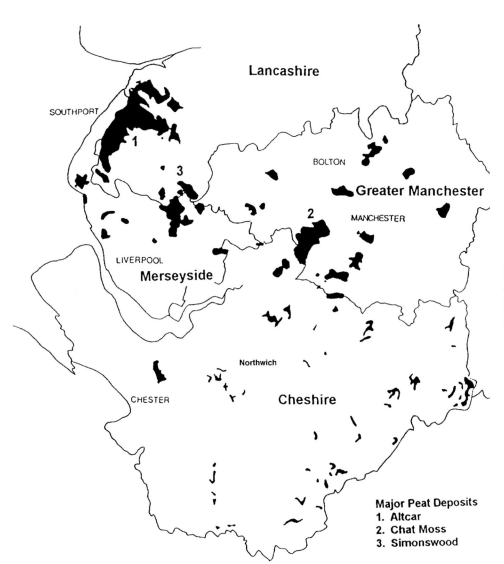

Figure 7.1. Extent of peat development in the Mersey Basin (after Fairhurst 1992 and English Nature 1992).

Major Peat Deposits
1. Altcar
2. Chat Moss
3. Simonswood

land occurs in Greater Manchester County outside the Chat Moss complex (Greater Manchester Countryside Unit 1989), of which 44.5ha of Red Moss, near Bolton, is threatened by landfill.

The situation in Cheshire is also reaching critical levels. The 1983/84 survey of semi-natural habitats (English Nature 1983/4), indicated that only 159.2ha of lowland raised mire and 20.4ha of basin mire remained relatively intact.

However, many of the remnant mosses now enjoy protection in unitary/local plans or as Sites of Special Scientific Interest (SSSI). Many sites could, with appropriate management, be made wetter and significant *Sphagnum* cover re-established.

Meres

The meres of the Mersey Basin are an internationally important feature of the lowland glaciated landscape, which extends into Lancashire, Merseyside, Greater Manchester, Clwyd, Shropshire and Staffordshire. The sites in Cheshire, form part of the North West Midland Meres, a group of generally fertile lakes, occupying hollows in the glacial drift surface which cover most of

Figure 7.2. Birch clearance on Abbots Moss, Cheshire.

the Shropshire and Cheshire plain. There are more than 60 open water bodies known as meres or pools, 32 of which lie within the Cheshire area of the Mersey Basin.

Meres originated from several causes. Many of the basins in which they occur may be kettleholes, created by ice blocks, which became separated from the retreating ice face, or were buried in the glacial outwash of clays and sands, some 12,000 years ago. More recently, subsidence of the underlying salt beds as a consequence of brine extraction, caused their formation. This is believed to be a factor in the creation of Rostherne Mere and Oakmere.

Most of the meres are to a greater or lesser extent fed and maintained by mineral rich ground water, with long retention times being typical. They vary in depth (1m–27 m), area (<1ha to 70ha) and water chemistry. The pools typically shelve steeply and are fringed with Reed swamp and Alder carr which provide additional habitats for wildlife.

Since their formation, they have been vulnerable to fluctuations in ground water level and supply, brought about by climatic variations and changes to trophic status. Good examples of hydroseral succession are also found, e.g., Quoisley Meres. However, some meres have probably changed little since the 13th century, as past records of fishing rights at Oakmere and Budworth Mere demonstrate. Alternatively, some mosses are shown as meres on early maps, so that the time taken to change from open water to peatland has varied enormously from one site to another. It is believed that most of these changes have occurred naturally.

Since the development of agricultural improvements in the 18th and 19th centuries, however, substantial changes have occurred. Land drainage has reclaimed many of the peat mosses, marshes and fens adjacent to the meres. This has greatly reduced the area of wetlands surrounding the meres and thus the later stages of hydroseral succession, e.g., fen pools with Alder carr, are particularly rare in the area.

There has also been an increase in the overall nutrient load to meres through changes in agricultural practice over the years, e.g., increases in cattle keeping and a switch from pasture to arable farming (see chapter thirteen, this volume for a more detailed account of factors affecting the meres). Beyond this, some meres have suffered further eutrophication due to pollution of the streams entering them, usually by farm wastes. The meres are, however, claimed as Britain's naturally eutrophic lakes, with evidence of blue-green algal blooms dating back to the last century and earlier. Nevertheless, there has been considerable anthropogenic eutrophication in recent decades. Trampling by cattle also inhibits the normal succession of vegetation, and there is evidence that the plants and animals surrounding the meres have been greatly affected. Nutrient control is urgently required on a number of meres within the Mersey Basin, but to do so requires an understanding of the complex factors involved and could be difficult to implement. Additionally, because

zooplankton grazing is an important control of algae in some meres, reduction of the existing fish stock to discourage zooplankton feeding fish may be necessary.

In conclusion, the meres are suffering from a variety of anthropogenic effects. The recent designation of a number of the midland meres as Ramsar sites in addition to their status as SSSIs and National Nature Reserves (NNR) may help emphasise the importance of this series of freshwater pools and encourage people to look after this valuable asset.

Lowland heath

Heathland is characterised by a pioneer community of limited diversity, in which the vegetation is dominated by ericaceous species. The Nature Conservancy Council (1990) defined it as vegetation dominated by dwarf shrubs, notably Heather (*Calluna vulgaris*), Cross-leaved Heath (*Erica tetralix*), Bell Heather (*Erica cinerea*), Bilberry (*Vaccinum myrtillus*) and Western and Dwarf Gorse (*Ulex gallii* and *U. minor*).

Although some heaths may have an ancient, natural origin much of what is now lowland heath was probably broadleaved woodland (Webb 1986). The trees were cleared by man for crop growing and grazing about 4,000 years ago (chapter four, this volume). Gradually and largely because there were no trees to bring up nutrients from deeper layers, the sandy soils became leached and nutrient poor under the more open conditions, and heathland became established. For example, the Royal Forest of Macclesfield was established in the 12th century, but where land was subsequently cleared heathland dominated the area, which was used for pastural purposes until the introduction of agrochemicals in the last 50 years improved soil fertility.

Despite its largely man-made origin (see also chapter eight, this volume) lowland heathland is a significant wildlife habitat and landscape feature. The location of heathland vegetation today reflects not so much its natural biogeographical range, but where it has survived through human activities or where newly created sites have been formed, e.g., the sand quarries. Thus, many of the remaining sites are where grazing is light, the soil too shallow for trees and the land too steep for arable agriculture. They do not necessarily represent in extent or location the nature of the former heathland of the Mersey Basin and those that do survive often have a depauperate flora to that known 100 years ago (de Tabley 1899) and despite active conservation measures, e.g., at Thurstaston on Wirral, species continue to be lost. The total area of heathland types in the Mersey Basin as recorded in 1993 is noted in Table 7.2.

In Cheshire, lowland heath is associated with the sandy soils in the Delamere and Goostrey areas and the sandstone outcrops of Thurstaston, Runcorn Hill and the mid-Cheshire ridge. According to the Cheshire Heathland Inventory (Clarke 1995), lowland dry heath covers 49ha of the county and lowland wet heath covers 10.85ha, a total of 59.85ha, distributed over 45 sites.

The table refers only to the areas of open heathland currently recorded on sites and not their total areas which may include other habitats.

Heathland type	Greater Manchester	Lancashire	Cheshire	Derbyshire	Merseyside
Dry heath	50.1	30.05	54.9	1.6	262.2
Wet heath	15.3	7.5	10.9	0	1
Total	65.4	37.55	65.8	1.6	263.2

Figures taken from the Lowland Heathland Inventory (English Nature 1993).

Table 7.2. The Total area (ha) of heathland types in the Mersey Basin.

22.7ha (14.5%) of the heathland area (both upland and lowland) lies within the boundaries of nine SSSIs and 115.15ha (73.8%) within 34 Sites of Biological Interest (SBI).

More detailed consideration of uplands is given elsewhere in this volume (chapter eleven). However, upland dry heathland was recorded in the 1983/84 survey (English Nature 1983/4) mainly in the eastern fringe of Cheshire, together with a small amount at the southern end of the mid-Cheshire ridge. Figures taken from the Cheshire Heathland Inventory (Clarke 1995), suggest that upland dry heath covers 86.1ha of the county, whereas wet heath covers 10.0ha, a total of 96.1ha, distributed over 26 sites.

In general, heathland has been lost through agricultural improvements, especially following the Enclosure Acts in the period from the 16th century. However, heathlands were still traditionally widespread in Europe until the early years of this century, often being used for common grazing. Since 1949, 40% of lowland British heath on acid soil has been lost by conversion to arable or intensive grazing, afforestation and building or succession to scrub, due to a lack of management (Nature Conservancy Council 1984). Additional pressures include those from recreation, uncontrolled fires, forestry and increased fertility from agricultural runoff.

Recognition of the problems associated with the conservation of heathland can only improve the chances of its survival and of its associated wildlife. Changes in agricultural policy and improvements in habitat creation, management and protection will go some way towards conserving this habitat for the future.

Ancient woodland

Forest once covered large areas of the Mersey Basin. In Roman and Saxon times the area remained extensively wooded, but by late medieval times woods were fragmented through deforestation for agriculture, construction and fuel, largely restricting them to river valleys.

The sites that have been continuously wooded since 1600 and are present today, may be fragments of this original 'wildwood' (ancient woodland), or they may have had their structure modified by past management (ancient semi-natural). The management undertaken, the relatively undisturbed soils and the length of time these sites have been continuously wooded has enabled rich communities of flora and fauna to develop and persist. However, the decline in this habitat continues today. Table 7.3 illustrates the overall trend in the change in area of ancient semi-natural woodland, throughout and beyond the boundaries of the Mersey Basin.

In Cheshire, ancient woodlands are mainly found on the steep-sided cloughs of the river valleys, especially those of the River Dane, River Bollin and River Weaver and their tributaries, and along the mid-Cheshire ridge in the vicinity of Peckforton. The other large areas of remaining woodland are found principally on private estates. Cheshire is deficient in woodland by national standards. Only 3.8% (8,640ha) of the county is covered

	Cheshire	Derbyshire	Lancashire	Manchester	Merseyside
Area of county	232,842	263,098	306,951	128,674	65,202
Area of ancient woodland	1,681	4,392	2,764	783	111
Area of semi-natural woodland	1,263	2,583	2,314	769	111
Area cleared since 1920	102	651	116	24	18

Figures taken from English Nature Research Report 177: amendments to the Ancient Woodland Inventory (July 1994 – February 1996).

Table 7.3. Area (ha) of ancient and semi-natural woodland of the counties of the Mersey Basin.

by woodland, although the area of woodland has remained broadly the same since the Second World War. The majority of remaining ancient woodlands are less than 11ha in area and 65% of them are less than 5ha.

In Rossendale and Blackburn, the two districts of Lancashire within the Mersey Basin, present woodland cover is 0.9% and 4.1% respectively. The loss of ancient woodland is probably more complete here than anywhere else in the county, a mere 19ha remain in Rossendale. There is little evidence that the remaining ancient woodland was systematically managed in the past and only a small minority of the woods in eastern Lancashire are now stockproof. The gradual attrition of woodland through grazing, can be seen perhaps most clearly in Rossendale.

In the Greater Manchester/Merseyside area of the Mersey Basin, most of the ancient semi-natural woodland lies on the acidic and neutral soils of the underlying coal measures and bunter sandstone. Very few are on the more calcareous soils of the keuper marls to the south, largely because these were reclaimed for agriculture. 29.5% of the woodland in Manchester and 6.6% of the woodland in Merseyside, is thought to be of ancient origin. Of this, some 98% is presently considered to be semi-natural. 62% of the woodlands are less than 5ha in size and only four are larger than 10ha. Unlike other areas, however, only a small proportion of ancient woodland found scattered in these urbanised counties, has been converted into plantation. This is because of their small size, their poor degree of accessibility and their urban setting.

Ancient woodland in blocks of 2ha constitutes about 2% of Derbyshire's area, although little of this is actually present in the part of the county covered by the Mersey Basin. The size and distribution of woodlands in the county is much the same as the rest of the Mersey Basin, and similarly there have been significant changes in the ancient woodland cover over the last 90 years. Clearance for agriculture has accounted for 75% of ancient woodland loss in Derbyshire, both as a result of over-grazing, the resulting lack of regeneration and direct grubbing out of sites.

The importance of semi-natural ancient woodlands is now widely acknowledged although despite this, the resource is declining rapidly. It is now a major priority to prevent any further reduction in the area or in the nature conservation value of remaining ancient woodland.

In addition to site protection changes in the policies of the Forestry Commission, improved grant schemes, initiatives such as the planting of New Native Woodlands and improved links between woodland owners and timber using bodies, will hopefully prove beneficial for the conservation of woodlands. However, financial incentives remain low when compared with agricultural incentives and potential climatic change may further influence the development of woodlands. A more detailed account of woodlands in the region is given elsewhere in this volume (chapter ten).

Rivers and riverine features

The Mersey Basin consists of a network of rivers and streams, draining 5,000km² of Merseyside, Greater Manchester, Cheshire, Derbyshire and Lancashire (Figure 7.3). All waters within the catchment eventually enter the Mersey Estuary and finally the Irish Sea. The Basin is made up of 1,725km of rivers and streams.

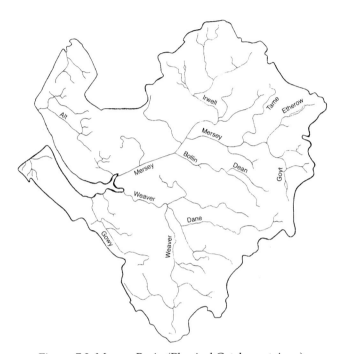

Figure 7.3. Mersey Basin (Physical Catchment Area).

Riverine systems exhibit an extremely diverse range of habitats. In a natural system, such habitats are found in a continuum and are constantly changing. Upland streams, e.g., the upper reaches of the River Goyt and River Etherow, are characteristically fast flowing bedrock streams. Both these rivers flow through an Environmentally Sensitive Area (ESA) designated for its 'high landscape, wildlife and historical value', containing two Special Protection Areas (SPA), as well as several SSSIs. In contrast, lowland rivers such as the River Dane, are often meandering systems with areas of erosion and deposition. 295.2ha of the River Dane have been designated SSSI, due to its important fluvial geomorphology, which includes clearly visible river terraces.

In-channel features, e.g., gravel bars, offer refuges for invertebrates and riffles provide spawning areas for fish. The River Medlock, although highly modified in parts, retains some areas of natural in-channel features, e.g., stony substrates, marginal gravel and some riffles. Eroding bankside cliffs found by the River Bollin and River Dane provide nesting sites for Kingfishers (*Alcedo atthis*) and Sand Martins (*Riparia riparia*).

A variety of natural habitats can also be found within the river flood plains. Backwaters, whether connected to the main channel or in the form of ox-bow lakes, offer refuges for wetland wildlife during periods of flooding and pollution. An excellent example of this feature can be found at Castle Hill on the River Bollin, where a 250m ox-bow lake is present. The ponds and ox-bows often have a diverse marginal flora, offering ideal sites for Warty or Great Crested Newt (*Triturus critatus*), a species protected under the Wildlife & Countryside Act (Department of the Environment 1981). Backwaters are also important spawning areas for fish. Additional significant wildlife habitats associated with these riverine systems, are the clough woodlands, Alder/Willow carr and flushes.

Unmodified rivers occasionally flood adjacent areas. In low lying river systems this can result in the formation of flood meadows, which have continuously high water tables and support a richly diverse flora and fauna. Flood meadows are no longer a common feature of the Mersey Basin, due mainly to agricultural changes, land drainage and flood defence measures. Less than 0.2% of unimproved, species rich wet grassland remains in Cheshire. The River Gowy drains a large and diverse ditch system, with associated unimproved acidic grassland over alluvial soils and deep acidic peat, which has formed under estuarine conditions, 90ha of which are a Site of Biological Importance (SBI). Stanley Bank Meadows SSSI covers 14.9ha of damp unimproved neutral grassland, which is now an extremely rare habitat in Merseyside.

A large majority of water courses within the Mersey Basin have been artificially straightened. The downstream section of the River Weaver, for example, was canalised for navigation purposes, resulting in extensive losses to the natural riverine features. The few remaining un-engineered stretches of this river offer a good insight into its natural course. Extensive areas of the flood plains have been modified to increase the area available for agriculture and to limit the risk of flooding.

In urban areas the land next to water courses has been developed to the bank top. The River Mersey itself has also been subject to increasing urban development pressure. Substantial flood bank protection has occurred, to form a fairly uniform area with little wildlife interest. Banks have also been reinforced by various means to limit natural erosion and the meandering nature of the river or stream.

Straightening, widening, deepening and embanking of rivers with low flow rates means that it is extremely difficult for these low energy systems to reassert their natural channel structure, after such engineering. Similarly, upland rivers have been subject to constraint, by the damming of the upper reaches to form reservoirs, e.g., River Croal and River Roach.

Very few waterways have suffered little or no human influence. Pollution, domestic and industrial, also puts pressure on the flora and fauna of the river and its associated river corridor. For example, leaching spoil heaps and direct discharges from extensive mining operations have left a legacy of water quality problems in the River Sankey/River Glaze catchment area. Urban run off via storm drains and surface water run off, contribute further to the pollution problems. If the wildlife of the riverine systems in the Mersey Basin is to be conserved in the long term, such problems need to be and indeed are being addressed.

Sand dunes/saltmarsh systems and mudflats

The coastal habitats of the Mersey Basin include the sand dunes of the Sefton coast and the saltmarshes and intertidal flats of the Mersey Estuary. These coastal ecosystems are valuable wildlife habitats, but as they are considered elsewhere (chapter sixteen, this volume) only a brief mention will be given here.

Only the southern half of the Sefton coast lies within the Mersey Basin area. However, it is impossible not to consider the system as a whole even though it extends beyond the Basin boundary. Approximately 2,100ha of dune survives, out of a total area of blown sand, which was probably once in excess of 3,000ha. In 1991 the dunes were estimated to extend for 17km and have an average breadth of 1.5km.

The dune vegetation on the Sefton coast currently stretches in a shallow crescent, from the north of Seaforth Dock to the north of Southport. At both the southern and northern ends of this area, building development and alteration of the natural landscape have confined the surviving dune vegetation to a very narrow strip at the top of the beach. Despite the loss of over 35% of the original dune area to development (Jackson 1979), the dunes are still a good example of a west-coast calcareous dune type, important on a European, national and local level, forming part of a complex of dune sites on the east Irish Sea Coast. It is the home of many rare and scarce species.

The importance of the Sefton coast for nature conservation is recognised by the designation of nearly the whole dune system as SSSI. This recognition is further strengthened by the presence of two National Nature Reserves, two Local Nature Reserves and National Trust land (Doody 1991).

The dunes of the Wirral Coast including Red Rocks SSSI (11.38 ha) and Wallasey dunes, also deserve a mention. Red Rocks Marsh lies between two parallel ridges of sand dune and until recently held a breeding colony of Natterjack Toads (*Bufo calamita*). A reintroduction programme is underway to re-establish these amphibians.

Within the Mersey Basin there are two estuaries, the Mersey Estuary and the much smaller Alt Estuary. The Mersey Estuary is 8,914ha, with an intertidal area of 5,607ha and a shore length of 102.6km. It comprises a number of valuable estuarine habitats including saltmarshes and intertidal mud flats, supporting a variety of associated species. Because of the importance of this area for wildlife, 24 SBIs have been designated, along with one NNR, three SSSIs, two SPAs and two Ramsar sites.

The Mersey catchment and its estuary continues to suffer, serious environmental degradation, receiving effluent from the major industrial sites and conurbations. The heavy pollution levels, have had a major impact on the wildlife, in addition to the effects of disturbance and loss of habitat. The Mersey Estuary Conservation Group and the Mersey Basin Campaign were established to promote the importance of the area and to safeguard and improve the environment. Following substantial investment, improvements in water quality and reduction in the pollution entering the estuaries are helping them recover from neglect and consequently support more wildlife (chapter fifteen, this volume).

Over the last 25 years, the importance of the Mersey Estuary for wildlife has increased substantially despite its pollution load. A key factor contributing to this was the increasing loss of the traditional European wintering grounds so that flocks of birds moved to the Mersey Estuary, as the best alternative site. During the mid-1960s, the treatment of effluent improved water quality with a consequent return of planktonic and benthic life. The high organic load of the estuary ensured maximum invertebrate productivity, which may be part of the reason why the Mersey Estuary continues to support higher densities of wildfowl and waders than neighbouring estuaries. Finally, there have been changes in the hydraulic regime which has allowed the return of a more dynamic system, making invertebrate food sources more available for some bird species. The area of saltmarsh available for feeding and roosting also seems to have increased around the same time. Hopefully, such positive changes will continue.

Wet flushes

The extent of wet flushes in the Mersey Basin is poorly documented. However, the majority of such sites are found in the upper reaches of the catchment. The surrounds of the River Goyt are typified by wet and dry heather moorland and acidic grassland with associated flushes, mires and blanket bogs. Flushes may be acidic or calcareous, both being found in the Mersey Basin, although the former type is more common.

The flushes are important in the way they alter the physical conditions of a site, allowing species with different ecological requirements to survive, increasing the diversity of the area. Longworth Clough SSSI exemplifies the complex transitions between vegetation communities related to drainage patterns and soil water conditions, which exist because of the base-poor flushes.

Lower Red Lees Pasture SSSI in south eastern Lancashire, is one of the few remaining examples of a herb-rich, unimproved neutral to slightly acidic pasture. Water seepages along a shallow tree- and scrub-invaded clough on the northern boundary of the site give rise to base-rich flush communities and areas of marshy grassland. Rushes, sedges and Yorkshire-fog (*Holcus lanatus*) dominate the flora, along with species such as Wild

Angelica (*Angelica sylvestris*), Marsh Thistle (*Cirsium palustre*), Ragged Robin (*Lychnis flos-cuculi*) and Cuckooflower (*Cardamine pratensis*).

In Cheshire, many of the flushes are associated with the river valleys. Dane-in-Shaw Pasture (SBI) is one of the largest, most botanically diverse areas of flushed neutral grassland remaining in lowland Cheshire, with springs which issue from the north-facing slopes.

Where calcareous springs create base-rich flushes, carpets of lime-loving bryophytes in which the moss *Cratoneuron commutatum* (Hedw.) Roth. is characteristic, occur. Greater Tussock sedge (*Carex paniculata*), often associated with calcareous fens, is also a feature.

This type of habitat was probably once more widespread, however, due to falling ground water levels, drainage and improvement of the land for agricultural purposes, it is becoming increasingly scarce. Current conservation measures are limited to the designation as SSSIs or SBIs of a few sites.

Conclusion

It was clear from our research that a considerable amount of information exists on the semi-natural habitats of the Mersey Basin. It is however, dispersed, mostly botanical, of variable age and in a variety of formats and therefore, not easily comparable. The information, when pieced together, forms a depressing picture of high quality habitat loss. However, several trends within the Basin, give cause for optimism:

1. the rapid increase of awareness within local government of the need to integrate environmental thinking within all policy areas;

2. national government's commitment to safeguarding biodiversity, through its endorsement of *Biodiversity: The UK Steering Group Report* (Department of the Environment 1995);

3. the increase in partnership solutions to wildlife conservation problems;

4. the protection received through national and international wildlife related designations and the development and implementation of agri-environment schemes, e.g., Countryside Stewardship.

However, there is no room for complacency. Statutory designations are not enough to safeguard the Basin's biodiversity, the network of non-statutory wildlife sites also need protection to allow them to provide for the enriching of the wider countryside and urban areas.

The challenge for us all is to work together and through the process of producing and implementing Local Biodiversity Action Plans commit ourselves to a programme of actions, which will help to develop a more sustainable future for the wildlife and natural habitats of the Basin and beyond.

References

Included here are many unpublished documents and publications used in compiling this chapter but not necessarily cited in the text.

Bevan, J.M.S., Robinson, D.P., Spencer, J.W. & Whitbread, A. (1992). *Derbyshire Inventory of Ancient Woodland*, NCC, Peterborough.

Carter, A. & Spencer, J. (1988). *Greater Manchester and Merseyside Inventory of Ancient Woodland (Provisional)*. NCC, Peterborough.

Cheshire Wildlife Trust and Cheshire County Council (1995). *Sites of Biological Importance Register*. Deposited at Cheshire Wildlife Trust, Grebe House, Reaseheath, Nantwich, Cheshire, CW5 6DG.

Clarke, S. A. (1995). *Cheshire Heathland Inventory*. Deposited at Cheshire Wildlife Trust, Grebe House, Reaseheath, Nantwich, Cheshire, CW5 6DG.

de Tabley, Lord (1899). *The Flora of Cheshire*. Longmans, Green and Co., London.

Department of the Environment (1981). *Wildlife & Countryside Act*. HMSO, London.

Department of the Environment (1995) *Biodiversity: The UK Steering Group Report*. HMSO, London.

Doody, J. P. (1991). Foreword. *The Sand Dunes of the Sefton Coast* (eds D. Atkinson & J. Houston), pp. v–vi. National Museums & Galleries on Merseyside in association with Sefton Borough Council, Liverpool.

English Nature (1983/4). *Cheshire Survey of Semi-natural Habitats*. Deposited at English Nature, Attingham Park, Shrewsbury, SY4 4TW.

English Nature (1992). *Distribution and Status of Lowland Peat in the Mersey Basin Area*. Deposited at English Nature, Attingham Park, Shrewsbury, SY4 4TW.

English Nature, RSPB (1993). *The Lowland Heathland Inventory*, English Nature, Peterborough.

English Nature. *Site of Special Scientific Interest Schedules, Cheshire, Derbyshire, Greater Manchester, Lancashire and Merseyside*. Deposited at English Nature, Attingham Park, Shrewsbury, SY4 4TW.

Environment Agency (1996). *Alt/Crossens Catchment Management Plan Consultation Report*. Deposited at Environment Agency, Richard Fairclough House, Knutsford Road, Warrington, WA4 1HG.

Environment Agency (1996). *Bollin Sub-Catchment Report*. Deposited at Environment Agency (as above).

Environment Agency (1996). *Goyt/Etherow Sub-Catchment Report*. Deposited at Environment Agency (as above).

Environment Agency (1996). *River Habitats in England and Wales: A National Overview*. Deposited at Environment Agency (as above).

Environment Agency (1996). *Sankey/Glaze Local Management Plan Consultation Report 2nd Draft*. Deposited at Environment Agency (as above).

Environment Agency (1996). *Upper Mersey Catchment Management Plan Consultation Report*. Deposited at Environment Agency (as above).

Fairhurst, J. (1992). *Cheshire State of the Environment Project: Technical Report Number 3, Habitats and Wildlife*. Cheshire County Council, Chester

Fairhurst, J. (1992). *Cheshire State of the Environment Project: Technical Report Number 4, Economic Landuse*. Cheshire County Council, Chester.

Greater Manchester Countryside Unit (1989). *The Mosslands Strategy: A Strategy for the Future of Chat Moss, Greater Manchester*. Greater Manchester Countryside Unit, Manchester

Isaac, D. & Reid, C. (1991). *Amendments to the Ancient Woodland Inventory for England, July 1994 – February 1996*. English Nature Research Reports, No. 177. English Nature, Peterborough.

Jackson, H.C. (1979). The decline of the sand lizard, *Lacerta agilis* L., population on the sand dunes of the Merseyside coast, England. *Biological Conservation*, **16**, 177–193.

Mersey Basin Campaign (1994). *Mersey Estuary Management Plan, Draft*. Deposited at Mersey Basin Campaign, Voluntary Sector Network, 111 The Piazza, Piccadilly Plaza, Manchester. M1 4AN.

Morries, G. (1986). *Lancashire's Woodland Heritage*. Lancashire County Council, Preston.

Moss, B., McGowan, S., Kilinc, S. & Carvalho, L. (1992). *Current Limnological Condition of a Group of the West Midland Meres that Bear SSSI Status*. Final Report of English Nature Research Contract Number F72-06-14, Department of Environmental and Evolutionary Biology, University of Liverpool.

National Rivers Authority (1994). *River Irwell Catchment Management Plan: Consultation Report, Chapter one – River Irwell introduction*. Deposited at Environment Agency (as above).

National Rivers Authority (1994). *River Irwell Catchment Management Plan: Water Quality Supplement*. Deposited at Environment Agency (as above).

National Rivers Authority (1995). *River Gowy Rapid Corridor Survey*. Deposited at Environment Agency (as above).

National Rivers Authority (1995). *River Irwell Catchment Management Plan: Action Plan*. Deposited at Environment Agency (as above).

National Rivers Authority (1995). *The Mersey Estuary A Report on Environmental Quality*. Water quality series, **No 23**. HMSO, London. Deposited at Environment Agency (as above).

National Rivers Authority North West (1995). *Stillwaters Project Summary*. Deposited at Environment Agency (as above).

Nature Conservancy Council (1984). *Nature Conservation in Britain*. NCC, Peterborough.

Nature Conservancy Council (1990). *Handbook for Phase 1 Habitat Survey*. NCC, Peterborough.

Newton, A. (1971). *Flora of Cheshire*. Cheshire Community Council, Chester.

Phase 1 Habitat Survey of Greater Manchester (1990/1992). Greater Manchester Ecology Unit. Deposited at GMEU, Council Offices, Wellington Road, Ashton-under-Lyne, Tameside, OL6 6DL.

Phillips, P.M. (1994). *Lancashire Inventory of Ancient Woodlands (Provisional)*. English Nature, Peterborough.

Robinson, D.P. & Whitbread, A. (1988). *Cheshire Inventory of Ancient Woodland (Provisional)*. NCC, Peterborough.

Ward, D., Holmes, H. & José, P. (eds) (1994). *The New Rivers and Wildlife Handbook*. RSPB, Sandy.

Savage, A.A. (1976). *The Nature and History of the Cheshire Meres*. News Bulletin of the Cheshire Conservation Trust, **2 (9)**, 1–2.

Webb, N. (1986). *Heathlands*. The New Naturalist. Collins, London.

CHAPTER EIGHT

Man-made habitats of the Mersey Basin: what is new?

H.J. ASH

Introduction

Mankind's activities in the Mersey Basin have destroyed much wildlife habitat. However, they have also created new opportunities, often accidentally. Looking at the Basin's areas, it is easy to be depressed by the lack of opportunity for species other than humans. A few forms have prospered at the expense of many others. In urban areas such as Knowsley and St Helens, half the urban greenspace (20% total urban area) is devoted to amenity grassland, dominated by Perennial Rye-grass (*Lolium perenne*) and a handful of other grasses and rosette species (Gilbert 1989; St Helens Wildlife Advisory Group 1986). This supports such a limited range of organisms, it has been dubbed 'green desert'. Another quarter of the greenspace (10% urban area) is occupied by rough, unmanaged grassland, dominated by False Oat-grass (*Arrhenatherum elatius*) and Cock's-foot (*Dactylis glomerata*) with a handful of tall herbs such as thistles (*Cirsium arvense*, *C. vulgare*), docks (*Rumex* spp.) and Rosebay Willow-herb (*Chamerion angustifolium*). This is floristically poor, but has some value to small mammals, including children, who are often short of informal play space close to home. However, hidden away among the debris of past industrial activity, there are sites supporting a surprising range of wildlife; not replacements for what has been lost, but different communities which would not be here without man's intervention.

Calcareous sites

The Mersey Basin has few naturally-occurring calcareous areas; just the coastal dunes and Pennine fringe. However, limestone is used in a range of industrial processes, in the chemicals and glass industries particularly, and forms the basis of a number of waste deposits:

1. blast furnace slag, e.g., Kirkless Lane, Wigan;

2. Leblanc waste from 19th-century manufacture of sodium carbonate and bleaching powder, e.g., Nob End, Bolton; St Helens; Bury; Widnes;

3. calcium sulphate from hydrofluoric acid production, e.g., Runcorn.

Examples of each of these have been left untreated since tipping ceased, often decades ago, so that although the substrate is unnatural, the colonisation is a natural primary succession. Such calcareous sites start with a pH of 8–10, weathering to a level not dissimilar to a limestone grassland (Ash 1983, 1991). Nutrient levels, especially nitrogen and phosphorus, are very low, with phosphorus fixed in insoluble forms by the calcium. Drought, extreme surface temperatures and, on blast furnace slag, very stony texture add to the problems of colonising plants. The nearest natural calcareous habitats are 30–40kms away, so these areas are 'islands' in a sea of acidic/neutral substrates. The colonising flora is a mixture of such widespread species as can cope with the edaphic conditions, e.g., Red Fescue (*Festuca rubra*), Cock's-foot, Colt's-foot (*Tussilago farfara*), and species which have succeeded in spreading from calcareous habitats, primarily the sand dunes (Ash, Gemmell & Bradshaw 1994). Prominent among the latter are the marsh and spotted orchids (*Dactylorhiza praetermissa*, *D. fuchsii* and hybrids, and *D. incarnata*) and sometimes Fragrant Orchid (*Gymnadenia conopsea*), but others include Common Centaury (*Centaurium erythraea*), Blue Fleabane (*Erigeron acer*) and Creeping Willow (*Salix repens*). One site, Nob End near Bolton, is now a Site of Special Scientific Interest (SSSI) for its flora (Table 8.1), and is a country park. Most of these long-distance colonisers have good dispersal mechanisms, e.g., wind-blown seeds. Other species from equivalent habitats will succeed if introduced artificially, e.g., Yellow-wort (*Blackstonia perfoliata*) and Autumn Gentian (*Gentianella amarella*) on Leblanc waste (Ash *et al.* 1994). On very infertile sites such as these the succession does not start with annuals – probably there are just not enough nutrients to complete a life cycle in one season. The primary colonisers are perennial herbs and grasses, and the succession proceeds from open grassland with herbs and bare ground, to a closed, usually species-rich, grassland community. The usual tree colonists of derelict land, Silver Birch (*Betula pendula*) and willows (*Salix caprea*, and *S. cinerea*) do not succeed on these calcareous wastes, but Hawthorn (*Crataegus monogyna*) will gradually invade if seed sources are available and grazing animals are absent. Growth rates are very slow, but some 115

Acer pseudoplatanus	+	Dactylorhiza praetermissa	3	Plantago lanceolata	4
Achillea millefolium	2	Dactylorhiza purpurella	3	Plantago major	+
Achillea ptarmica	+	Equisetum arvense	2	Pohlia nutans	1
Agrostis capillaris	1	Euphrasia nemorosa	1	Potentilla anglica	1
Agrostis stolonifera	3	Festuca arundinacea	2	Potentilla reptans	+
Angelica sylvestris	2	Festuca ovina	2	Ranunculus acris	+
Anthoxanthum odoratum	1	Festuca rubra	5	Ranunculus repens	1
Arabidopsis thaliana	+	Gymnadenia conopsea	4	Rubus fruticosus	3
Arrhenatherum elatius	1	Heracleum sphondylium	3	Rumex acetosa	+
Bellis perennis	4	Hieracium sabaudum	1	Sambucus nigra	+
Carex flacca	2	Hieracium vulgatum	2	Senecio jacobaea	1
Carlina vulgaris	2	Holus lanatus	1	Sisyrinchium bermudiana	2
Centaurea nigra	6	Hypochoeris radicata	2	Solidago canadensis	1
Centaurium erythraea	1	Lathyrus pratensis	2	Succisa pratensis	5
Cerastium fontanum	1	Leontodon autumnalis	+	Taraxacum agg.	1
Cirsium arvense	2	Linum catharticum	4	Tragopogon pratensis	4
Cirsium vulgare	+	Lolium perenne	1	Trifolium pratense	4
Crataegus monogyna	4	Lotus corniculatus	7	Trifolium repens	1
Dactylis glomerata	4	Molinia caerulea	1	Tussilago farfara	2
Dactylorhiza fuchsii	2	Ophioglossum vulgatum	1	Urtica dioica	+
Dactylorhiza fuschii x D.praetermissa	2	Orobanche minor	2	Orchis morio recorded 1970s, but not since.	
Dactylorhiza incarnata	3	Pilosella officinarum	3		

Table 8.1. Plant species list for Leblanc waste area of Nob End SSSI, near Bolton. 7.5ha, pH 7.9–8.2. Waste tipping ceased 1881. Large orchid populations noted by 1954. Estimated DOMIN cover values.

years after abandonment, Nob End is having to be managed to control scrub and maintain the species-rich grassland.

Acidic sites

Man has been unintentionally creating habitats on acidic substrates in Merseyside for centuries. Wirral still has considerable areas of lowland heath at Thurstaston Common, Heswall Dales, Caldy Hill and several smaller sites. Although most heaths have been created by human activity (Webb 1986) it is likely that at least some have an ancient and natural origin. Thus, some of the coast-facing slopes of Wirral heaths may be truly 'natural', but most are the product of the activities of man and his animals on thin, acid sandy soils and Cowell (chapter four, this volume) shows that humans have lived in the area continuously since Mesolithic times. Most of the areas are dry maritime heath, with Western Gorse (*Ulex gallii*) and Bell Heather (*Erica cinerea*) abundant among the Heather (*Calluna vulgaris*). Small damp areas support Cross-leaved Heath (*Erica tetralix*), Deergrass (*Trichophorum cespitosum*), Bog Asphodel (*Narthecium ossifragum*) and even relict populations of Round-leaved Sundew (*Drosera rotundifolia*) and Marsh Gentian (*Gentiana pneumonanthe*). Grazing and controlled burning faded from these heaths in the first half of this century, as development reduced and fragmented them. Birch, Scots Pine (*Pinus sylvestris*) and Oak (*Quercus* spp.) have invaded, converting some sites, e.g. Irby Common,

to secondary woodland. Considerable conservation effort has been expended on the larger sites in recent years, removing scrub and controlling Bracken (*Pteridium aquilinum*). On Thurstaston Common, the National Trust has re-instated grazing in three large paddocks, using a small flock of Herdwick wethers (sheep). Once initial problems with some dogs and their owners had been overcome, these have proved successful managers.

Man's activities have also served to change the nature of existing soils, especially through the effects of acid rain. In the early 19th century a Leblanc works was established in Liverpool (Jarvis & Reed, chapter six, this volume), the hydrochloric acid from whose chimney had such devastating effects on the farmers of Everton that the owner was forced to move to Earlestown (where his waste heap, known as Mucky Mountains, is extant adjacent to the Sankey Canal, supporting an interesting flora including Pyramidal Orchid (*Anacamptis pyramidalis*) and Quaking-grass (*Briza media*)). More generally, the acidity from coal-burning has lowered the pH of the soils of the older parks such as Sefton Park, Liverpool, especially where soils were thin or sandy, rather than clay. Common Bent (*Agrostis capillaris*) and Red Fescue dominate in such grasslands, with Common Bird's-foot-trefoil (*Lotus corniculatus*), Autumn Hawkbit (*Leontodon autumnalis*) and Heath Bedstraw (*Galium saxatile*). Where limewash was used to line out the sports pitches, a narrow band of less calcifuge plants provides more permanent markings: better grass growth and the presence of species such as Creeping Buttercup (*Ranunculus repens*) and Daisy (*Bellis perennis*) (Bradshaw 1980).

Coal fuelled industry in the Mersey Basin for around 200 years. Its extraction left many waste deposits, which have been the major target of four decades of reclamation programmes. The shales associated with the Lancashire coalfield usually have a high pyrites content, which under natural weathering releases acids (Bradshaw & Chadwick 1980). As a consequence most of the Basin's colliery wastes are acidic (pH 3–5), and reclamation techniques have centred on treatment with large quantities of limestone, followed by grass and tree establishment. However, there remain sufficient examples of colliery shale, and other acidic wastes such as clinker, cinders and sandstone, to illustrate the succession.

When new, these materials are severely deficient in nutrients, especially nitrogen (at least in available form) and phosphate. They are subject to extreme temperatures, drought, poor physical structure, erosion, and, on recent deposits, compaction by heavy machinery. Whereas calcareous wastes weather to a neutral or slightly alkaline substrate, acidic ones only ameliorate as an organic layer accumulates on the surface. However, they are somewhat less isolated from potential colonisers.

As with calcareous wastes, the first colonisers are perennials: Common Bent, Yorkshire-fog (*Holcus lanatus*), Rosebay Willow-herb, Colt's-foot and hawkweeds (*Hieracium* section *Sabauda*). Pioneer trees (*Betula* spp. and *Salix* spp.) often colonise at this early stage of very open vegetation, especially if seed sources are close, but grow very slowly (Curtis 1977). Usually the vegetation slowly closes over 50 years or more, becoming grassland with scrub and eventually Birch/Willow woodland with a species-poor, grassy understorey. If oaks (usually *Quercus robur*) are nearby, and carriers for their acorns such as Jays (*Garrulus glandarius*) or people, they will establish at the grassland stage and eventually dominate the woodland. On some less hostile spoils Sycamore (*Acer pseudoplatanus*) also invades as the organic layer accumulates.

Sometimes the succession takes other routes (Figure 8.2). On very acid spoils Wavy Hair-grass (*Deschampsia flexuosa*) can dominate to the virtual exclusion of other species. Elsewhere, especially on small rural heaps, various species characteristic of moorland occur: Matgrass (*Nardus stricta*), Purple Moor-grass (*Molinia caerulea*), Heather and Crowberry (*Empetrum nigrum*). Other acidic wastes develop similarly, but sometimes become colonised by Gorse (*Ulex europaeus*), which may form impenetrable stands; Gorse is rare on colliery shales. The ecology of these different routes is little understood, although seed sources and (for Gorse) phosphate levels may be presumed to play a part. A better understanding would be worthwhile, not least to help those trying to develop alternative reclamation techniques, which do not involve digging up environmentally sensitive limestone areas to reclaim wastes!

Figure 8.1. Nob End Country Park, Bolton. SSSI on Leblanc waste (*Photo: H. Ash*).

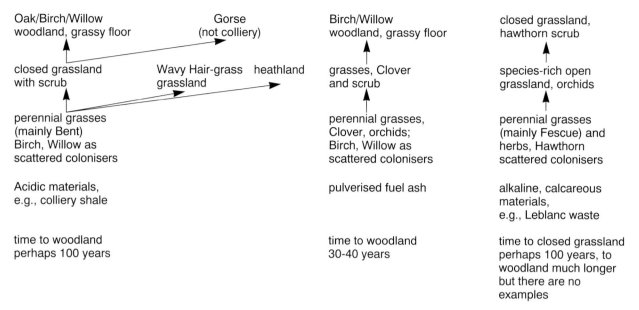

Figure 8.2. Typical plant successions on industrial wastes in the Mersey Basin.

Other unusual substrates

Calcareous wastes bear some resemblance to limestone soils, and colliery wastes to natural acidic substrates, but not all man-made substrates have such analogues. One of the few industrial wastes still to be deposited in large amounts is pulverised fuel ash (PFA), from coal-burning power stations. Unlike colliery shale or chemical wastes, it is fairly benign, with a silty texture, some phosphate (though no nitrogen) and a pH just above neutral, gradually reducing under weathering (Ash 1983). However, it is liable to form cemented layers which impede root growth, and when fresh it has toxic levels of boron. These leach sufficiently to allow plant growth within 5–10 years (less if the ash is lagooned). Some sites are initially saline and are colonised by plants tolerant of brackish conditions: Spear-leaved Orache (*Atriplex prostrata*) and Red Goosefoot (*Chenopodium rubrum*). On less saline sites, or as the salts leach out, perennial grasses and herbs establish, often the same suite of species, capable of survival on very low nutrients, that appear on other wastes: Yorkshire-fog, bents, Red Fescue, Colt's-foot and Field Horsetail (*Equisetum arvense*). Clovers (*Trifolium* spp.) are at a double advantage, being boron-tolerant as well as nitrogen-fixing, and can dominate a site in early years. Frequently there are swarms of marsh and spotted orchids with spectacular hybrids. Birch and Willow also colonise while the vegetation remains open, and proceed to turn the area into woodland. This succession can be relatively swift; PFA tipped into a subsidence flash at Wigan Power Station took 20 years to develop spectacular marsh orchid colonies, and within another 10 years was dense Willow/Birch woodland, very good for birds especially Short-eared Owls (*Asio flammeus*), but needing management to retain an interesting ground flora. With such management Marsh Helleborine (*Epipactis palustris)* and Round-leaved Wintergreen (*Pyrola rotundifolia*)

have flourished and some marsh orchids have been retained (S. Crombie pers. comm.).

Often industries are or were associated with a particular area. St Helens' glass production resulted in several large heaps of Burgy waste, a mixture of sand and jeweller's rouge once used for polishing plate glass. It is rarely found elsewhere in the country. Similarly Leblanc waste is almost confined to north-western England, and Prescot has an area of copper-contaminated soils. Burgy waste was tipped behind bunds, rising to 15m and

Figure 8.3. Marsh Helleborine on PFA, Wigan (*Photo: H. Ash*).

covering several hectares. It is largely dominated by Tall Fescue (*Festuca arundinacea*), a species frequently found on alkaline soils of industrial origin. One heap, topped with sand washings and war-time sandbags as well as Burgy, supports sand dune species and well developed Willow scrub with many birds, and is listed as of major ecological importance to the town: that did not stop it being given permission on appeal for housing. It had not been built on by July 1996, but the planning permission remained in force. Such industrial wastes have links to local history, and are often the only remaining relics of the industry on which the development of the community was based, and thence have cultural as well as biological significance. These sites are valuable to wildlife and to people. They often include some of the best sites for wildlife in urban areas, and the most available opportunities for recreation in semi-natural surroundings. The problem is that it is still difficult to defend them from proposals for redevelopment or even reclamation. I believe they must be defended, just as much as ancient woodlands and grasslands, if they and the plants and animals they support are to survive.

Another common substrate in urban areas, especially where demolition occurs, is brick rubble. However, the issues associated with this substrate are considered by Bradshaw (chapter twelve, this volume).

Wetlands

Such considerations do not just apply to dry land. Man's activities have decimated the region's natural wetlands, but created a variety of new ones; canals, starting with the Sankey Navigation in 1757, reservoirs, subsidence flashes and many small marshes where drainage has been impeded. Ponds are treated in detail by Boothby & Hull (chapter fourteen, this volume). In water bodies isolated from land drainage, such as some industrial reservoirs and canals, water quality can be good and a range of aquatic species flourish – duckweeds (*Lemna* spp.), pondweeds (*Potamogeton natans, P. crispus, P. pectinatus*), waterweeds (*Elodea* spp.), water-starworts (*Calitriche* spp.), but also unusual plants. Fringed Water-lily (*Nymphoides peltata*) grows along considerable stretches of the Leeds-Liverpool Canal in Liverpool, Floating Water-plantain (*Luronium natans*) in Manchester, White Water-lily (*Nymphaea alba*) at Top Dam, Eccleston, and Water-soldier (*Stratiotes aloides*) is embarrassingly successful since introduction to a number of Wirral marl pits. Many such sites are popular with anglers, whilst the large area of subsidence flashes south of Wigan are a major locality for migrating and resident birds.

Marshes colonise quickly, and can become valuable wildlife habitats in a few decades. Kraft Fields at Knowsley Industrial Estate was stripped of topsoil in the 1960s. Within 20 years it had a plant list of over 100 species (Table 8.2), including Common Cottongrass (*Eriophorum angustifolium*), Common Club-rush (*Schoenoplectus lacustris*), marsh orchids and large areas

Achillea millefolium	R	*Juncus conglomeratus*	O
Achillea ptarmica	O	*Juncus effusus*	O
Agrostis capillaris	R	*Juncus inflexus*	R
Agrostis stolonifera	O	*Lapsana communis*	R
Alchemilla xanthochlora	R	*Lathyrus pratensis*	O
Angelica sylvestris	F	*Lathyrus sylvestris*	R
Anthoxanthum odoratum	R	*Lemna minor*	R
Apium nodiflorum	R	*Leontodon autumnalis*	O
Arrhenatherum elatius	F	*Leucanthemum vulgare*	O
Artemisia vulgaris	R	*Listera ovata*	R
Aster novi-belgii	O	*Lotus corniculatus*	O
Bellis perennis	R	*Lotus pedunculatus*	O
Calystegia sepium	R	*Melilotus officinalis*	R
Carex demissa	R	*Mentha arvensis*	R
Carex flacca	F	*Oenanthe crocata*	R
Carex hirta	O	*Ononis repens*	R
Carex ovalis	R	*Ophrys apifera*	R
Carex pseudocyperus	R	*Phalaris arundinacea*	F
Carex remota	R	*Phleum pratense*	R
Centaurea nigra	R	*Plantago lanceolata*	O
Centaurium erythraea	R	*Plantago major*	R
Cerastium fontanum	R	*Potentilla anserina*	F
Chamerion angustifolium	F	*Potentilla reptans*	F
Cirsium arvense	R	*Prunella vulgaris*	O
Cirsium vulgare	R	*Pulicaria dysenterica*	R
Convolvulus arvensis	R	*Quercus robur*	R
Crataegus monogyna	R	*Ranunculus repens*	F
Cynosurus cristatus	O	*Rubus caesius*	R
Cytisus scoparius	R	*Rubus fruticosus*	O
Dactylis glomerata	F	*Rumex acetosella*	R
Dactylorhiza fuchsii	O	*Rumex crispus*	R
Dactylorhiza fuschii x *praetermissa*	O	*Rumex obtusifolius*	R
Dactylorhiza incarnata	R	*Salix cinerea*	R
Dactylorhiza praetermissa	O	*Salix repens*	R
Deschampsia flexuosa	R	*Salix viminalis*	R
Dipsacus fullonum	R	*Schoenoplectus lacustris*	R
Eleocharis palustris	F	*Senecio erucifolius*	R
Elytrigia repens	O	*Senecio jacobaea*	R
Epilobium hirsutum	F	*Sisyrinchium bermudiana*	R
Epilobium montanum	R	*Solidago canadensis*	R
Epilobium palustre	O	*Sonchus arvensis*	R
Epilobium parviflorum	O	*Stachys palustris*	O
Equisetum arvense	O	*Stachys sylvatica*	R
Equisetum palustre	O	*Tragopogon pratensis*	R
Eriophorum angustifolium	R	*Trifolium hybridum*	R
Fallopia japonica	R	*Trifolium medium*	O
Festuca rubra	R	*Trifolium pratense*	O
Galium palustre	R	*Trifolium repens*	O
Geranium dissectum	R	*Tussilago farfara*	O
Glyceria fluitans	R	*Typha latifolia*	F
Glyceria notata	R	*Ulex europaeus*	R
Heracleum sphondylium	O	*Vicia cracca*	O
Hieracium sabaudum	R	*Vicia hirsuta*	R
Holcus mollis	O	*Vicia sativa*	R
Holus lanatus	F	*Vicia sepium*	R
Hypochoeris radicata	O	*Viola arvensis*	R
Juncus acutiflorus	R		
Juncus articulatus	F		

Table 8.2. Flora of Kraft Fields, Knowsley Industrial Estate, Merseyside. Former farmland, topsoil stripped 1960s leaving sands and clays. 10ha. pH 6.5–7.3. DAFOR abundances.

of rushes (*Juncus* spp.), Bulrush (*Typha latifolia*) and Great Willowherb (*Epilobium hirsutum*). A boardwalk was installed and the site used for education. At Dibbinsdale in Wirral, a dam installed on the River Dibbin in the 1860s appears to have caused the formation of a large reed-marsh (*Phragmites australis*) on the flood plain upstream, which now has breeding Sedge Warbler (*Acrocephalus schoenobaenus*) and Reed Warblers (*A. scirpaceus*) and is an important part of its SSSI. Marshes can be attractive communities to people, especially where tall herbs give good floral displays, e.g., an area at Eastham, Wirral, which has suffered impeded drainage since the M53 was built, where Southern Marsh-orchid (*Dactyloriza praeter-missa*) and Common Fleabane (*Pulicaria dysenterica*) flourish, or the bright fringe to many polluted waterways provided by Indian or Himalayan Balsam (*Impatiens glandulifera*). Leachate from tipped waste can have inter-esting effects; the alkaline leachate from a tip at Runcorn supports Sea Club-rush (*Bolboschoenus maritimus*), while saltmarsh has come as far inland as St Helens where Reflexed Saltmarsh-grass (*Puccinellia distans*) has colonised the leachate from the Burgy banks.

More saline-tolerant plants may become obvious in future. The effects of spreading salt on our roads are now visible every spring as a band of pale lilac along the very edge of main road verges. The plant responsible is Danish Scurvygrass (*Cochlearia danica*), a common species around the coasts on walls, banks, sand and rock

(chapter nineteen, this volume). Countrywide a number of other maritime species have been reported from road-sides (Scott 1985), such as saltmarsh-grasses (*Puccinellia* spp.) and Lesser Sea-spurrey (*Spergularia marina*).

Urban wetlands are just as vulnerable to drainage and natural succession as their traditional counterparts. They are unlikely to be drained for agriculture, but changing industrial needs, new development and failure of old dams can all change water levels. Once again, decisions have to be made over what is to be retained and how to do it.

Management

There is another problem in relation to wildlife on derelict land. All these sites are undergoing succession, either primary or secondary. Some of them have reached a stage which is particularly valuable to wildlife, such as species-rich grassland, of which semi-natural versions have suffered drastic declines in recent decades. Or they may be particularly attractive to people, like the orchids on PFA. If such sites are to be kept in a desired state, they will need management, just as semi-natural sites do, and this will require time and resources. Most of these sites have been totally neglected since they were formed, so no-one is in the habit of taking care of them. It has been suggested for PFA that new sites should be created to replace existing ones as they follow their natural succes-

Figure 8.4. New Ferry Butterfly Park, Wirral: school visit to former railway goods yard, now community nature reserve (*Photo: H. Ash*).

sion (Shaw 1994); that may be extreme, but thought is needed as to whether the best waste sites should be conserved, and how, and we should be prepared to take action where necessary. Many such sites are irreplaceable, since industries have changed, and fortunately modern pollution legislation does not allow some past practices. Even if new sites are created, and people are prepared to live with an ugly view for decades while succession proceeds, new sites may never become as rich as the older ones, because in the modern, depauperate landscape the seed sources are much sparser and poorer in species than 50 or 100 years ago.

One way to address both these problems is to get the local community involved. At Bebington railway station on Wirral, the goods yard became derelict in the 1960s, and in 1993 was turned into 'New Ferry Butterfly Park'. It has a variety of soils – coal from the staithes, lime from a water treatment works which softened water for steam engines, ballast from old sidings, building rubble and clay. Over the years a wide variety of plants and invertebrates moved in; 23 species of butterfly have been recorded, at least 12 breeding. None is rare, but the variety is remarkable for a small area of 1.4ha in a densely-built, urban area. A local resident tried for 20 years to get recognition and protection for the site. The railway company, who owned the site, had planning permissions, but failed to attract a buyer, because of poor ground conditions and inadequate road access. Eventually the last planning permission lapsed and Cheshire Wildlife Trust were able to lease the site and now help a group of local people to run it. Extensive management work was needed, to erect fencing, lay paths, clear mounds of rubbish, thin scrub, dig a pond, control bramble, start grassland management; all done by volunteers and grants to maintain and enhance its wildlife value. The Park was officially opened in July 1995 by Lyndon Harrison, MEP for the area and since then local schools have used it for field studies. All this has not stopped the railway company from applying to have part of the site zoned for housing in Wirral's Unitary Development Plan: needless to say there have been vociferous objections, and in July 1996 the results of a public inquiry were awaited hopefully! [In August 1997 the Inspector refused the zoning for housing.]

If such 'man-made', or at least 'man-started' sites are considered valuable to our wildlife and culture, the community will have to work to maintain and defend such areas, just as much as traditional wildlife sites.

New-made communities

Most of the habitats so far considered have arisen by natural processes on man-made sites, with little or no interference in their successions. In recent years conscious 'habitat creation' has begun to make an impact on the towns of the Mersey Basin, with deliberately planned landscapes, intended primarily to improve the amenity value of urban areas, but also increasing the opportunities for wildlife. Knowsley Borough Council, among other efforts to improve the Borough's image, has worked with the Groundwork Trust and Landlife over a decade, on both practical projects and research into effective techniques. At Tobruk Road, Huyton, the Alt valley used to be a canalised stream bordered by mown grassland. The stream still runs in its canal, but the banks have been landscaped with tree and shrub plantings, in substantial blocks, and areas of wild flower grassland (Luscombe & Scott 1994). Management is something of a problem; mowing gangs do not always get the boundaries to the wildflower areas correct, leaving areas of amenity grassland unmown which become unsightly as the season progresses. This is at least better than the problem at Tower Hill, Kirkby, where wildflower areas were mown early in 1995 instead of being left to flower. Disposal of long cuttings after the autumn cut is a problem, until large-scale composting schemes are established. Stadtmoers Country Park, Whiston is a large area established on former derelict land – claypits and collieries, some subsequently filled with domestic refuse. A wildflower grassland sown on sandy subsoil in 1983 shows that this habitat can be sustained, given the correct soil conditions and management. Other parts of the Park have extensive grasslands and experimental areas, used to test methods for enriching both managed and unmanaged grasslands (Ash, Bennett & Scott 1992). The improvements of these plantings to the visual landscape are obvious; systematic work on their effects on wildlife are still lacking, although casual observation indicates that insect life certainly increases in the short term, and sustainable bird populations would be expected to increase as time goes on. Usually these landscapes replace mown amenity grassland or newly-demolished buildings, and can only increase the local opportunities for wildlife, and for human contact with that wildlife. Re-creation of lost natural habitats, such as ancient woodlands and meadows, is not possible, but these new habitats can enrich urban areas, offer new homes for some species, and by allowing urban people contact with nature, improve their lives and the value placed on the other species with which we share our planet. Landlife's wildflower seed production grounds in Knowsley are visually spectacular in flower; they also allow Skylarks (*Alauda arvensis*) to breed.

Conclusion

Man's activities in the Mersey Basin have left a range of habitats of wildlife interest. They demonstrate plant succession, offer homes to a wide range of species and can be attractive to people. If the best are to be retained, they will need active protection and management, just as much as more natural habitats. Many are in urban areas and, along with deliberately-created habitats, can improve people's lives and contribute to persuading them to value the natural world in which we all live.

References

Ash, H.J. (1983). *The Natural Colonisation of Derelict Industrial Land*. PhD thesis, University of Liverpool.

Ash, H. J. (1991). Soils and vegetation in urban areas. *Soils in the Urban Environment* (eds P. Bullock & P.J. Gregory), pp. 153–172. Blackwell Scientific Publications, Oxford.

Ash, H.J., Bennett, R. & Scott, R. (1992). *Flowers in the Grass*. English Nature, Peterborough.

Ash, H.J, Gemmell, R.P. & Bradshaw, A.D. (1994). The introduction of native plant species on industrial waste heaps: a test of immigration and other factors affecting primary succession. *Journal of Applied Ecology*, **31**, 74–84.

Bradshaw, A.D. (1980). Mineral Nutrition. *Amenity Grassland: An Ecological Perspective* (eds I.H. Rorison & R. Hunt), pp. 101–18. Wiley, London.

Bradshaw, A.D. & Chadwick, M.J. (1980). *The Restoration of Land*. Blackwell Scientific Publications, Oxford.

Curtis, M. (1977). *Trees on Tips*. MSc dissertation, University of Salford.

Gilbert, O.L. (1989). *The Ecology of Urban Habitats*. Chapman & Hall, London.

Luscombe, G. & Scott, R. (1994). *Wildflowers Work*. Landlife, Liverpool.

Shaw, P. (1994). Orchid Woods and Floating Islands – the Ecology of Fly Ash. *British Wildlife* **5**, 149–57.

Scott, N.E. (1985). The updated distribution of maritime species on British roadsides. *Watsonia*, **15**, 381–86.

St Helens Wildlife Advisory Group (1986). *A Policy for Nature*. Available from: The Land Manager, Community Leisure Department, Century House, Hardshaw Street, St Helens, Merseyside.

Webb, N. (1986). *Heathlands*. The New Naturalist. Collins, London.

CHAPTER NINE

Recent landscape changes: an analysis post-1970

C.J. BARR AND G.J. STARK

Introduction to landscape change

Landscape change means different things to different people and good, consistent definitions are needed so that the statistics of change may be properly understood. Landscape change may represent changes in land use from, say, food production to housing, shifts which are driven by the changing needs and priorities of an ever more affluent society. Some changes are permanent and irreversible, such as the conversion of permanent pasture to arable crops, while others represent part of a normal agricultural rotation. While landscape change is often measured quantitatively (e.g., the length of hedgerow lost), it is important to note that more subtle, qualitative changes may also be taking place, e.g., in the species composition of grasslands.

In the natural sciences, landscape change may be analysed in terms of (in descending order of spatial organisation): land cover (e.g., crops, woodland, and urban land); landscape elements and habitats (e.g., trees, hedges and ponds); and flora and fauna, often represented by populations which fluctuate over a long time-scale. Other interest groups might be concerned with visual amenity (e.g., changing seasonal patterns); cultural changes (which are often irreversible); and socio-economic aspects (clear driving forces behind both urban and rural change). While landscape change is very important in terms of its consequential effects on the flora and fauna, it is a large, complex, and involved area for study and not one which the biologist should tackle in isolation.

Scale and definitions

Analysis of change at the regional scale, and its relationship to the national picture, is an important task for a wide range of organisations with responsibilities for planning and resource assessment. Care must be taken, however, in the interpretation of regional data and especially when analysing change. Differences in the definitions of categories used by different data gatherers, or differences in the geographical extent of an area, may be far more significant than any 'real' changes detected by comparison of data sets. For example, the Forestry Commission North West Conservancy stretches from Cumbria to Warwickshire (inclusive), while the Department of the Environment (DOE) 'Standard Region' classification considers the North West to comprise Lancashire, Greater Manchester, Merseyside and Cheshire only. For the best estimates of change, landscape surveys must be undertaken in the same geographical region, using identical methods and definitions. The results of such surveys are rare indeed, and so compromises have to be made and the relationships between different surveys properly understood.

Sources of information

Many sources of information are available from different agencies to describe landscapes in the Mersey Basin. Statistics published by the Forestry Commission (FC) and based on their 1980 woodland census, estimate that there was a total of 1,677ha of woodland in the county of Merseyside and a woodland cover density of 2.1% for the county of Greater Manchester (Locke 1987). The Ministry of Agriculture, Fisheries and Food (MAFF) Digest of Agricultural Census Statistics for 1994 indicates the area of grassland in Cheshire to have been 132,000ha, 79% of agricultural land in the county (MAFF 1991 to 94). According to the 1990 Land Cover Map of Great Britain, created from an analysis of satellite imagery (e.g., Fuller & Parsell 1990), there was a mean density of 21.6ha per km square of urban and suburban land in the Mersey Basin in that year. From these and other statistics we can describe the landscape of the Mersey Basin as one with a high density of urban areas, little woodland and the majority of agricultural land being under grass.

However, many of these sources do not adequately convey landscape change. Although there were relatively early surveys on land use, e.g., the First and Second Land Utilisation Surveys (Stamp 1937–47; Coleman 1961), consistent information on change at the national or regional level has not become available until relatively recently. There are several examples of regional change statistics in both the published and the unpublished literature. Data on agricultural crops and uses are published annually (e.g., MAFF. 1992) showing, for example, a 25%

Greater Manchester

	1974	%	1979	%	1984	%	1989	%	1994	%
Total Agr. land (major)	48777	%	45604	%	43321	%	42158	%	40647	%
Arable	8198	17	7674	17	7299	17	7841	18	6238	15
Wheat	632	1	566	1	850	2	1674	4	2094	5
Barley	5971	12	5834	13	4986	12	4231	10	2654	7
Crops for stockfeed	195	0	155	0	212	0	233	1	389	1
Potatoes	703	1	720	2	807	2	668	2	523	1
Rape	0	0	38	0	277	1	428	1	323	1
Sugar beet	0	0	45	0	42	0	44	0	41	0
Other arable crops	697	1	316	1	125	0	203	0	214	0
Horticulture	992	2	987	2	966	2	915	2	752	2
Vegetables	858	2	890	2	844	2	797	2	642	2
Orchards and small fruit	8	0	5	0	26	0	22	0	15	0
Glass	57	0	17	0	17	0	18	0	19	0
Bulbs, flowers and HNS	69	0	75	0	79	0	78	0	76	0
Grass and other land	39630	81	36940	81	35056	81	33762	80	33720	83
Under 5 yrs	3828	8	2757	6	3553	8	3387	8	3252	8
Over 5 yrs and other land	35801	73	34183	75	31503	73	30375	72	30468	75

Cheshire

	1974	%	1979	%	1984	%	1989	%	1994	%
Total Agr. land (major)	175326	%	174810	%	175166	%	173298	%	167468	%
Arable	37312	21	37242	21	34402	20	35849	21	33941	20
Wheat	3753	2	4351	2	7121	4	9830	6	9416	6
Barley	24684	14	24807	14	19446	11	17089	10	11764	7
Crops for stockfeed	1636	1	2302	1	1085	1	2009	1	5995	4
Potatoes	3944	2	4325	2	4599	3	4235	2	4148	2
Rape	8	0	55	0	1158	1	1626	1	1512	1
Sugar beet	34	0	310	0	121	0	154	0	161	0
Other arable crops	3253	2	1902	1	872	0	906	1	945	1
Horticulture	1458	1	1648	1	1512	1	1299	1	1076	1
Vegetables	873	0	981	1	714	0	574	0	536	0
Orchards and small fruit	279	0	351	0	423	0	345	0	168	0
Glass	26	0	30	0	35	0	43	0	42	0
Bulbs, flowers and HNS	280	0	286	0	340	0	337	0	330	0
Grass and other land	136558	78	135920	78	139252	79	136150	79	132408	79
Under 5 yrs	39967	23	32741	19	31961	18	28158	16	29008	17
Over 5 yrs and other land	96591	55	103179	59	107291	61	107992	62	103400	62

Table 9.1. Agricultural statistics for the counties of Greater Manchester, Merseyside, Cheshire and Lancashire for the period 1974 to 1994. Areas are in hectares (Source: MAFF 1973 to 79; MAFF 1980 to 89 and MAFF 1991 to 94).

increase in the planted area of oilseed rape between 1989 and 1990. County figures are available although not published routinely. Similarly, the Forestry Commission (e.g., Forestry Commission 1983) publishes data at intervals, but with regional summaries; for example non-woodland trees increased in volume by about 57% in the North West Conservancy between 1951 and 1980.

Other sources of information on landscape change include: the 'Monitoring Landscape Change' project funded by the DOE and the Countryside Commission (Hunting Surveys and Consultants Ltd 1986); the former Nature Conservancy Council's 'National Countryside Monitoring Scheme', completed for only a few counties in England, but including Cumbria (Budd 1989); the National Park Monitoring project (Countryside Commission 1991); the Phase I and Phase II habitat surveys of the national conservation agencies (e.g., Moreau 1990); and the monitoring undertaken by the Ordnance Survey on behalf of the DOE (Department of the Environment 1992). The County Councils in the North West are in the process of producing Green Audits (e.g., Lancashire County Council 1991), and further new initiatives include the Countryside Commission's 'New Map of England'. It is unclear how these initiatives will contribute to landscape change statistics in the future.

The Institute of Terrestrial Ecology (ITE) has carried out a series of national surveys of land cover, landscape features, and vegetation, using a sampling approach. Since the methods and definitions are constant between surveys, reliable estimates of change (together with statistical error terms) can be generated. The latest of these surveys was 'Countryside Survey 1990' (CS1990) which involved not only a detailed field survey of 508 representative 1km squares in Great Britain (GB), but also the construction of a Land Cover Map of the whole land surface, from remotely-sensed data (Barr *et al.* 1993).

To find out what other information would be avail-

Merseyside

	Year									
	1974	%	1979	%	1984	%	1989	%	1994	%
Total Agr. land (major)	**21094**	%	**20345**	%	**20002**	%	**20606**	%	**19528**	%
Arable	**12112**	57	**11483**	56	**11702**	59	**12383**	60	**10111**	52
Wheat	1225	6	836	4	1688	8	3528	17	3453	18
Barley	7874	37	8676	43	7668	38	6012	29	3543	18
Crops for stockfeed	119	1	166	1	314	2	895	4	991	5
Potatoes	1291	6	1228	6	1251	6	925	4	776	4
Rape	0	0	24	0	312	2	505	2	740	4
Sugar beet	111	1	290	1	219	1	202	1	224	1
Other arable crops	1492	7	263	1	250	1	316	2	384	2
Horticulture	**1634**	8	**1616**	8	**1945**	10	**1335**	6	**929**	5
Vegetables	1556	7	1508	7	1736	9	1160	6	822	4
Orchards and small fruit	23	0	39	0	127	1	98	0	27	0
Glass	13	0	15	0	12	0	16	0	15	0
Bulbs, flowers and HNS	42	0	54	0	70	0	61	0	65	0
Grass and other land	**7348**	35	**7246**	36	**6355**	32	**6888**	33	**8488**	43
Under 5 yrs	2662	13	1811	9	1615	8	1344	7	1507	8
Over 5 yrs and other land	4686	22	5435	27	4740	24	5544	27	6981	36

Lancashire

	Year									
	1974	%	1979	%	1984	%	1989	%	1994	%
Total Agr. land (major)	**239887**	%	**235570**	%	**223997**	%	**222841**	%	**220441**	%
Arable	**24434**	10	**27439**	12	**24487**	11	**24620**	11	**2118**	10
Wheat	3266	1	2411	1	4147	2	6821	3	7910	4
Barley	16769	7	18117	8	15147	7	11616	5	6232	3
Crops for stockfeed	893	0	815	0	846	0	1862	1	2366	1
Potatoes	3009	1	3031	1	3207	1	2657	1	2683	1
Rape	0	0	24	0	456	0	834	0	827	0
Sugar beet	64	0	338	0	296	0	307	0	338	0
Other arable crops	432	0	2703	1	388	0	523	0	824	0
Horticulture	**4844**	2	**5625**	2	**6058**	3	**5192**	2	**5367**	2
Vegetables	4493	2	5322	2	5679	3	4828	2	4935	2
Orchards and small fruit	58	0	34	0	103	0	72	0	26	0
Glass	158	0	178	0	169	0	184	0	184	0
Bulbs, flowers and HNS	134	0	91	0	107	0	108	0	222	0
Grass and other land	**210609**	88	**202506**	86	**193452**	86	**193029**	87	**193895**	88
Under 5 yrs	10668	4	12704	5	15019	7	13821	6	16169	7
Over 5 yrs and other land	199941	83	189802	81	178433	80	179208	80	177726	81

able to study recent landscape change in the Mersey Basin, the authors consulted local organisations such as County Councils and Universities. The response demonstrated that whilst some local information might be available, e.g., as a set of records or aerial photographs, it would be difficult to compile consistent data sets for the Mersey Basin as a whole and almost impossible to draw conclusions about change from these.

Using county-based data

For information on the Mersey Basin, it is possible to extract data from sources of national statistics published by county. The counties of Cheshire, Merseyside and Greater Manchester are almost exclusively in the Mersey Basin, accounting for approximately 50%, 15% and 30% of the region respectively. Lancashire occupies a further 5% of the northern part of the Basin.

June agricultural census returns are published annually by MAFF and county figures are included for some topics (MAFF 1973 to 79; MAFF 1980 to 89; MAFF 1990; MAFF 1991 to 94). County figures for Cheshire,

Merseyside, Greater Manchester and Lancashire during the period 1974 and 1994 are shown in Table 9.1. The chosen categories are those which have more or less constant definitions over the period.

Using the countryside information system

The Countryside Information System (CIS) is a computer-based package which gives a user the ability to define particular areas of GB (e.g., Scotland, National Parks, land over 100m, Bedfordshire, or the Mersey Basin) and to compute countryside data for the specified region. Where change data exist, then changes can be computed for the same areas. Information can be overlaid so that the area of coniferous woodland on land over 150m in Wales could be calculated simply. The CIS was developed as a tool for analysing, interrogating and presenting data about the countryside in a way which was responsive to policy questions (Haines-Young, Bunce & Parr 1994) and, although developed originally to present Countryside Survey data to a wide user community, it is not limited to the use of such data and allows any spatially referenced database to be incorporated.

The initial development was through extensive evaluation and testing of prototype systems by users in government departments and agencies (Howard *et al.* 1994). CIS is able to provide access to a wide range of information about the British countryside by adopting the 1km squares of the National Grid as a standard by which to summarise data. Many kinds of information about the countryside can be presented as either 'presence/absence in', or 'area occupied within', a 1km square. In CIS terminology 'presence/absence' datasets are called region files (an example would be all 1km squares which contain part of the Mersey Basin) and 'area occupied' datasets are known as census files (for example, the number of hectares of woodland in all 1km squares).

An additional feature of CIS is the ability to view data collected on a sample basis. CIS integrates sample data with other data sources through the ITE Land Classification (Bunce *et al.* 1996). This is a classification of all 1km squares in GB into 32 Land Classes, based on a multivariate analysis of a range of environmental parameters (such as geology, climate and physiography). A sample dataset in CIS consists of a mean and associated variance for each Land Class. These values can be used to estimate the amount of a feature for any region defined by the user, based on the occurrence of the ITE Land Classes. Sample data for change of land use are available for the period between CS1990 and an earlier ITE survey of land use conducted using the same methodology in 1984. These data can be used to make estimates for the Mersey Basin region, though the CS1990 field survey was designed for calculation of national statistics and estimates for small regions should be treated with caution. Estimates of statistical error are given and areas such as the Mersey Basin are at the lower end of the acceptable range. However, in the absence of any other data, the CIS does provide consistent estimates of the stock and change of landscape features in the Mersey Basin.

Some specific examples of landscape change in the Mersey Basin

Agriculture

Between 1974 and 1994, MAFF data show that there was an overall decline in the total area of agricultural land in the four counties of 37,000ha, amounting to 8% of the total area of agricultural land (Figure 9.1). Considerable fluctuations in the number of agricultural holdings in the four counties were recorded over the period (Figure 9.2). At least part of the explanation for this is likely to be the periodic reassessment of holdings included in the census. Overall there were 874 fewer holdings at the end of the period. Figure 9.3 contrasts land use on agricultural holdings at the beginning and end of the period for each county. Change has been slight with the most

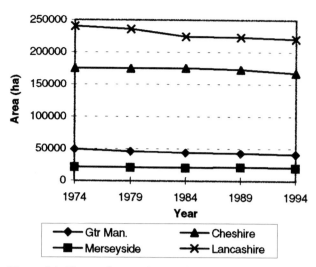

Figure 9.1. The total area of agricultural land in the counties of Cheshire, Merseyside, Greater Manchester and Lancashire between 1974 and 1994 (Source: MAFF 1973 to 79; MAFF 1980 to 89 and MAFF 1991 to 94).

noticeable change being from arable land to grass and other land uses, in all four counties.

Changes in land use are recorded for the DOE by the Ordnance Survey (OS) as map revisions are undertaken. Statistics are published by DOE as the 'Land Use Change in England' series and county figures for change of land use to urban and residential are available for the period

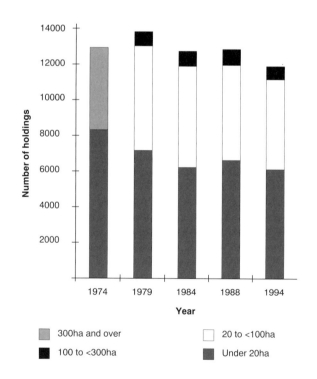

Figure 9.2. The number of agricultural holdings in Cheshire, Merseyside, Greater Manchester and Lancashire between 1974 and 1994 (Source: MAFF 1973 to 79; MAFF 1980 to 89 and MAFF 1991 to 94). Figures for the larger size classes in 1974 are not available, since earlier farm size ranges quoted in acres have no equivalent in the current hectare ranges.

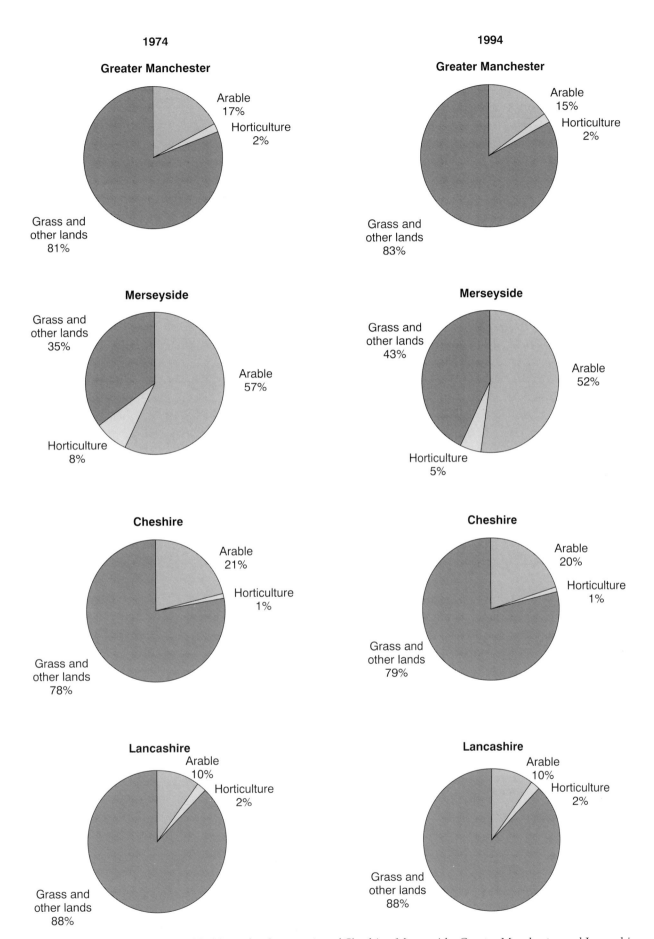

Figure 9.3. Land use on agricultural holdings for the counties of Cheshire, Merseyside, Greater Manchester and Lancashire for 1974 and 1994 (Source: MAFF 1973 to 79 and MAFF 1991 to 94).

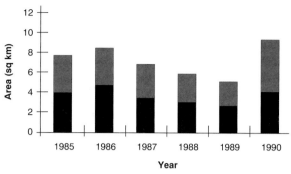

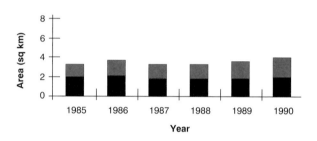

Figure 9.4. Changes in land use to urban and residential for the counties of Cheshire/Lancashire and Merseyside/Greater Manchester between 1985 and 1990 (Source: DOE 1992 to 95).

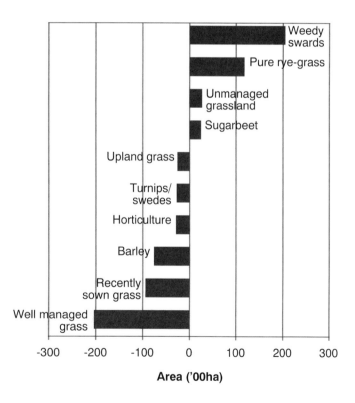

Figure 9.5. Estimated land cover changes in the Mersey Basin region between 1984 and 1990. Figures were calculated in CIS using data from CS1990. Only land covers with over 2,000ha of change have been included.

1985 to 1990 (Department of the Environment 1992 to 95). Figure 9.4 shows the area of land which was recorded as changing to urban or residential for each year, apportioned according to whether the previous land use had been rural or urban. Over the six-year period for which statistics are available, 32ha of land previously classed as rural was recorded as converted to urban and 17ha to residential, in the four Mersey Basin counties. These figures demonstrate a gradual conversion from rural to urban land use, though they do not account for the larger losses of agricultural land indicated by the MAFF statistics.

CS1990 results for land cover are summarised under 58 categories. Land cover categories for which a change of greater than 2,000ha is estimated are shown in Figure 9.5. The largest changes are increases of 11,600ha of pure Rye-grass and 20,300ha of weedy swards and decreases of 20,100ha of well managed grass and 9,400ha of recently sown grass. These shifts mirror those seen nationally and are a result of changing grassland management practices towards less frequent re-seeding. The consequences of these changes are the more frequent occurrence of weedy swards and the use of non-seeding Rye-grass species.

The landscape impact of these changes to grassland management may be slight. More significant for the landscape of the Mersey Basin may be the shift away from arable crops. Barley in particular has declined over this period, an estimated 7,400ha for the Mersey Basin between 1984 and 1990 according to the CS1990, and a recorded 8,299ha for the counties of Cheshire, Merseyside, Greater Manchester and Lancashire between 1984 and 1989 according to MAFF statistics (MAFF 1980–89).

Woodland and forests

Figures from the 1980 FC woodland census are available by county (Locke 1987). The density of woodland in all four counties (Cheshire 3.8%, Greater Manchester 2.1%, Merseyside 2.6% and Lancashire 3.7%) is low by comparison with other counties in England. There is a lack of comparable data at the beginning and end of the period under consideration and, anyway, changes in the method of recording woodlands have made the detection of real change difficult (Peterken 1983).

Figures for the area of woodland on agricultural holdings are available by county for the period 1973 to 1979 and then 1990 from the MAFF statistics (MAFF 1973–79

and MAFF 1990). The amount of woodland increased by 1,908ha between 1973 and 1979 and had increased by a further 1,712ha by 1990 (Figure 9.6).

Landscape features

CS1990 included the recording of landscape features such as field boundaries. Figure 9.7 shows changes to field boundaries between the 1984 and 1990 surveys within the region. Overall there has been a decrease of 1,730km (5%) in the length of boundaries in the region. Within this total the length of hedgerow has declined by 20% which is comparable with the national estimate of 23%. In GB as a whole, it has been shown that much of this change is due to lack of hedgerow management leading to hedges growing into lines of trees, or degenerating into scattered shrubs, rather than outright removal (Barr, Gillespie & Howard 1994).

General trends in the landscape, post-1970

This cursory analysis of available statistics points to a period of relative stability in the landscape of the Mersey Basin since 1970. There was an overall reduction in the area of agricultural land and a commensurate loss in the number of agricultural holdings. Within these holdings there have been few changes in the use of the land although there is a reduction in the area of arable crops. There have been modest increases in the area of woodland on farmland and, as elsewhere in GB, hedgerows have been lost, probably due to a lack of management.

The apparent general stability in landscape change, post-1970, may however disguise changes in the quality of landscape elements, as indicated by the subtle changes in grassland types detected within CS1990. These might be due to intensification of land use in some areas and relaxation of management (or even set aside) in other parts of the region. Current work at ITE is looking at changes, and causes of change, in the botanical composition of landscape elements.

Conclusions

In gathering information for this paper on landscape change, it has become clear that it is difficult to find relevant data which have been collected in a consistent way for the Mersey Basin as a whole, and therefore it is difficult to compute change statistics with confidence. Potentially, the CIS is a useful tool in this respect as it allows data from different sources to be compared for any specified region.

Overall, there appears to be little net change in land use and landscape features in the basin in the period post-1970. It remains predominantly an agricultural area, mostly under grassland systems, but also with a relatively large proportion of urban land. However, some results suggest that there may be changes in the quality of the landscape and this provides scope for more research.

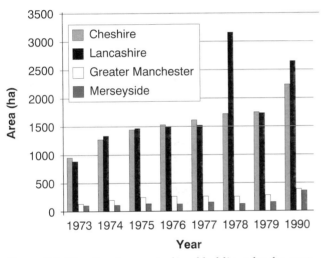

Figure 9.6. Woodland on agricultural holdings for the counties of Cheshire, Merseyside, Greater Manchester and Lancashire for the period 1973 to 1979 and 1990 (Source: MAFF 1973 to 79 and MAFF 1990).

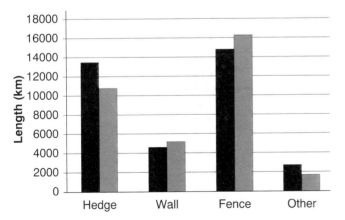

Figure 9.7. Changes to boundary features in the Mersey Basin region between 1984 and 1990. Figures were calculated in CIS using data from CS1990. Where several boundary types were present in a single boundary preference was given in the order: hedge, wall, fence then other.

References

Barr, C.J., Bunce, R.G.H., Clarke, R.T., Fuller, R.M., Furse, M.T., Gillespie, M.K., Groom, G.B., Hallam, C.J., Hornung, M., Howard, D.C. & Ness, M.J. (1993). *Countryside Survey 1990 Main Report*. Volume 2 in the Countryside 1990 series. Department of the Environment, London.

Barr, C.J., Gillespie, M.K. & Howard, D. (1994). *Hedgerow Survey 1993 stock and change estimates of hedgerow lengths in England and Wales, 1990–1993*. Department of the Environment, Bristol.

Budd, J.T.C. (1989). National Countryside Monitoring Scheme. *Rural Information for forward planning* (ITE symposium no. 21) (eds R.G.H. Bunce & C.J. Barr). Institute of Terrestrial Ecology, Grange-over-Sands.

Bunce, R.G.H., Barr, C.J., Clarke, R.T., Howard, D.C. & Lane A.M.J. (1996). Land classification for strategic ecological survey. *Journal of Environmental Management*, **47**, 37–60.

Coleman, A. (1961). The second land use survey: progress and prospect. *Geographical Journal*, **127**, 168–86.

Countryside Commission. (1991). *Landscape change in the national parks*: summary report of a research project carried out by Silsoe College. Countryside Commission, Cheltenham.

Department of the Environment. (1992). *Land Use Change in England*. DOE Statistical Bulletin (92)3. DOE, London.

Department of the Environment. (1992 to 95). *Land Use Change in England* No. 7 to 10. DOE Statistical Bulletin, London.

Forestry Commission. (1983). *Census of Woodlands and Trees 1979–82 (North West England)*. Forestry Commission, Edinburgh.

Fuller, R.M. & Parsell, R.J. (1990). Classification of TM imagery in the study of land use in lowland Britain: practical considerations for operational use. *International Journal of Remote Sensing*, **11**, 1901–17.

Haines-Young, R.H., Bunce, R.G.H. & Parr, T.W. (1994). Countryside Information System: an information system for environmental policy development and appraisal. *Geographical Systems*, **1**, 329–45.

Howard, D.C., Bunce, R.G.H., Jones, M. & Haines-Young, R.H. (1994). *Development of the Countryside Information System*. Volume 4 in the Countryside 1990 series. Department of the Environment, London.

Hunting Surveys and Consultants Ltd. (1986). *Monitoring Landscape Change*. Huntings, Borehamwood.

Lancashire County Council. [1991]. *Lancashire – a green audit: summary*. Lancashire County Council, Preston.

Locke, G.M.L. (1987). *Census of Woodlands and Trees 1979–82*. Forestry Commission Bulletin 63. Forestry Commission, Edinburgh.

MAFF. (1973 to 79). *Agricultural Statistics England and Wales*. Volumes 1973 to 1978/1979. HMSO, London.

MAFF. (1980 to 89). *Agricultural Statistics United Kingdom*. Volumes 1980 and 1981 to 1989. HMSO, London.

MAFF. (1990). *Final Results of the June 1990 Agricultural and Horticultural Census: England and Wales, Regions and Counties*. MAFF, London.

MAFF. (1991 to 94). *The Digest of Agricultural Census Statistics: Volumes 1991 to 1994*. HMSO, London.

MAFF. (1992). *Agriculture in the United Kingdom: 1991*. HMSO, London.

Moreau, M. (1990). *The Phase I Habitat Survey in Bedfordshire*. Nature Conservancy Council, Letchworth.

Peterken, G. (1983). Woodland surveys can mislead. *New Scientist*, **100** (1388), 802–03.

Stamp, L.D. (1937–47). *The Land of Britain: The Final Report of the Land Utilisation Survey of Britain*. Geographical Publications, London.

Changing Habitats 1

E.F. GREENWOOD

In the previous two sections the changing landscape of the Mersey Basin since the retreat of the last glaciation, which began some 15,000 years ago, has been described and the increasing effect of human impact noted.

In this and subsequent sections a more detailed look is taken at selected habitats, plants and animals that live or have lived in the region. In the first section, on changing habitats, an analysis of three of the most important terrestrial ones is undertaken. Chapter ten shows what happened to the woodland that covered so much of the Mersey Basin. At the time of Doomsday over a thousand years ago there was plenty of woodland, but even then less than 30% of Cheshire was woodland. In the following centuries much of this woodland was lost through human impact, but from about the end of the 18th century plantations and parkland trees were planted. Nevertheless, the Mersey Basin is today one of the least woodland areas in Europe with only 4% cover or less. A determined effort is now being made to increase the amount of woodland with the creation of the Mersey and Red Rose Forests.

Much of the east of the Mersey Basin is upland and even here woodlands once covered a large part of the area (chapter eleven). From the earliest times human intervention has reduced this woodland to remnants. In many places bogs developed with *Sphagnum* spp. being a major component of the vegetation. But the effects of the industrial revolution were highly damaging and most of the *Sphagnum* was killed off. Today the uplands continue to endure severe human impact, especially from overgrazing by sheep, resulting in a degraded landscape with erosion in many places.

Chapter twelve demonstrates what is happening to plants and animals in the large urban habitats created by human intervention. Here chance and the adaptability of plants and animals to changed and sometimes hostile environments is shown. These positive reactions, if allowed, result in the development of woodland once again.

These three chapters all show the considerable and sometimes alarming deleterious effect of human intervention over not hundreds, but thousands of years. Yet they also show hope and promise through a combination of the considerable resilience and adaptability of many plants and animals to changed circumstances, and more recently efforts by humans themselves to improve the environment, so creating new opportunities for wildlife to flourish.

CHAPTER TEN

A history of woodland in the Mersey Basin

D. ATKINSON, R.A. SMART, J. FAIRHURST, P. OLDFIELD
AND J.G.A. LAGEARD

Introduction

From prehistoric times to the present day, the history of woodland in the Mersey Basin, like that in much of industrial north-west Europe, has been dominated by deforestation and damage with relatively modest re-planting. The present-day counties of Cheshire, Greater Manchester and Merseyside, whose boundaries define the geographical limits of this review, are among the least wooded in Europe; each with 4% woodland cover or less (see chapter nine, this volume). This compares unfavourably with the *c*.10% cover in Britain as a whole (Forestry Commission 1997). Woodland appears always to have provided a resource. But how people have viewed it and therefore exploited or managed it has changed many times and often quite dramatically, even in recent decades. Details of these changes since the Norman Conquest, and their effects especially on the landscape contained within the present-day boundaries of Cheshire and Merseyside, form the subject of this chapter. We conclude with an assessment of the wood-land resource for the 21st century: this provides some hope for the future, but also points to problems that need to be addressed if ecological and economic sustainability are to be achieved.

Domesday woodland and the medieval Royal Forests

Despite the difficulties in assessing the extent of wood-land cover at the time of Domesday, two estimates for Cheshire based on a variety of evidence suggest that the county was well-wooded with perhaps 25–27% cover (Rackham 1986, p. 78; Yalden 1987). Since much more land was capable of supporting woodland, it appears that the majority had been cleared for agriculture before Domesday. Administrative areas (hundreds) in the west of the county, including Wirral, which was both densely populated (Table 10.1) and adjacent to two navigable river estuaries, contained little taxable woodland. Yet to the south and south-east (e.g., in the Macclesfield hundred) woodland cover was extensive (Table 10.1, Figure 10.2). In the hundred of West Derby to the north of the River Mersey, it seems that except for pre-

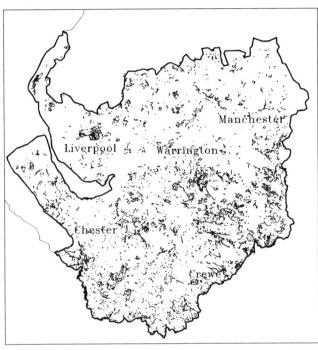

Figure 10.1. Woodlands in the Mersey Basin. Woodland of more than 2ha area in the counties of Cheshire, Greater Manchester and Merseyside. Data obtained from aerial surveys flown in 1993.

Hundred	Woodland (km²)	Population (per km²)
Wirral	5.78	1.43
Bucklow West	21.44	0.66
Bucklow East	37.05	0.23-0.27
Broxton	39.02	0.39-1.24
Eddisbury North	25.50	1.0
Eddisbury South	35.01	0.27-1.27
Nantwich	129.08	0.39-0.69
Northwich	86.95	0.39-0.58
Macclesfield	579.93	0.15
	959.76	

(Calculated from Yalden 1987)

Table 10.1. Cheshire Domesday woodland recorded by Hundreds.

Forest	Boundaries	Earliest reference	Changes in extent	Year of disafforestation
Delamere	Rivers Mersey (N), Gowy (W), Weaver (E) and Weaver tributary (S). Precise boundaries unclear (Green 1979, p. 172).	Indirectly in Domesday. By name in about 1129 (Green 1979, p. 172).	By 1600 almost all of old forest of Mondrem excluded (Green 1979, p. 172).	1812
Macclesfield	Rivers Mersey (N), Goyt (E), and Dane (S). Western boundary not clearly defined (Green 1979, p. 179).	Indirectly in Domesday (Morris 1978 1:25, 1:26). By name in a Charter of 1153–60 (Green 1979, p. 178).	The forest of Leek detached in the 13thC (Green 1979, p. 178).	1684 (Green 1979, p. 184).
Wirral	The Wirral peninsula (Green 1979, p. 184).	1194–1228 Charter (Green 1979, p. 185).	Stanney Grange disafforested by Ranulph III (1181–1232) (Green 1979, p. 187).	1376 (Green 1979, p. 187).
West Derby	West Derby Hundred south west of a line running approximately from Southport to Warrington (James 1981, p. 76).	By 1199 (Farrer & Brownhill 1990, pp. 1–2).	Toxteth disafforested about 1593 (Shaw 1956, p. 466).	Uncertain. A 1716 survey omits any mention of West Derby Forest (Shaw 1956, p. 466).
Rossendale (area between Accrington and Bacup only)	Southern and eastern boundary the River Irwell as far as its source on Thieveley Pike (James 1981, p. 76).			About the beginning of the 16thC (James 1981, p. 76).

Table 10.2. Medieval Royal Forests in the Mersey Basin (disafforestation refers to the removal of Forest Law).

Conquest nucleated settlements, such as at Kirkby and Eccleston, much of the area was wooded including considerable tracts of mossland and heath (Cowell & Innes 1994, p. 133). Studies of place names in West Derby have given an indication of past woodland cover and landscape. For instance, the derivation of Akenheaved, a minor name in Aintree township, suggests the top of an oak headland (Russell 1987, pp. 173–74).

The Royal Forests of Delamere and Mondrem, Macclesfield, Wirral, West Derby, and part of Rossendale were extensive tracts of unenclosed land in the region designated as royal hunting reserve in the 11th century. They contained not just woodlands, but also villages and areas of cultivation, pasture, and wasteland owned partly by the king as demesne and partly by his subjects (James 1981, Chapters 1–3). The Forest Law, which helped to preserve the woodland, including wood-pasture, probably helped to conserve biodiversity for up to several hundred years before disafforestation (Table 10.2).

Imparkment or enclosure of land, woodland or forest for sporting purposes occurred on an increasing scale from the 13th century onwards (e.g., Aston Wood, Cheshire (Warburton Muniments); see also Table 10.3). Medieval parks were often compartmented in order that deer, pannage and woods could be managed efficiently (Lasdun 1991, p. 7; Ives 1976, p. 51).

Woods regenerated naturally and were managed as coppice or woodpasture in which occasional trees were left to grow on for timber. Coppice mainly of Oak

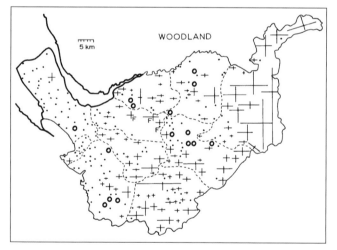

Figure 10.2. The distribution of woodland in Domesday Cheshire. The lines indicate the length and breadth of woodland. Places named in Domesday but with no mention of woodland are indicated by dots. Places with small amounts of woodland – less than 144ha ($\frac{1}{2}$ league × $\frac{1}{2}$ league) – are shown as hollow circles. Manors associated with Delamere Forest are marked F (after Yalden 1987).

(*Quercus* sp.), Ash (*Fraxinus excelsior*), Alder (*Alnus glutinosa*), Willow (*Salix* spp.), Hazel (*Corylus avellana*), and Wych Elm (*Ulmus glabra*) provided the basic needs of fuel, fencewood and poles (Rackham 1990, pp. 55, 65). Pollards in wood pastures provided similar material.

Of the indigenous British trees, Beech (*Fagus sylvatica*),

Location	Year	Reference
Neston	1258	Tait 1923, p. 302.
Knowsley	1292	Farrer & Brownhill 1990, p. 158.
Tarvin	1299	Green 1979, p. 178.
Shotwick	1327	Stewart-Brown 1912, p. 100.
Arley	by 1383	Arley Charters 1866. Lease of 11th Nov, 7. Ric II 1302, Peter de Werbeton to Wm.de.Wermyncham.
Lyme	after 1388	Elizabeth Banks Associates (1993) Restoration Management Plan 2.2.
Adlington	1462	Green 1979, p. 179.

Table 10.3. Some examples of medieval imparkment

Hornbeam (*Carpinus betulus*) and Scots Pine (*Pinus sylvestris*) were probably not present in Mersey Basin woodlands in 1086, though Scots Pine had been a dominant species in the distant past (chapter three, this volume). An introduced tree, the English Elm (*Ulmus procera*) was almost certainly present while it is possible that the Romans had brought Sweet Chestnut (*Castanea sativa*) and Walnut (*Juglans regia*) to the area, but these trees were most likely to have been planted round settlements rather than in woodlands (Evans 1984, pp. 168, 204, 208; Mitchell 1984).

Woodland losses

Domesday to the Industrial Revolution

Pressures on the forest increased with the rise in population until the mid-14th century and many assarts or woodland clearances took place (Green 1979; Farrer & Brownhill 1990; Newbigging 1868; Shaw 1956; Cowell & Innes 1994, also Table 10.2). This was seen in West Derby, for instance, in the 13th century where agricultural expansion was also associated with a large increase in rectorial corn tithes and manorial grants (Berry 1980, p. 8). By the early 16th century about half of the forest of Macclesfield was occupied by settlers with farms (Davies 1961, p. 44).

Dwindling wood supplies prompted a 1482 Act to protect coppice regrowth (22 Edw. IV, Cap. 7), and later the 1543 Preservation of Woods Act (35 Henry VIII, Cap. 15). Tudor alterations to Bramall Hall were one victim of the shortage of good oaks, and had to be carried out using old timbers.

The Civil War also affected the woodland resource. Overcutting and devastation of the Deer populations during that war altered the forest composition, and by 1661 many of the 2,200 oaks that were estimated to be present in Delamere Forest at the start of the war had been lost. Only enough trees for fuel remained (Green 1979, p. 176). During that period the forest areas at Delamere became predominantly heathland grazed by 20,000 sheep (Palin 1843, p. 19).

Wood demands increased and shortages occurred as industries developed. Tree-ring evidence suggests that

Oak was used from the 12th century in the salt industry infrastructure (Leggett 1980; see also Table 10.9). Wood or charcoal provided fuel for the salt, glassmaking, iron and copper industries, but were gradually replaced by coal in the 17th and 18th centuries (King 1656; Awty 1957; Vose 1977; Carlon 1979, 1981; see Table 10.4). Until the 17th century, supplies of fuelwood for the salt industry in Nantwich were partly met from woods at Awsterton, Cheshire, which were managed on twenty-year coppice rotations (King 1656, p. 66). A fivefold increase in Lancashire's coal production between the time of Elizabeth I (1558–1603) and 1700 was partly attributed to an increasing shortage of timber which pushed up wood prices and caused people to search for a substitute (Challinor 1972, pp. 10–11). But coal-mining in Lancashire and Cheshire also required ever-increasing supplies of pit timber as production rose further in the 18th century and mines became deeper.

Large quantities of bark, especially Oak, were required for tanning hides (Table 10.4), and often generated more income than the timber beneath it. Early in the 18th century bark was being exported to Ireland (Davies 1960, p. 12), but by the beginning of the 19th century Cheshire tanners were having great difficulty finding sufficient supplies of bark (Holland 1808, p. 197).

Many wooden naval, as well as other ships, were built in Liverpool after 1739 (Northcote Parkinson 1952, p. 105). Woodland exploitation was preferentially near to supply routes, where shortages were therefore most acutely felt, as John O'Kill indicated in 1763: 'as to the decrease of timber fitting for His Majesty's use in Lancashire, Cheshire, and North Wales I believe fifteen parts out of twenty are exhausted within these fifty years. I mean what was growing near any navigable river' (Fisher 1763, pp. 46–47). Indeed, it was the need for Oak for shipbuilding which was the driving force behind forest development until the change from wooden ships to ironclads in the mid-19th century (James 1981, p. 189).

The Industrial Revolution and two World Wars

The effects of industrialisation are illustrated by the Garston Hall estate, south Liverpool. All the 40 woods and shelter belts planted between 1800 and 1840 had been destroyed by about 1930 following the development of railway marshalling yards associated with Garston Docks and subsequent residential development (Berry & Pullan 1982). Other examples are listed by Handley (1982).

Timber imports increased considerably following industrialisation and the development of the railways in the 19th century (Forestry Commission 1921). At the start of the First World War home-grown timber barely met 10% of the nation's requirements and extensive fellings were made during the war to make up for the deficit in imports (Forestry Commission 1921). In Cheshire 1,310ha were felled (Forestry Commission 1928). The Forestry Commission was established in 1919 partly in response to these national shortages (James 1981). Grants

Species		Uses	Period	Reference
Oak				
	25–100 yrs	Building timber	Medieval – 19thC	Rackham (1991)
	Larger trees	Bridge bearers, mill posts	Medieval – 19thC	Bagley (1968)
	Unspecified	Shipbuilding	Particularly from 17thC	Smart (1992)
		Fencing, furniture, gates, wooden pipes.	Medieval – 13thC	Leggett (1980)
	Bark	Tanning, the most favoured species.	Medieval largely until early 20thC	Edlin (1949, p.87) Hodson (1978, p.138)
Ash		Shafts, tools, handles, carts, rails, wheels, bentware.		Smart (1992, p.116)
		Cooper timber	18thC	Bagley (1968)
Elm		Wheelwrights timber. Furniture, coffins, wooden pipes, vessels and storage ships. Underwater work. Keels and bottoms of flats.		Smart (1992, p.116).
Poplar		Flooring, boarding in carts and boxes, low-grade furniture and fittings.		Smart (1992, p.116). Holland (1808, p.206)
		Rebuilding bridge over Mersey at Stretford.	1745	Jarvis (1944).
Alder		Turnery, clogs		Edlin (1949, p.23–24)
	Poles	Drying cotton yarn	18thC	Holt (1795, p.85)
	Bark	Tanning		Edlin (1949, p.87)
	Bark	Dyes	18thC, 19thC	Holland (1808, p.206)
Birch		Turnery		Edlin (1949, p.42)
		Bobbins, spools, reels	19thC	Edlin (1949, p.42)
		Lancashire cotton industry		
	Twigs	Besoms	Medieval onwards, 1713	Bagley (1968) Edlin (1949, p.42)
	Bark	Tanning	Medieval on	Edlin (1949, p.87)
Willow		Basket work		Holt (1795, p.85)
		Baskets for salt	18thC	Holland (1808, p.206)
	Bark	Tanning		Edlin (1949, p.87)
Unspecified		Mining timber	Mainly from late 17thC	Shercliff, Kitching & Ryan (1983)
		Fuel for salt industry	Medieval to 1690	King (1656)
		Fuel for glassmaking		
		in Vale Royal	1284–1309	Vose (1977)
		in West Derby	c.1600	Vose (1995)
		Charcoal – iron forges	From 1619 at Tib Green, Doddington to the 18thC	Awty (1957)
		Copper smelting (e.g., Alderley Edge and Gallantry Bank, Cheshire (small scale)	Up to 18thC	Carlon (1979)

Table 10.4. Some historical uses of timber prior to the 20th Century

were made available to encourage planting and restocking of both conifers and hardwoods, but take-up was slow. By 1930 there had been little success in either arresting the deterioration of the home woodlands in private ownership or restoring the pre-war position (James 1981). However, locally there was more activity. On the Peckforton Estate in Cheshire, for example, where there were widespread war-time fellings, 86ha were reforested between 1922 and 1927 (Peckforton Estate Papers). Of the extensive fellings during the Second World War in the Mersey Basin, 960ha were from Cheshire of which 138ha were in Delamere Forest (Forestry Commission 1952; Forestry Commission Delamere Records). On the Sefton coast, net losses from all causes were estimated at about 20% for the period 1925–45 (Joint Countryside Advisory Service 1990).

Post-war losses

From about 1960 to 1982 Cheshire lost an estimated 660ha of woodland, Greater Manchester 190ha, and Merseyside 60ha – a total of 910ha due mainly to urban development and associated land use and agriculture (Forestry Commission 1984). There was some compensation by planting along motorways and trunk roads, and the natural colonisation of disused railway tracks.

Areas of coppice have constantly declined. In the 1924 census Cheshire had 134ha of coppice and 157ha of coppice-with-standards. By the time of the 1947–49 census there were 21ha of coppice and no coppice-with-standards, and in the 1979–82 census just 13ha of coppice-with-standards. At that time there was only 1ha of coppice recorded in Greater Manchester and none in Merseyside. It should be noted that Cheshire county boundaries changed in 1974 and the figures are not strictly comparable.

Pathogens and pollution

Pathogens have always played an important role in tree health, but their impact will have been altered by the widespread importation of timber; the development of near-monoculture plantations; improved control measures for pests and pathogens; and the increase in other stress-factors, especially pollutants which may alter the balance of the relationship between a tree and its pathogens (e.g., Innes 1987). The most catastrophic effect of a pathogen in recent decades was the outbreak of Dutch Elm disease which spread through the area during the 1970s and 1980s. The first official record of Dutch Elm disease in Merseyside was from Birkenhead in 1973. Whether or not this arrived as a direct import from Canada, as suggested by Marshall & Dawkins (1981), or was simply part of the general northward spread of the disease is not clear. By 1975 the disease was a major problem. However, a lack of political and financial commitment resulted in many diseased trees being left standing for well over a year (Greig & Gibbs 1983). The consequences of this were substantial. Of an estimated 76,000 highway trees in Merseyside in the mid-1970s about 18,600 were elms. Between 1975 and 1981 an estimated 10,800 of these elms were felled. Since highway trees may comprise only about a third of total trees in the county (Marshall & Dawkins 1981) and since the disease continued well into the 1980s, the total number of elms lost in the county is likely to have been several tens of thousands. The argument that pollutants or other stressors were necessary before the disease could overcome the defences of the trees was countered by Heybroek, Elgersma & Scheffer (1982) who noted that it was the trees showing strong growth that were the first to fall prey, while trees that had practically stopped growing suffered less. Instead, they said the disease could be regarded simply as an ecological accident resulting from the importation of an aggressive form of a fungal pathogen from one continent (North America) to another (Europe). A further impression of the impact of this 'accident' in the Mersey Basin area can be seen from the proportion of dead trees noted in the 1979–82 census that were Elms. In Merseyside, 55% of dead trees were elms; in Cheshire, 52%; and in Greater Manchester, 10% (Forestry Commission 1984). Given the history of this 'accident' it is appropriate to ask if lessons can be learned that are relevant for other potential pathogens, e.g., Oak Wilt and Ash Dieback?

Dramatic losses of trees caused by pollution were observed in the 1880s when emissions from chemical factories at Weston near Runcorn killed trees within a distance of 13km (8 miles), including 5,000 trees on the Norton estate (Dodd 1987). It was also suggested (Farrar, Relton & Rutter 1977) that the distribution of Scots Pine in the industrial Pennines is limited by concentrations of sulphur dioxide in the air.

Impacts of pollutants are sometimes more subtle, as was observed in a wood in Prescot, Merseyside, following a hundred years of metal processing in the town (Dickinson *et al.* 1996). Despite the uptake of the metals by the trees, which were mainly Sycamore (*Acer pseudoplatanus*), only a very limited effect was observed on tree growth before 1970. With emissions declining, this effect became even less after 1970. However, toxicity symptoms were observed in tree seedlings and the decay of leaf litter was inhibited at the site.

Other sub-lethal effects of pollutants were observed at Delamere Forest, where the goldening of Pine needles and needle loss in year three were associated with high levels of sulphur in the foliage which may be related to high levels of sulphur dioxide in combination with other pollutants (Inman & Reynolds 1995).

Development of plantations

Landscape, timber and game cover

Most of the present-day woodland cover comprises plantations dating from the 18th and later centuries. However, woodlands are known to have been planted since at least the first part of the 17th century (e.g., close to Simonswood; Shaw 1956, p. 468).

The great Halls were being landscaped in the 17th century, e.g., at Knowsley during the period 1651–72 (Sholl 1985, p. 14) and Dunham Massey, where a Kip engraving of 1697 shows established avenues and trees in formal gardens.

At the Royal Society in 1662, when there was great concern at the depletion of Britain's timber stocks, particularly shipbuilding Oak, John Evelyn presented his paper 'Sylva' which encouraged the planting of trees and care of woodlands. By the beginning of the 18th century trees were being planted for timber and underwood to meet estate needs (Table 10.5). At the beginning of the 19th century, Dunham Massey was considered to hold the largest stock of Oak timber in Cheshire (Holland 1808, p. 197).

Following the 1756 Enclosure Act opportunities arose for areas of moss and common to be planted as at Hill Top, Appleton which was fenced, drained and planted in 1765 (Warburton Muniments 1758–70). The Royal Society for the Arts gave awards to encourage the establishment of plantations from 1757 until 1835, e.g., the planting of 54ha of waste moorland near Delamere in 1795, and 847,650 trees at Taxall, Cheshire in 1796 (Royal Society of Arts 1801–02).

An indication of land use and extent of woodland, including new plantations can be seen on the first reasonably accurate maps of Lancashire (by Yates in 1787) and Cheshire (by Burdett in 1777). Though mixed plantations were favoured a 1792 report stated that Oak was still much planted in Cheshire and other species were cut to make way for it. By contrast the Lancashire report stated that fewer Oak were planted than any other timber (Commissioners 11th Report 1792), and Holt (1795, p. 84) described the woods and plantations in Lancashire as 'embellishments for gentlemen's seats, cover for game, or shelter from the blast rather than with a view to supplying the country with timber, and preventing importation'.

The Act of Enclosure for Delamere in 1812 was part of a parliamentary programme to use Crown lands for much-needed timber production. Although half of the land was set out in allotments in four new townships, some 1,557ha was to be planted as Crown forest (Simpson 1967) which is by far the largest planting scheme carried out in the Mersey Basin area (Smart 1992, pp. 73–86 and Fairhurst 1988). By 1823 all but 81ha of this land had been planted with Pedunculate Oak (*Quercus robur*), although initial difficulties led also to the planting of Scots Pine and European Larch (*Larix decidua*) as nurse species (Palin 1848; Simpson 1967, p. 270). Interestingly, Sessile Oak (*Q. petraea*), rather than Pedunculate Oak had probably been the main Oak constituent of the natural woodland (Smart 1992, p. 74). Extensive marling was undertaken between 1856 and 1864 (Grantham 1864) to reclaim 498ha of forest for agriculture leaving the Crown woods to be converted into predominantly coniferous plantation at the beginning of the 20th century (Popert 1908).

Game cover shrubs, including Rhododendron (*Rhododendron ponticum*), were included in many of the plantations established at Knowsley by the 12th Earl of Derby (1770–1830), and coverts became widely established on the estates in the 19th century, e.g., Fox Covert, (Lower Peover, 1832, Smart 1992, p. 62; Table 10.5). Later in the century pheasant shooting became an important country pursuit and mixed planting of broadleaves and conifers predominated, the latter providing shelter and cover.

Between 1840 and 1910 at a time of increasing prosperity, 246 new woods larger than 0.2ha were planted on private estates in what is now Merseyside, e.g., Halsnead in St Helens and in new public parks, e.g., Birkenhead and Sefton (Berry & Pullan 1982). But at the same time urban, industrial and agricultural activity led to the destruction of 156 Merseyside woods: the net change was an increase from 1,371ha to 1,651ha (Berry & Pullan 1982).

The choice of tree species planted changed with time (Table 10.5). The European Larch was planted in great quantity at Tatton and other estates in the early 19th century (Caldwell archives 1828-34). Later in the century western American conifer species were introduced including Sitka Spruce (*Picea sitchensis*), (Smart 1992, p. 61). Another successful introduction, Corsican Pine (*Pinus nigra* ssp. *laricio*) first used in Delamere in 1894 is now the main species planted there. Widely used on the estates, it has also proved invaluable in sand dune stabilisation on the Sefton coast. Oak was (and remains) the most planted hardwood in the mixed stands favoured on most estates and Beech has been widely planted on the lighter soils on which it is most successful.

Protection woodlands

Coastal plantations

Trees have been planted adjacent to the coast between Liverpool and Southport since at least the early 18th century, and probably much earlier, to protect sensitive

Figure 10.3. Austrian Pine woods planted on sand dunes at Ainsdale, Merseyside (*Photo: D. Atkinson*).

areas immediately inland of the dunes. In February and March 1712, for instance, Nicholas Blundell planted Horse-Chestnut (*Aesculus hippocastanum*), willows and poplars (*Populus* spp.), as well as a variety of shrubs adjacent to his ditches to protect them from encroachment by sand (Bagley 1968, 1970). Conifer plantations behind areas of planted Marram (*Ammophila arenaria*) on the most seaward dunes can stabilise further the partly vegetated surfaces (Macdonald 1954) and help to reduce erosion by wind and blown sand (Lehotsky 1941). By the end of the 19th century no technical barriers remained to the creation of large pine plantations – both techniques and species from continental Europe had been employed

Table 10.5. (right) History of plantations 1650–1900

Species used

A = *Alnus glutinosa*	N = *Castanea sativa*
B = *Fraxinus excelsior*	O = *Acer pseudoplatanus*
C = *Fagus sylvatica*	P = *Juglans regia*
D = *Betula* spp.	Q = *Salix* spp.
E = *Ulmus* spp.	R = Broadleaved species
F = *Carpinus betulus*	S = *Abies alba*
G = *Aesculus hippocastanum*	T = *Populus balsamifera*
H = *Tilia* spp.	U = *Larix decidua*
I = *Quercus* spp.	V = *Pinus nigra* spp. *laricio*
J = *Quercus ilex*	W = *Pinus sylvestris*
K = *Platanus* spp.	X = *Pinus strobus*
L = *Populus tremula*	Y = *Picea abies*
M = *Populus nigra, P.alba*	Z = Conifer species

Table: Species used with mixture percentages when known (see Key opposite)

Date	Location	A	B	C	D	E	F	G	H	I	J	K	L	M	N	O	P	Q	R	S	T	U	V	W	X	Y	Z	Note	Reference
1650s	The Mere, Alderley	X		X																								Seed source Worcestershire	Ormerod 1882, pp. 305–06
1702–28	Little Crosby	X	X	X	X	X	X	X	X	X	X	X	X	X		X	X	X	X	X	X			X				Nursery and plantations	Bagley 1968, 1970, 1972
1710–50	The Park, Dunham Massey			X		X				X																		100,000 trees for landscape and timber	National Trust, 1991, Dunham Massey archives
1748	Tatton Plantation													X															Tatton Muniments
1749–56	Peover – Nursery		X	X		X	X		X	X	X	X			X	X			X	X	X	X		X		X			Mainwaring Collection
1756	– Circular Plantation			14		13	14	3							14				1			18		18		5			Mainwaring Collection
1759	– Plantations													X				X										Farm poplar and willow plantations	Mainwaring Collection
1792	Higmere					15	9							15								30		31					Caldwell Archives
1795	Near Delamere		1	4		1				3						3			1			37		50				133 acres heathland site	RSA 1801–02
1796	Taxall	7	14	4		18				3						11			2			35		4			2	Upland site 847,650 trees	RSA 1801–02
1797	Alderley Edge			26		5								5								12			26	26			Caldwell Archives
1813	Delamere Forest									X																		First Delamere planting 100% Oak	Smart 1992
1820	Delamere Forest									X					X									X				Oak with Sweet Chestnut and Scots Pine	Smart 1992
1822	Delamere Forest									X					X							X		X				Larch introduced	Smart 1992
1825	Howies Plantation, Knowsley		2							2					2	8						44		25		17			Sholl 1985
1828	Arley – Covert		10							10												40		20		20			Caldwell Archives
1828–34	Tatton Plantations		1																2	2	1	84		6		6		May have included hardwoods from other sources	Caldwell Archives
1829	Adlington									45												22		19		14			Caldwell Archives
1832	Lower Peover		10							9												41		30		10			Caldwell Archives
1894	Delamere Forest				X																	X	X	X	X				Smart 1992
1900	Delamere Forest																						X	X	X				Smart 1992

1707: Nicholas Blundell plants 'witherns' [willows] and other [unspecified] species on the dunes.	Bagley (1968)
By 1795: Rev. Formby successfully established Sycamore, Ash, Alder, 'platanus' [Plane] and 'fir' [probably Scots Pine] at Firwood, Formby.	Holt (1795)
From 1837: Large-scale planting of mainly Scots Pine and Larch at Culbin Sands, north-east Scotland, using techniques which had been employed as early as 1789 along the Gulf of Gascony, France.	Ross (1992)
From c.1850: Large-scale Pine planting at Holkham, Norfolk and the first extensive use of Corsican Pines on British dunes.	Macdonald (1954)
1887: Experimental tree-planting on Sefton coast by Charles Weld Blundell.	Jones, Houston & Bateman (1993)

Table 10.6. Events preceding the large-scale tree planting on the Sefton coast between 1894 and 1925

successfully in the creation of plantations on other British dunes (Table 10.6). Afforestation was also possible on a large scale at this time because all the coastal land between Ainsdale and Ravenmeols (at the southern edge of Formby) was effectively under the ownership of just two estates. A major programme of planting was therefore carried out between 1894 and 1925 by these estates – mainly of Austrian (*Pinus nigra* ssp. *nigra*) and Corsican Pines. Before this there had been only small-scale planting (Table 10.6). The development of these plantations is described by Ashton (1920), Gresswell (1953), Joint Countryside Advisory Service (1990), Jones, Houston & Bateman (1993) and Wheeler, Simpson & Houston (1993). By 1925 the basic pattern of woodland that is seen today on the coast was in place.

Even by 1925 some of the more seaward planting had been overwhelmed by sand or eroded by the sea (Joint Countryside Advisory Service 1990), and Sea-buckthorn (*Hippophae rhamnoides*) was introduced to protect the seaward edge of the plantations (Clements & Lutley 1987). Nonetheless losses of seaward plantations continue to the present day.

Since 1965 a greater variety of landowners have provided more varied management (Table 10.7). Although these plantations now serve many purposes (Wheeler *et al*. 1993), their role as protection woodlands – providing a valuable shelterbelt for Freshfield and Formby – remains paramount. Even the ongoing phased clear-felling of up to 40ha of frontal woodland at Ainsdale Sand Dunes National Nature Reserve is seaward of a large area of rear dune woodlands which is managed for continuity of cover, and forms an important shelter belt. This bold approach to dune management aims to restore and rejuvenate a dynamic semi-natural sand-dune system following the deterioration in nature conservation value caused at least partly by the dense plantations of non-native pines and a reduction in grazing following myxomatosis in the mid-1950s (Sturgess 1993, Sturgess & Atkinson 1993; Wheeler *et al*. 1993). In 1991 the total estimated area under tree-cover on the Sefton dune coast was 289ha; this represents about 14% of all the woodland in Merseyside (Wheeler *et al*. 1993). Pines still dominate these areas. The amount of deciduous tree-cover has also increased since the original period of conifer planting, but still accounts for less than 5% of the coastal woodland (Wheeler *et al*. 1993).

Catchment plantations

By far the largest plantation in a reservoir catchment in the Mersey Basin is Macclesfield forest. The 390ha of woodland comprises mainly Japanese Larch (*Larix kaempferi*) and Sitka Spruce with smaller amounts of other softwoods and some hardwoods including Beech and Sycamore. Its origin is intimately associated with the increasing demand for water, and the construction of reservoirs since the Industrial Revolution. The population of Macclesfield increased rapidly from about 1850 to its peak in 1881 when there were nearly 4,000 more

ACTIVITY	EXAMPLES
Felling and re-planting of small pockets of trees to regenerate Pine woods.	National Trust, Formby Point.
New planting on exposed western edge of plantation.	Sefton MBC, Lifeboat Road, Formby.
Thinning of mature stands of Pine to obtain advanced natural regeneration before felling.	Formby Golf Club (R.A. Smart, pers. obs.).
Thinning of pines and underplanting with deciduous species such as Beech and Oak rear woodlands.	National Trust, Formby and English Nature, Ainsdale NNR.
Coppicing of Alder, Sycamore and Birch to yield timber for brushwood or poles.	National Trust, Formby.
Clear-felling: (a) to create fire-breaks (b) to remove frontal woodland	Ainsdale NNR.
No management; natural regeneration of pines and other species.	Some private land.

Table 10.7. Some woodland management activities on the Sefton coast since the late 1970s (from Wheeler, Simpson & Houston 1993 and R.A. Smart, personal observations)

people than there are today. The demand for water led to the construction of the three Langley reservoirs, yet domestic demand continued to increase, prompting Macclesfield Corporation to construct the Trentabank Reservoir in 1929. After completion in 1930, trees were planted in its catchment to protect the water supply from pollution and to produce marketable timber (Smart 1992, p. 86).

Post-war forestry

A report on post-war forest policy (Forestry Commission 1943) set a national target of 5 million acres (*c*.2 million ha) of managed and developed forest, and proposed that private woodlands should be managed primarily for timber production. The 1947–49 census of woodlands of 2ha and over established the current situation in both Cheshire and Lancashire (Forestry Commission 1952). A Dedication Scheme announced in 1943, which became the main subject of the 1947 Forestry Act, was finally accepted by woodland owners in 1950. It gave them long-term confidence and provided incentives in the form of grants to regenerate, manage, and extend woodlands (James 1981). Other schemes and grants were also introduced to encourage management and timber production (Table 10.8).

In 1974 a new Dedication Scheme, Basis III, was introduced because Government took the view that the provision of employment and environmental gain should be the primary purpose of grant aid rather than the encouragement of timber production. Subsequent grant schemes swung the emphasis further from the needs of a strategic timber reserve and reduction of imports, towards multi-purpose forestry and environmental benefits. However, the many changes in grant scheme (Table 10.8), and conditions attached, combined with changes in taxation served as a disincentive to some estate and woodland owners.

A woodland resource for the 21st century

The woodland and timber resource comprises not just the present and planned areas of tree cover, but also ancient tree and pollen grains preserved by peat accumulation, which provides a valuable resource for research (for environmental and climatic reconstructions). This lends further argument for the conservation of wetlands which enable these woodland relicts to be preserved. Table 10.9 and Figure 10.5 list and locate prehistoric and historic tree-ring

1943	Post-war Forestry Policy and Dedication Scheme announced.
1945	Forestry Act passed.
1947	Forestry Act to deal with matters arising from the Dedication Scheme.
1947	Town and Country Planning Act giving local authorities powers to make Tree Preservation Orders.
1949–55	Thinning grants available.
1950	Dedication Scheme finally accepted by woodland owners.
1950	Small Woods Scheme announced.
1950–55	Poplar planting grants available.
1951	Forestry Act included a replanting commitment as a condition of granting a felling licence.
1953	Approved woodland scheme planting grants introduced.
1953–62	Scrub clearance grants available.
1967	Forestry Act.
1972	Review of forest policy following cost-benefit analysis.
1973	A new Dedication Scheme announced which would replace existing dedication, approved, and smallwoods schemes.
1974–81	The new Dedication Scheme, Basis III, contained incentives to plant hardwoods and improve the environment.
1981–88	Forestry Grant Scheme with further incentives for hardwood planting.
1985–88	Broadleaved Woodland Grant Scheme gave added protection to semi-natural ancient woodland.
1988 on	Woodland Grant Scheme with subsequent changes.
1988 on	Farm Woodland Scheme to encourage diversification into forestry followed in 1992 by Farm Woodland Premium Scheme.
1988–89	Provisional County Inventories of ancient woodlands published by the Nature Conservancy Council.
1991	Mersey Forest establishment started.
1992	Red Rose Forest establishment started.
1994	Government statement 'The UK Programme and Sustainable Forestry', a follow up of the Earth Summit in Rio. Government's proposals for forestry put forward in 'Our Forests – The Way Ahead'. Further changes to the Woodland Grant Scheme which included considerable cuts in restocking grants.
1995	EU Agricultural Council agreed that tree planting including short rotation coppice, would be allowed on set-aside land.

Table 10.8. Some post-war Forestry Policies, Acts, Grants and Schemes
Compiled from: James (1981); Forestry Commission Scheme Leaflets; Institute of Chartered Foresters News

Table 10.9. Tree-ring chronologies from in and surrounding the Mersey Basin region.

PREHISTORIC PINE

	Site	Chronology Name/code	Radiocarbon years before present (BP)	NGR	Source
1.	Church Moss, Davenham (date from one tree)	–	7810±40 (B-82584)	SJ 665 714	University of Manchester Archaeological Unit 1995
2.	Lindow Moss, Wilmslow	LM1	5190±50 (GU-5567)	SJ 822 807	Lageard, Chambers & Thomas 1999
	Lindow Moss	LM2	5260±70 (GU-5568)	"	"
	Lindow Moss	LM3	5150±50 (GU-5569)	"	"
	Lindow Moss	LM4	5330±80 (GU-5570)	"	"
3.	White Moss, Alsager	Chron 1	4505±40 (SRR-4500)	SJ 774 552	Lageard 1992; et al. 1999
			4335±40 (SRR-4501)	"	
	White Moss	WM4	4160±40 (SRR-3941)	SJ 774 552	Calendar age see Chambers et al. 1999
			4125±50 (SRR-3942)	"	
			4115±40 (SRR-3943)	"	
			4090±50 (SRR-3944)	"	
			4015±45 (SRR-3945)	"	
			4055±45 (SRR-3946)	"	
4.	Day Green Farm		–	SJ 781 573	Lageard, samples collected

PREHISTORIC OAK

	Site	Chronology Name/code	Calendar age	NGR	Source
5.	Meols, Liverpool	–	Undated	SJ 238 909	Hillam unpub.
6.	White Moss	WM1	2190–1891 BC	SJ 777 552	Lageard et al. 1999.
	White Moss	WM2	3228–2898 BC	"	Lageard et al. 1999.
7.	Ashton Lane		4465–3929 BC	SD 412 436	Brown pers. comm. for all subsequent prehistoric oak chronologies.
					Sources for these include: Pilcher et al. 1984 Baillie and Brown 1988 Hillam et al. 1990 Baillie 1995
8.	Balls Farm		4433–4165 BC	SD 408 220	
9.	Berry House Farm		4922–4623 BC	SD 425 158	
10.	Broad Lane		3516–2986 BC	SD 405 444	
11.	Clay Brow Farm		4770–4601 BC	SD 426 149	
12.	Croston Moss 1		3198–1682 BC	SD 472 170	
	Croston Moss 2		1584–970 BC	SD 472 170	
13.	Curlew Lane		Undated	SJ 441 141	
14.	Eskham House Farm 1		3601–3109 BC	SD 440 440	
	Eskham House Farm 2		5012–4604 BC	SD 440 440	
15.	Hill Farm 1		3807–3494 BC	SD 314 018	
	Hill Farm 2		3519–3282 BC	SD 314 018	
16.	Leyland		1553–1032 BC	SD 523 236	
17.	Lower House Farm		4433–4224 BC	SD 3512	
18.	Meanygate Farm		2976–2698 BC	SD 404 173	
19.	New Eskham House Farm		3199–3006 BC	SD 418 437	
20.	New House Farm		Undated	SD 450 136	
21.	New Lane Farm		4940–4748 BC	SD 4213	
22.	North Wood's Hill Farm 1		4800–4647 BC	SD 449 457	
	North Wood's Hill Farm 2		3516–2986 BC	SD 449 457	
23.	Rougholm Farm		3445–3135 BC	Unknown	

24.	South Wood's Hill Farm 1	3931–3238 BC	SD 450 455
	South Wood's Hill Farm 3	4371–4113 BC	SD 450 455
	South Wood's Hill Farm 4	3145–2717 BC	SD 450 455
25.	Tinsley's Lane 1	4286–3862 BC	SD 406 437
	Tinsley's Lane 2	3851–3572 BC	SD 406 437
	Tinsley's Lane 3	3489–3160 BC	SD 406 437
26.	Whams Farm	Undated	SD 412 162
27.	Wild Goose Slack Farm	4935–4569 BC	SD 449 008
28.	Wood Moss Lane	Undated	SD 388 143

HISTORIC OAK

	Site	Chronology Name/code	Calendar age	NGR	Source
29.	Baguley Hall		AD 1037–1290	SJ 817 887	Leggett 1980
30.	Nantwich		AD 930–1330	SJ 650 523	"
31.	Peel Hall		AD 1378–1481	SJ 833 873	"
32.	Staley Hall		AD 1365–1554	SJ 975 997	"
33.	Eccleston Hall		AD 1121–1301	SJ 488 950	Groves unpub.
34.	The Falcon Inn, Chester		AD 991–1234	–	Groves & Hillam forthcoming
35.	36 Bridge Street, Chester		AD 1073–1317	SJ 405 662	Groves & Hillam forthcoming
36.	Bowers Row Car Park, Nantwich		AD 920–1208	–	Hillam unpub.
37.	Lightshaw Hall, nr. Wigan		AD 1106–1270 AD 1414–1552	SJ 614 996 "	Groves forthcoming "
38.	Sefton Fold, nr. Manchester		AD 1507–1601	SD 648 097	Groves & Hillam unpub.
39.	Willaston, nr. Nantwich		AD 917–1205	SJ 671 525	Groves 1990
40.	Wigan – 2 timber posts		AD 1029–1205		Groves 1987
41.	Lydiate Hall		AD 1369–1541	SD 364 049	Leggett 1984–85
42.	The Scotch Piper Inn, Lydiate		AD 1366–1531	SD 365 048	Leggett 1984–85
43.	21–23 Eccleston Road, Prescott (one timber, date for latest tree-ring)		AD 1513	–	Leggett pers. comm. in Cowell & Chitty 1982–83
44.	Old Abbey Farm, Risley		–	SJ 662 935	Dendrochronological analyses pending 1996, Lancaster University Archaeology Unit
45.	Kersall Cell, Salford		AD 1367–1510	SD 810 955	Howard pers. comm.
46.	Staircase Café, Stockport		AD 1389–1458	SJ 898 904	ibid
47.	Speke Hall		AD 1387–1598	SJ 410 820	ibid
48.	Brook Farm, nr Knutsford		AD 1402–1585	SJ 791 764	ibid
49.	Little Moreton Hall, nr Congleton		AD 1393–1538	SJ 833 589	ibid
50.	Ordsall Hall, Salford		AD 1385–1512 AD 1076–1345	SJ 815 973	ibid
51.	Morleys Hall, nr Leigh		AD 1386–1463	SJ 689 992	ibid

MODERN OAK

	Site	Chronology Name/code	Calendar age	NGR	Source
52 and 53.	Peckforton		AD 1780–1976	SJ 533 582 SJ 536 578	Leggett 1980

DENDROCHRONOLOGY LABORATORIES

Dendrochronology Laboratory, Palaeoecology Centre, Queen's University of Belfast, BT7 1NN. (Mr D. Brown.)
Tree Ring Dating Laboratory, Department of Archaeology, University of Nottingham, NG7 2RD. (Mr R. Howard.)

Dendrochronology Laboratory, Archaeology Research School, University of Sheffield, West Court, 2 Mappin Street, Sheffield, S1 4DT. (Dr J. Hillam & Ms C. Groves.)

Figure 10.4. Aerial view of planted Pine woods on the sand dunes of the Sefton Coast (*Photo: J. Houston*).

chronologies in and adjacent to the Mersey Basin.

The development of the Mersey Forest and Red Rose Forest provide important mechanisms to increase significantly the extent of woodland in the 21st century and involve local people in its use and management (the Mersey Forest Plan 1994 and Red Rose Forest Plan 1994; Table 10.9). Some 50 woods in the Mersey Forest area, covering 220ha, are now managed for public access, landscape and wildlife by the Woodland Trust. Whilst this includes outstanding sites such as Floodbrook Clough SSSI, in Runcorn, (which suffered particularly loss of canopy through Dutch Elm disease) many of the sites are secondary woodland and indeed the Trust seeks to acquire land for new woodland planting. The significance of both existing and projected Mersey Basin woodlands (Figure 10.1) can be considered against both global and local needs, which have been summarised in four major international agreements made at the United Nations Conference on Environment and Development 1992 (The Earth Summit, Rio) (Quarrie 1992, see Table 10.11). In addition to supporting landscape and recreational amenity three major functions of woodland for the future are:

1. A source of biological diversity (biodiversity)
The distribution of the significant number of ancient woodland fragments (continuously wooded since AD 1600) across the Mersey Basin is displayed in Figure 10.6, and can be compared with that for total woodland in

Figure 10.1 (see also chapter seven, this volume). Whilst more recent plantations are often associated with poorer agricultural soils, ancient woodland tends to be confined to areas where access for management has long been difficult, particularly in the incised cloughs of river valleys (Figure 10.7).

Since most of the woodland across the region has been planted or replanted since the 18th century, many sites have few plant species. Thus, 70% of woods surveyed in north Merseyside contained fewer than 21 species of flowering plant (Berry & Pullan 1982). In addition most of the extant woodlands shown in Figure 10.1 are small with 74% of the 679 woodland sites in Merseyside extending to less than 2ha. Moreover, Rhododendron was present in about 60% of woods and Sycamore was regenerating in half of them, altering their composition (Berry & Pullan 1982; see also Roberts 1974).

The proposed national Biodiversity Action Plans include the requirement to sustain ancient semi-natural woodland. The first local Biodiversity Action Plans for woodlands in the Mersey Basin were formulated in 1996/7 as a component of Local Agenda 21.

2. A component in the carbon cycle
Increasingly, woodland is recognised as a way of offsetting the effects of practices that produce carbon dioxide (CO_2).

In addition to restocking, an average area of 90.4ha per year was newly planted in the Mersey Basin in the

Figure 10.5. Location of tree-ring chronologies in and surrounding the Mersey Basin. Numbers refer to Table 10.9 which gives further information of chronologies.

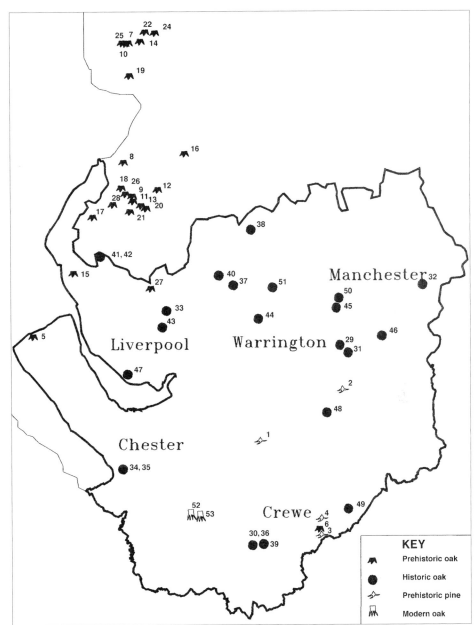

Figure 10.6. Ancient woodland sites in the Mersey Basin. These sites have been continuously wooded since at least AD 1600. Data from ancient woodland inventories for Cheshire, Greater Manchester and Merseyside (English Nature), and selected background sources.

period 1992–95 (average block size, 2.2ha), 92% of which was with broad-leaved species. Also, new temporary plantations, including coppices, have been established: these include *c.*70ha that were planted on vacant industrial land in Knowsley Metropolitan Borough in the 10-year period to April 1996.

3. A sustainable source of timber

In 1980 the region had only about 12,622ha of high forest (Forestry Commission 1984). The current low restocking rates (29.3ha per year across the region) are cause for considerable concern. On an assumed rotation of 100 years 126ha would be regenerated annually in a normal forest. The considerable reduction in restocking grants in 1994 could lead to even lower restocking, woodland neglect and hence reduced sustainability.

The sustainable timber resource must also include the important non-woodland trees (isolated trees, clumps and linear features, e.g., in hedgerows and parks). The 1980 census demonstrated an ageing population

Figure 10.7. Aerial view of woodlands in the Weaver valley east of Calton Hall, Kingsley, Cheshire (*Photo: R.A. Philpott*).

Guiding principles
improve the landscape and protect high quality areas
increase opportunities for access, sport and recreation
protect the best agricultural land from irreversible development
regenerate the environment within green belt and equivalent areas
protect sites of nature conservation
ensure community forests can be used for environmental education
improve the economic well-being of towns
encourage a high level of community involvement
seek private sector support to implement the forests
Six central themes for the Mersey Forest
convert wasteland to woodland
create networks of wooded greenways
green key transport routes
return farmland to forestry
weave woodland into new development
capitalise on the existing woodland assets
(The Mersey Forest Team 1994 and Red Rose Forest Team 1994)

Table 10.10. The Mersey and Red Rose Forests

numbering about 4,627,400 (Forestry Commission 1984, Table 10.12). The maintenance of non-woodland trees was seen to be no longer integrated into economic land use at a time when agricultural productivity was at a premium, and that retention and restoration of hedgerows with trees can only be achieved through financial incentives. Given an assumed average life span of 100 years, a minimum planting programme of some 46,000 trees per annum is required.

More encouragingly, United Utilities (formerly North West Water) which manages most of the water supply catchment plantations in north-west England, including Macclesfield, propose to maintain a sustained timber yield. They intend also, in the five years from 1995, to increase their planting and replanting programme in north-west England about threefold to a million trees a year, and to adopt a policy of maintaining continuous woodland cover (North West Water 1995).

The Sustainable Development Panel, set up by UK government as part of its commitment to Agenda 21 (Table 10.11), recommends that a national strategy should integrate forestry with other land uses recognising that the distinction between forestry and agricultural products is now becoming blurred. Such a National Forestry Strategy supported by regional strategies such as those for the Mersey Basin, which include the Mersey Forest and Red Rose Forest, should identify the incentives needed to meet specific agreed targets.

The **Convention on Biological Diversity** requires an audit of the biological resource with a commitment to sustaining the diversity of species. In the woodland context this dramatically changes the status of the largely neglected Ancient Woodlands.

The **Convention on Climate Change** requires particular consideration of the forest/woodland ecosystems in maintaining carbon sinks and climatic stability.

The **Declaration of Forest Principles** requires a fundamental review of our consumption of timber-based products to ensure the sustainable management of forests world wide.

Agenda 21 is the global strategy engaging all levels of society, including scientists, educationalists, local authorities and local communities to each play their part in shifting from an exploitative culture to a sustainable culture which respects both human needs and life supporting ecosystems.

Table 10.11. Four international agreements from the Earth Summit (1992) with direct relevance to the Woodland Resource of the Mersey Basin (Quarrie 1992)

	Density per sq km			Total Non-Woodland Tree Population
	Isolated Trees	*Clumps*	*Linear Features*	
Cheshire	207	46	1.47km	2,030,450
Greater Manchester	399	85	0.9km	1,876,510
Merseyside	267	49	0.8km	720,440
Total				4,627,400

All figures from 1979–92 census Forestry Commission.
Clumps – small woods less than 0.25ha
Linear features – strips less than 20m mean width and more than 25m length.

Table 10.12. Non-woodland Tree Census (1979–82)

Economic sustainability now needs to be integrated with ecological sustainability.

Acknowledgements

David Brown, Cathy Groves, Jennifer Hillam and Robert Howard kindly provided details of tree-ring chronologies. Other information was generously provided by I. Briscoe, S. Freeman, G. Heddon, C.J. Henratty, Dr D.P. O'Callaghan, P. Rawlinson and P. Russell. Hayley Atkinson cheerfully helped to extract all references to trees and woodland from the Great Diurnal of Nicholas Blundell. JGAL acknowledges the support of a NERC research studentship, grants from Cheshire County Council and the British Ecological Society (Small Ecological Project Grant No. 1145). Finally to the Word Processing Centre, Cheshire County Council for the table layouts and deciphering varying degrees of scrawl.

References

The location of archives and unpublished sources is given in parenthesis at the end of the reference citation. These locations include Cheshire Record Office (Cheshire CRO); Joint Countryside Advisory Service (JCAS) at Bryant House, Liverpool Road, Maghull and the John Rylands Library, Manchester (JRULM). To help find the documents, reference numbers are also often given.

Arley Charters, (1866). Calendar p. 11. (Cheshire CRO).

Ashton, W. (1920). *The Evolution of a Coast-line.* Stanford Ltd, London.

Awty, B.G. (1957). Charcoal ironmasters of Cheshire and Lancashire. *Transactions of the Historical Society of Lancashire and Cheshire*, **109**, 71–124.

Bagley, J.J. (1968). *The Great Diurnal of Nicholas Blundell of Little Crosby, Lancashire, Volume One, 1702–1711.* The Record Society of Lancashire and Cheshire, Manchester.

Bagley, J.J. (1970). *The Great Diurnal of Nicholas Blundell of Little Crosby, Lancashire, Volume Two, 1712–1719.* The Record Society of Lancashire and Cheshire, Manchester.

Bagley, J.J. (1972). *The Great Diurnal of Nicholas Blundell of Little Crosby, Lancashire, Volume Three, 1720–1728.* The Record Society of Lancashire and Cheshire, Manchester.

Baillie, M.G.L. (1995). *A Slice through Time, Dendrochronology and Precision Dating.* Batsford, London.

Baillie, M.G.L. & Brown, D. (1988). An overview of oak chronologies. *British Archaeological Reports, British Series*, **196**, 543–48.

Beck, H. (1969). *Tudor Cheshire.* Cheshire Community Council, Chester.

Berry, P.M. (1980). *An Evaluation of Woodlands on North Merseyside.* PhD thesis, University of Liverpool.

Berry, P.M. & Pullan, R.A. (1982). The woodland resource: management and use. *The Resources of Merseyside* (eds W.T.S. Gould & A.G. Hodgkiss), pp.101–18. Liverpool University Press, Liverpool.

Caldwell Archives. (Cheshire CRO DDX 363).

Carlon, C.J. (1979). *Alderley Edge Mines.* John Sherratt & Son Ltd, Altrincham.

Carlon, C.J. (1981). *The Gallantry Bank Copper Mine, Bickerton, Cheshire.* British Mining No. 16, Northern Mine Research Society, Sheffield.

Challinor, R. (1972). *The Lancashire and Cheshire Mines.* Frank Graham.

Chambers, F.M., Lageard, J.G.A., Boswijk, G., Thomas, P.A., Edwards, K.J. & Hillam, J. (1997). Dating prehistoric fires in northern England to calendar years by long-distance cross-matching of pine chronologies. *Journal of Quaternary Science.* **12**, 253–256.

Clements, D. & Lutley, W. (1987). *National Trust Biological Survey: Formby.* Report of the NT Biological Team. Unpublished report. The National Trust, Cirencester. (JCAS database).

Commissioners 11th Report (1792). Enquiring Into the State of The Woods and forests of the Crown, Appendix 11, Question 13. J. Debrett, London. (Forestry Commission Library, Alice Holt, Hants).

Cowell, R.W. & Chitty, G.S. (1982–83). A timber framed building at 21–23 Eccleston Street, Prescot (site 30). *Journal of the Merseyside Archaeological Society*, **5**, 23–33.

Cowell, R.W. & Innes, J.B. (1994). *The Wetlands of Merseyside.* Lancaster Imprints, Lancaster.

Davies, C.S. (1960). *The Agricultural History of Cheshire 1750–1850.* The Chetham Society 3rd Series, 10, Manchester.

Davies, C.S. (1961). *History of Macclesfield.* Manchester University Press, Manchester.

Dickinson, N.M., Watmough, S.A. & Turner, A.P. (1996).

Ecological impact of 100 years of metal processing at Prescot, N.W. England. *Environmental Reviews*, **4**, 8–24.

Dodd, J.P. (1987). *A History of Frodsham and Helsby*. Privately published (J.P. Dodd), Frodsham.

Dunham Massey. Accession List Records, Windblown Trees in Dunham Massey Park, Box, 15. (JRULM.)

Earwaker Collection, City Record Office, Chester. (CR 6311–2.)

Edlin, J.T. (1949). *Woodland Crafts in Great Britain*. Batsford, London.

Elizabeth Banks Associates (1993). *Lyme Park Restoration Management Plan 2.2*. Unpublished report. (Stamford Estate Office, Altrincham.)

Evans, J. (1984). *Silviculture of Broadleaved Woodland*. Forestry Commission Bulletin No. 62, HMSO, London.

Fairhurst, J.H. (1988). *A Landscape Interpretation of Delamere Forest*. Unpublished dissertation, University of Liverpool (Continued Education Library, University of Liverpool).

Farrar, J.F., Relton, J. & Rutter, A.J. (1977). Sulphur dioxide and the scarcity of *Pinus sylvestris* in the Industrial Pennines. *Environmental Pollution*, **14**, 63–68.

Farrer, W. & Brownhill, J. (1990). West Derby. Vol. 3 *The Victoria History of the Counties of England. A History of Lancaster* (eds W. Farrer & J. Brownhill). Published by Archibald Constable & Co. 1907, and reprinted photographically by William Dawson & Sons, Folkestone.

Fisher, R. (1763). *Heart of Oak, The British Bulwark*. Johnson, London.

Forestry Commission (1921). *The First Annual Report of the Forestry Commissioners, Year ending September 1920*. HMSO, London.

Forestry Commission (1928). *Report on Census of Woodlands and Census of Home Grown Timber 1924*. HMSO, London.

Forestry Commission (1943). *Post War Forestry Policy*. HMSO, London.

Forestry Commission (1952). *Census of Woodland 1947–49: Woodlands of 5 acres and over*. HMSO, London.

Forestry Commission (1984). *Census of Woodland and Trees 1979–82, Counties of Cheshire, Greater Manchester and Merseyside*. Forestry Commission, Edinburgh.

Forestry Commission (1997). *Forestry Commission: Facts and Figures 1996–97*. Forestry Commission, Edinburgh.

Grantham, R.B. (1864). A description of the works for reclaiming and making parts of the late forest of Delamere in the County of Cheshire. *Journal of the Royal Agricultural Society*, **35**, 369–80.

Green, J.A. (1979). Forests. *A History of Chester*, Vol. II *The Victoria History of the Counties of England* (ed. B.E. Harris). Published for the Institute of Historical Research by Oxford University Press, Oxford.

Greig, B.J.W. & Gibbs, J.N. (1983). Control of Dutch Elm Disease in Britain. *Research on Dutch Elm Disease in Europe* (ed. D.A. Burdekin), pp. 10–16. HMSO, London.

Gresswell, R.K. (1953). *Sandy Shores in South Lancashire; the Geomorphology of South-West Lancashire*. Liverpool University Press, Liverpool.

Groves, C. (1987). *Tree-ring dating of two timber posts from Wigan, 1984*. Ancient Monuments Laboratory Report Series 133/87. (Ancient Monuments Laboratory, English Heritage, London.)

Groves, C. (1990). *Tree-ring analysis of medieval bridge timbers from Willaston moated site, near Nantwich, Cheshire*. Ancient Monuments Laboratory Report Series 29/90. (Ancient Monuments Laboratory, English Heritage, London).

Handley, J. (1982). The land of Merseyside. *The Resources of Merseyside* (eds W.T.S. Gould & A.G. Hodgkiss), pp. 83–100. Liverpool University Press, Liverpool.

Heybroek, H.M., Elgersma, D.M. & Scheffer, R.J. (1982). Dutch elm disease: an ecological accident. *Outlook on Agriculture*, **11**, 1–19.

Hillam, J., Groves, C.M., Brown, D.M., Baillie, M.G.L., Coles, J.M. & Coles, B.J. (1990). Dendrochronology of the English Neolithic. *Antiquity*, **64**, 210–20.

Hodson, J.L. (1978). *Cheshire 1660–1760*. Cheshire Community Council, Chester.

Holland, W.B. (1808). *General View of the Agriculture of Cheshire*. Phillips, London.

Holt, J. (1795). *General View of the Agriculture of the County of Lancaster*. G. Nicol, London.

Inman, M. & Reynolds, S. (1995). *Physiological health check monitoring of (Pinus sp.) in Delamere*. Unpublished report. University of Wolverhampton. (Planning Department, Cheshire County Council, file reference 4143.)

Innes, J.L. (1987). *Air Pollution and Forestry*. Forestry Commission Bulletin 70. HMSO, London.

Ives, E.W. (1976). Letters and Account of William Brereton 1490–1536. *Record Society of Lancashire and Cheshire*, **116**, 51.

James, N.D.G. (1981). *A History of English Forestry*. Blackwell, Oxford.

Jarvis, R.C. (1944). The rebellion of 1745: the turmoil in Cheshire. *Transactions of the Lancashire and Cheshire Antiquarian Society*, **57**, 43–70.

Joint Countryside Advisory Service (1990). *A Working Plan for Woodlands on the Sefton Coast*. (JCAS database 080 Me 006.)

Jones, C.R., Houston, J.A. & Bateman, D. (1993). A history of human influence on the coastal landscape. *The Sand Dunes of the Sefton Coast* (eds D. Atkinson & J. Houston), pp. 3–18. National Museums & Galleries on Merseyside in association with Sefton Metropolitan Borough Council, Liverpool.

King, D. (1656). *Vale Royal of England*. Daniel Webb, London.

Lageard, J.G.A. (1992). *Vegetational history and palaeoforest reconstruction at White Moss, south Cheshire, UK*. PhD thesis, Keele University.

Lageard, J.G.A. (1998). Dendrochronological analysis and dating of subfossil *Pinus sylvestris* L. at Lindow Moss, Cheshire. *Bulletin of the British Ecological Society*, **29 (2)**, 31–32.

Lageard, J.G.A., Chambers, F.M. & Thomas, P.A. (1999). Climatic significance of the marginalisation of Scots Pine (*Pinus sylvestris* L.) circa 2500 BC at White Moss, south Cheshire, UK. *The Holocene*. (In press.)

Lasdun, S. (1991). *The English Park*. André Deutsch, London.

Leggett, P.A. (1980). *The use of tree-ring analyses in the absolute dating of historic sites and their use in the interpretation of past climatic trends*. PhD thesis, Liverpool Polytechnic.

Leggett, P.A. (1984–85). Dendrochronological study of timbers from the Scotch Piper Inn. *Journal of the Merseyside Archaeological Society*, **6**, 69–73.

Lehotsky, K. (1941). Sand dune fixation in Michigan. *Journal of Forestry*, **39**, 998–1004.

Macdonald, J. (1954). Tree planting on coastal sands in Great Britain. *Advances in Science*, **11**, 33–37.

Mainwaring Collection, Box 42. Foresters Notebook 1749–1768. (JRULM).

Marshall, S.A. & Dawkins, R.D.H. (1981). The financial impact of the Dutch Elm Disease on a local authority. *Arboricultural Journal*, **5**, 256–62.

Mersey Forest (1994). *Mersey Forest Plan*. (Mersey Forest Project Office, Risley Moss, Warrington.)

Mitchell, A.F. (1984). Native British Trees. *Research Information Note 53.80.SILS*, Forestry Commission.

National Trust (1991). *Dunham Massey Guide*. National Trust, London.

Newbigging, T. (1868). *History of the Forest of Rossendale*. Simpkin Marshall, Bacup.

North West Water, (1995). *North West Water woodland strategy statement*. North West Water, Warrington.

Northcote Parkinson, C. (1952). *The Rise of the Port of Liverpool*. Liverpool University Press, Liverpool.

Ormerod, G. (1882). *History of Cheshire*. Routledge and Son, London.

Palin, W. (1843). The Farming of Cheshire. (Cheshire CRO X630.)

Peckforton Estate Papers. Reafforestation. (Cheshire CRO

DTW 2477/E/4.)

Pilcher, J.R., Baillie, M.G.L., Schmidt, B. & Becker, B. (1984). A 7,272 year tree-ring chronology from W. Europe. *Nature*, **312**, 150–52.

Popert, E.H. (1908). *Report on the Crown Woods at Delamere, Cheshire.* HMSO, London.

Quarrie, J. (ed.) (1992). *Earth Summit '92. The United Nations Conference on Environment and Development, Rio de Janeiro 1992.* Regency Press, London.

Rackham, O. (1986). *The History of the Countryside.* Dent, London.

Rackham, O. (1990). *Trees and Woodland in the British Landscape.* Revised Edition. Dent, London.

Red Rose Forest Team (1994). *Red Rose Forest Plan.* (Community Forest Centre, Salford Quays.)

Roberts, J. (1974). *The Distribution and Vegetation Composition of Woodland on the Wirral Peninsula, Cheshire.* PhD thesis, University of Liverpool.

Ross, S. (1992). *The Culbin Sands – Fact and Fiction.* Centre for Scottish Studies, University of Aberdeen.

Royal Society of the Arts (1801–02). Agricultural Minutes C10/60/F2 and C10/154/F2.

Russell, P. (1987). *The Nomenclature of the West Derby Hundred.* MPhil thesis, University of Liverpool.

Shaw, R.C. (1956). *The Royal Forest of Lancaster.* The Guardian Press, Preston.

Shercliff, W., Kitching, D. & Ryan, J. (1983). *Poynton A Coal Mining Village.* Publisher and place not stated. (Chester CRO.)

Sholl, A. (1985). *Historical Development of Knowsley Park.* Compilation of source material and historical notes. Groundwork Trust for St Helens and Knowsley. (Earl of Derby's Library, Knowsley.)

Simpson, E.S. (1967). The reclamation of the Royal Forest of Delamere. *Liverpool Essays in Geography – a Jubilee Collection* (eds R.W. Steel & R. Lawton), pp. 271–91. Longmans, London.

Smart, R.A. (1992). *Trees and Woodlands of Cheshire.* Cheshire Landscape Trust, Chester.

Stewart-Brown, R. (1912). Royal Manor and Park of Shotwick. *Historical Society of Lancashire and Cheshire*, **64**, 104.

Sturgess, P. (1993). Clear-felling dune plantations: studies in vegetation recovery on the Sefton coast. *The Sand Dunes of the Sefton Coast* (eds D. Atkinson & J. Houston), pp. 85–93. National Museums & Galleries on Merseyside in association with Sefton Metropolitan Borough Council, Liverpool.

Sturgess, P. & Atkinson, D. (1993). The clear-felling of sand-dune plantations: soil and vegetational processes in habitat restoration. *Biological Conservation*, **66**, 171–83.

Tait, J. (1923). *Chartulary of Register of the Abbey of St Werburgh, Chester.* The Chetham Society, New Series, **82**, Manchester.

Tatton Muniments. Egerton Family Correspondence 2/1/41. (JRULM.)

University of Manchester Archaeological Unit (1995). Davenham Bypass: Archaeological evaluation. Unpublished report. (University of Manchester Archaeological Unit.)

Vose, R.H. (1977). *Glassmaking at Kingswood, Delamere, Cheshire.* Winsford Local History Society and the Michaelmas Trust, Winsford.

Vose, R.H. (1995). Excavations at the *c.*1600 Bickerstaffe glasshouse, Lancashire. *Journal of the Merseyside Archaeological Society*, **9**, 1–24.

Warburton Muniments (1758–1770). Peter Harpers Account Books, 1758–1770 and 1302 Lease. (JRULM), Box 26 Folder 3.

Wheeler, D.J., Simpson, D.E. & Houston, J.A. (1993). Dune use and management. *The Sand Dunes of the Sefton Coast* (eds D. Atkinson & J. Houston), pp. 129–50. National Museums & Galleries on Merseyside in association with Sefton Metropolitan Borough Council, Liverpool.

Yalden, D.W. (1987). The natural history of Domesday Cheshire. *Naturalist*, **112**, 125–31.

CHAPTER ELEVEN

The uplands: human influences on the plant cover

J.H. TALLIS

Introduction

Geologically and geographically, the Mersey Basin uplands are part of 'highland Britain' (Figure 11.1). Consequently, the plant and animal life there is exposed to the harsh environmental conditions characteristic of all the highland zone: a cold wet climate, a prevalence of shallow and infertile soils, and a high proportion of steeply sloping and often rocky ground (McEwen & Sinclair 1983). Only the most hardy plants and animals can survive. The same hostile combination of climate and geology normally deters human settlement. However, if humans do invade the uplands, either as visitors or as settlers, then additional pressure is placed on a plant cover that is already stressed by the intrinsic harshness of the upland environment. Irremediable damage to that fragile cover can result. Nowhere in Britain is this more apparent than in the Mersey Basin uplands. Here there is a record of human presence extending back over 8,000 years, and a current population in the surrounding lowlands of more than 5 million people who have easy access (< 30km distance) to the uplands (Figure 11.1). For many of those people who visit the uplands for recreation and pleasure (at least 100,000 each week in summer – Shimwell 1981), the bleak peat-covered moorlands, the poor-quality pastures and the open treeless vistas of the Mersey Basin uplands may appear an epitome of 'highland Britain' – a landscape perhaps as Nature intended. The reality is probably closer to 'a man-made desert', with 'the creation of the moorland environment... as much a social failure as a natural phenomenon' (Spratt 1981).

The distinctive character of the Mersey Basin uplands

At least three characteristics of the plant cover of the Mersey Basin uplands can be highlighted which are not typical of other upland areas of Britain, and which in combination make it unique. These are:

1. A notable scarcity of particular plant groups in the vegetation:

   (a) mosses (and particularly bog moss, *Sphagnum*);

   (b) upland pasture grasses such as the fescues and bents (*Festuca* and *Agrostis* spp.);
   (c) grassland herbs such as Daisy (*Bellis perennis*) and Tormentil (*Potentilla erecta*); and
   (d) bog plants such as Bog-rosemary (*Andromeda polifolia*), Cranberry (*Vaccinium oxycoccos*) and Round-leaved Sundew (*Drosera rotundifolia*).

As a result there are extensive areas of species-poor grassland dominated by Wavy Hair-grass (*Deschampsia flexuosa*) or Mat-grass (*Nardus stricta*), and species-poor moorland dominated by cotton-grasses (*Eriophorum* spp.) or Purple Moor-grass (*Molinia caerulea*).

2. A lower-than-average representation of Heather (*Calluna*) moorland, and a higher-than-average representation of either *Molinia* grassland or *Eriophorum* moorland. This feature is apparent if comparison is made with the overall vegetation cover of the twelve upland parishes in England and Wales surveyed for the 'Upland Landscapes Study' in the early 1980s. The results of that survey (Allaby 1983) are given in Table 11.1, together with equivalent data for the Anglezarke Moors of Rossendale, and for the four large Mersey Basin parishes of the North Peak (Charlesworth, Hayfield, Saddleworth

	A	B	C
Smooth grassland (fescues, bents)	17.1	0	
Coarse grassland (*Molinia*)	11.1	**46.7**	35.8
Coarse grassland (*Nardus, Deschampsia*)	13.3	19.1	
Bracken	8.5	0.2	1.9
Dwarf shrubs : *Calluna*, etc.	32.6	**7.3**	**16.7**
Sedge & rush moorland	16.6	**26.7**	**41.8**

Table 11.1. The plant cover (as % total area) of (A) twelve upland parishes of England and Wales (combined results of Allaby 1983); (B) the Anglezarke moorlands, Rossendale (data of Bain 1991); (C) the combined parishes of Charlesworth, Hayfield, Saddleworth and Tintwistle (data of Phillips, Yalden & Tallis 1981). Noteworthy differences in columns B and C as compared with column A are shown in bold.

(a)

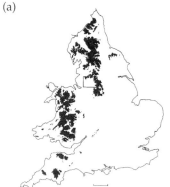

Figure 11.1. (a) The position of the Mersey Basin uplands in relation to upland areas of England and Wales as a whole; ground above 244m altitude is shown black. (b) The Mersey Basin Uplands, and location of sites mentioned in the text. The upland areas are shaded above the 244m and 427m contours – the notional lower limits of the uplands and of deeper blanket peat, respectively. Urban centres are shown in black, with populations in thousands superimposed.

(b)

and Tintwistle – total area 130km²). The lower-than-national representation of Heather moorland and the higher-than-national representation of *Molinia* grassland and *Eriophorum* (sedge) moorland in both these areas is clearly shown.

3. Widespread erosion of the plant cover, with patches or larger areas of bare soil and peat exposed. The peat cover that blankets the flatter ground above 450m altitude, in particular, is often massively eroded. In the four parishes of Charlesworth, Hayfield, Saddleworth and Tintwistle, about 12% of the land (15.8km²) was bare or partly bare in 1981 (Phillips, Yalden & Tallis 1981). Here there are probably 'greater expanses of deep and heavily eroded peat than can be found in any other mountain region of the British Isles' (Conway 1954). This eroded peat landscape has been aptly called a 'wet desert'.

The record of human presence

Human presence in the Mersey Basin uplands is documented in a number of ways. Settlement features (farm buildings, enclosed fields, etc.) from both the historic and prehistoric periods afford direct evidence of presence, as do written records from more recent centuries (maps, stock numbers, litigation proceedings, etc.). Burial mounds and various components of the hunting or farming 'tool-kit' (flint arrowheads, pottery vessels, etc.) afford further evidence of prehistoric presence. The growth of industrial towns and cities in the surrounding lowlands is documented directly in population censuses, and indirectly in the soot and heavy metal deposits on the upland vegetation (Livett, Lee & Tallis 1979).

These varied lines of evidence highlight five major periods of heightened human impact on the Mersey

Basin uplands:

1. from *c*.8,600 to 5,300 BP (the Mesolithic period) – when parts of the uplands were probably used as summer hunting grounds (Wymer 1977; Barnes 1982; Williams 1985);

2. from *c*.3,900 to 3,200 BP (the Early Bronze Age) – when the first upland farms were established (Barnes 1982; Vine 1982; Bain 1991);

3. from *c*.AD 1250 to 1507 (the Medieval period) – when local cattle and sheep ranches were set up as Forest Law was relaxed (Montgomery & Shimwell 1985; Bain 1991);

4. from AD 1507 to 1675 – the expansion of upland farming in the post-disafforestation period (Montgomery & Shimwell 1985); and

5. from AD 1675 to present – agricultural intensification and industrial expansion (Montgomery & Shimwell 1985).

The different upland resources utilised in these five impact-periods are summarised in Table 11.2.

A model of vegetation change

The cold wet climate of all upland areas of Britain results almost inevitably in long-term deleterious soil changes. The prevailing downward movement of water in the soil

	1	2	3	4	5
Food materials:					
Wild animals and plants	+	+			
Domesticated animals & plants		+	+	+	+
Raw materials:					
Wood		+	+	+	
Peat			+	+	
Stone				+	+
Water					+
Recreation:					
Sport			+		+
Walking					+
Tourism					+

Table 11.2. The upland resource and its utilisation; 1–5 are the human impact-episodes mentioned in the text.

produced by the annual excess of precipitation over evaporation leads inexorably to leaching, acidification and podsolisation. On flatter ground, where drainage is impeded, water ponds up in the soil, resulting in water-logging and ultimately the accumulation of peat. The potential productivity of the ecosystem is thus inevitably reduced, as indicated schematically by the line AB in Figure 11.2. Additional environmental stress (which could be climate change, but is more likely to be human-induced) is represented in Figure 11.2 by the displacements S_1, S_2 and S_3. These give successively lower

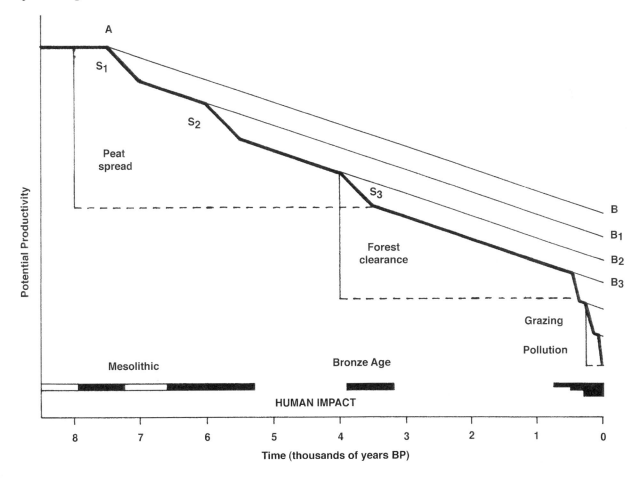

Figure 11.2. Schematic chart of major impacts on the Mersey Basin uplands.

positions of the overall trend line – to AB₁, AB₂ and AB₃ – and hence a less productive vegetation cover.

Six such displacements, representing responses to periods of accentuated environmental stress, can in fact be recognised in the documented history of the plant cover of the Mersey Basin uplands. Recognition is based on lines of evidence derived solely from the plant cover itself, without recourse to the evidence on human impact summarised above. These lines of evidence are considered in the following Sections.

The long-term perspective

Twenty thousand years ago, at the height of the last cold stage, the Mersey Basin uplands were a wasteland of snow and ice – a wind-blasted polar desert overlooking the ice sheets of the Lancashire and Cheshire lowlands. A single pollen diagram from an upland site just outside the Mersey Basin catchment (Middle Seal Clough, below the north face of Kinder Scout at 490m altitude), records the early stages of vegetation recovery between 9,800 and 8,900 years ago. During that period, scrub of Willow (*Salix*) and Birch (*Betula*), and later Hazel (*Corylus avellana*), spread rapidly upwards (Johnson, Tallis & Wilson 1990). By 8,600 years ago, Hazel was widespread over a considerable altitudinal range, as witnessed by the pollen diagrams at two further sites within the Mersey Basin catchment itself: Robinson's Moss (Tallis & Switsur 1990), and Soyland Moor (Williams 1985). These two sites also record the subsequent spread of forest trees, such that by 7,500 years ago the hillslopes were forested up to about 525m altitude (Figure 11.3); above

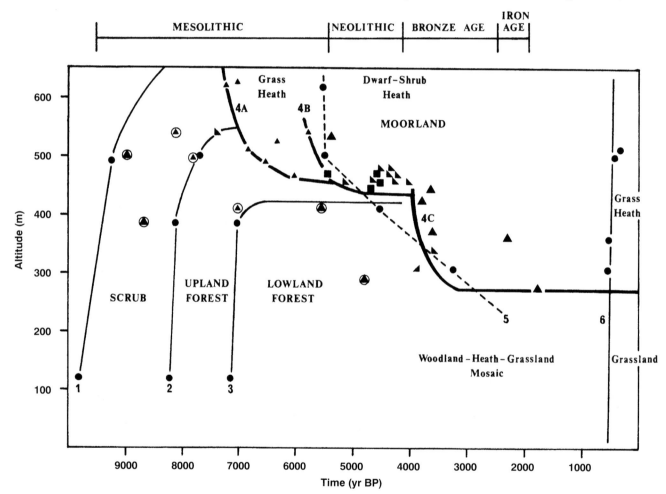

Figure 11.3. Chart of vegetation changes in the Mersey Basin uplands, based on radiocarbon dates and pollen analyses at various sites (see Appendix for details). The following codings are used for datings at individual sites:

▲ radiocarbon date of the basal peat
◢ radiocarbon date of tree remains below the peat
● other radiocarbon dates
○ sites with early peat formation

■ radiocarbon date of tree remains at the peat base
◣ radiocarbon date of tree remains above the peat base
▲ pollen-analytical date for the basal peat.

Lines 1–6 show particular features of the vegetation history:
1, the upper limit of scrub (marked by a rise in birch pollen and a fall in juniper pollen);
2, the upper limit of upland forest (marked by a rise in pine pollen);

3, the upper limit of lowland forest (marked by a rise in alder pollen);
4, the onset of peat formation;
5, soil podsolisation (marked by a rise in Ericaceae pollen);
6, spread of grassland (marked by a rise in grass pollen).

that level, scrub covered all but the most exposed ground. The pollen influx to Robinson's Moss then consisted of 80–90% tree and shrub pollen, as compared with < 20% at the present day.

Evidence of this former more extensive forest cover over the southern Pennines and Rossendale is also provided by the numerous roots, trunks and branches of trees embedded and preserved at the base of the upland peats (Tallis & Switsur 1983; Bain 1991). These tree remains date variously from about 7,500 to 3,500 years ago (Appendix). They show that Pine and Oak were common components of the forest up to 525m altitude, but at higher altitudes a scrub of Birch and Willow (and also Hazel, on the basis of the pollen evidence) was predominant (Table 11.3).

The upland forest and scrub established itself initially at these higher altitudes in a warm, dry and relatively benign climate. Beginning about 7,500 years ago, however, the climate became substantially wetter. That change led gradually but inexorably to waterlogging of the soils on the flatter ground, and to leaching and acidification of the soils on the hillslopes. As a result, the uplands became increasingly less favourable for tree and shrub growth. The progress of these soil changes is charted in Figure 11.3, where the initiation and spread

Altitudinal interval (m)	Total number of sites	Number of sites with:			
		Pine	Birch	Oak	Willow
<375	9	5	3	4	
380–425	10	8	4	1	2
430–475	15	7	8	5	3
480–525	16	7	7	5	6
530–575	1	1	1		1
580–625	2		1		2

Table 11.3. The distribution of peat-preserved tree remains in relation to altitude in the Southern Pennines (data of Tallis & Switsur 1983).

of blanket peat and the expansion of Heather are used as recognisable consequences of soil waterlogging and of soil acidification, respectively. Peat initiation and spread occurred between *c*.7,500 and 3,500 years ago, and expansion of Heather between *c*.5,500 and 3,500 years ago.

In the face of these changes, the forest and scrub receded – first from the higher flatter ground, where its former presence is recorded in the preserved tree remains, and later from the hillslopes. Its retreat from the hillslopes is shown in the five pollen diagrams summarised in Figure 11.4, from sites within, or close to,

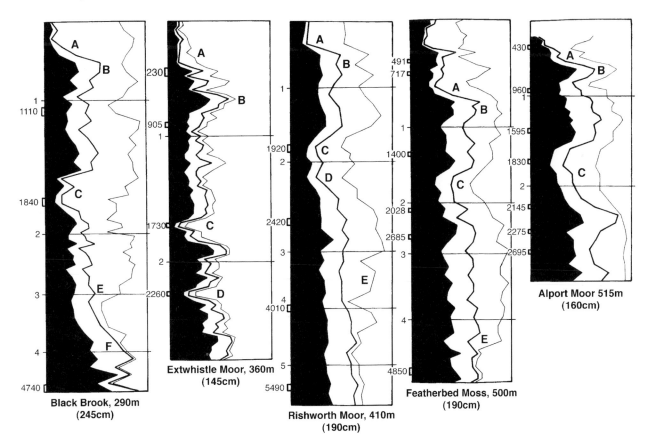

Figure 11.4. Summary pollen diagrams from: Black Brook, Rossendale (Bain 1991); Extwistle Moor (Bartley & Chambers 1992); Rishworth Moor (Bartley 1975); Featherbed Moss (Tallis & Switsur 1973); and Alport Moor (Tallis & Livett 1994). Components of each diagram, from left to right, are: total tree pollen, shrub pollen, Ericaceae, and other non-arboreal pollen (all as % total land pollen, TLP). Dates of radiocarbon-dated peat samples (yrs BP) are shown at the left of each diagram. The horizontal lines show approximate dates at 1 ka (thousand year) intervals. A, B, C, D, E and F are pollen features referred to in the text. Depths of the peat column analysed are shown below each site name.

the Mersey Basin uplands ranging in altitude from 290m to 515m. What is very apparent in Figures 11.3 and 11.4 is the discontinuous nature of the forest retreat. Three major episodes of peat spread are highlighted in Figure 11.3: at c.7,500–7,000, 6,000–5,500 and 4,000–3,500 BP. Marked reductions in tree and shrub pollen (APC) occur in some or all of the pollen diagrams in Figure 11.4 at the levels labelled F, D, C and A: at c.4,000 BP, c.2,260 BP, c.2,000–1,600 BP and c.500–400 BP. In between are periods of forest recovery.

The spread of peat

Numerous finds of small worked flints (microliths) buried below the peat (Wymer 1977; Barnes 1982; Williams 1985) testify to a regular summer population of Mesolithic hunters in the Mersey Basin uplands during the early stages of peat spread (7,500–5,500 BP). The pollen evidence for this timespan suggests a complex vegetation mosaic, of high forest, scrub, grassland, heath and bog, with the emphasis shifting perceptibly from forest towards bog through time. Long-continued episodes of reduced Hazel pollen values, accompanied by increased values of herbaceous and ericaceous plants, are recorded in both the Soyland Moor and Robinson's Moss pollen diagrams: from 7,900 to 7,250 and 6,600 to 4,965 BP at Soyland Moor (Williams 1985), and from 9,000 to 8,500, 7,700 to 7,500 and 5,500 to 4,900 BP at Robinson's Moss (Tallis & Switsur 1990). All the episodes at Robinson's Moss are accompanied by high concentrations of carbonised plant material in the peat (Figure 11.5). Accordingly it has been suggested that

these episodes record phases of repeated burning (either accidental or deliberate) of the high-altitude scrub (Tallis & Switsur 1990). At Holme Moss, the pollen evidence indicates that Hazel scrub occupied the plateau up to 5,400 BP, and was then destroyed by fire immediately before the onset of peat accumulation (Livett & Tallis 1994).

A plausible pattern of land-use (Jacobi, Tallis & Mellars 1976; Williams 1985) is of long-continued exploitation of the higher moorlands for wild game (perhaps Red Deer), through regular burning of the encroaching Hazel and Birch scrub, to provide more nutritious regrowth. This practice led eventually to soil impoverishment and also to the spread of peat, as removal of the tree cover could have been sufficient to raise the water balance of the soil to a level where waterlogging and the accumulation of peat occurred inevitably (Moore 1975, 1993).

It is likely that the Mesolithic hunters utilised only the zone of scrub and grassland/heath above the forests. Before 8,000 BP a range of altitudes could have been exploited, and at lower-altitude sites (such as Soyland Moor, and others in the Saddleworth – Marsden area) regular annual visits would have kept back the encroaching forest for many centuries. Less heavily-exploited, lower-altitude sites would eventually have been abandoned to the forest, however, so that from 8,000 to 7,000 BP perhaps only the highest ground was being visited regularly. As the climate became wetter after 7,500 BP, peat spread would eventually have led to the abandonment of these sites also. Progressive soil deterioration might then have resulted in an opening up

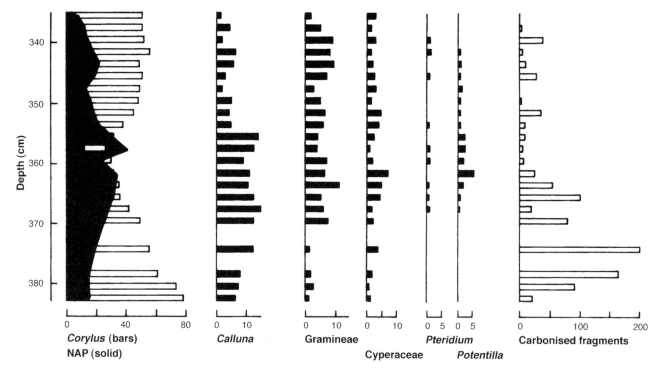

Figure 11.5. Changes in selected pollen and spore components at Robinson's Moss between c.7,700 and 7,500 BP; all values are expressed as % total land pollen, except for Corylus, which is shown as % total tree + shrub pollen. The frequency of burning is shown as number of carbonised fragments per slide-traverse. Re-drawn from Tallis & Switsur (1990).

of the uppermost levels of the forest below the peat blanket, and an extension of scrub, so that by 6,000 BP new hunting grounds at slightly lower altitudes were becoming available and being exploited. These, too, were eventually abandoned as peat continued to spread. On the lower-altitude Rossendale uplands, forest and scrub probably persisted until at least 4,000 BP (Bain 1991), but was then cleared, almost certainly as a result of Bronze Age activity in the uplands (level F in the Black Brook pollen diagram, Figure 11.4). Again, peat formation ensued.

In the southern Pennines peat formation began at a much earlier date than in most other upland regions of Britain. Thus in south-west, central and northern Scotland and the Outer Hebrides, the major period of peat spread was after 5,500 BP; in Orkney and Shetland, the Inner Hebrides, western and northern Ireland, mid- and north Wales and the northern Pennines, it was delayed until after 4,500 BP (Tallis 1995a). By this time, the present-day extent of blanket peat in the southern Pennines had nearly been achieved. In all these regions of Britain, much of the peat spread was over ground that was previously covered by forest and scrub (Birks 1988). The differences in timing of peat initiation between regions can conceivably be related to differing levels of prehistoric activity, and associated scrub clearance, in the uplands.

The clearance of the hillslope forests

Limited episodes of clearance of the forest cover from the Mersey Basin hillslopes over the time-period from 3,000 to 700 BP are recorded in all five of the pollen diagrams in Figure 11.4. In between the clearance episodes the forest returned to varying degrees. As the values for tree and shrub pollen at level B (*c.*700 years ago) were substantially the same as those at 3,000 BP, however, these forest clearance episodes can have had little long-term effect on the upland environment.

The forest regeneration episode at level B is almost certainly associated with the emplacement of large parts of the Mersey Basin uplands under Forest Law in the early Medieval period. The pollen evidence indicates that the composition of the upland forest then differed in detail from that of earlier forests (more Oak and Birch, less Hazel, for example), but that forest and scrub was nonetheless extensive. Subsequent clearance, in contrast to earlier episodes, was effectively permanent. Where documentary evidence exists, it indicates that substantial inroads were being made into the upland forests by AD 1300 (Tallis & McGuire 1972). The pace of clearance quickened after disafforestation in AD 1507, and in parts of Rossendale at least, clearance was almost complete by AD 1600 (Tallis & McGuire 1972). The uplands were then given over almost entirely to management for sheep and cattle, and locally later (post-AD 1800) for Red Grouse (*Lagopus lagopus*) and water catchment.

The grazing factor

A detailed picture of the upland landscape at the beginning of the 20th century can be derived from the maps of the Peak District by Smith and Moss (1903) and Moss (1913). In that landscape, Heather moor was more widespread than it is today and grassland less extensive. Table 11.4 summarises the differences for the four parishes of Charlesworth, Hayfield, Saddleworth and Tintwistle.

	1913	1979
Dwarf-shrub moorland (*Calluna* or *Vaccinium* dominant or co-dominant)	25.2	16.7
Eriophorum and eroding blanket bog	45.5	41.8
Grassland and bracken	29.3	37.7

Table 11.4. The extent of three major vegetation types (as % total area) in the four upland parishes of Charlesworth, Hayfield, Saddleworth and Tintwistle in 1913 and 1979 (data of Anderson & Yalden 1981).

The balance between Heather moor and grassland in upland areas is known to be influenced by grazing (Yalden 1981a). The western slopes of Kinder Scout were covered by Heather (*Calluna vulgaris*) and Bilberry (*Vaccinium myrtillus*) in the early 1900s (Moss 1913), but by 1982 the plant cover had degenerated to a discontinuous turf of Wavy Hair-grass interspersed with patches of Mat-grass, Bilberry and bare ground; Heather was virtually absent. Under National Trust management, sheep numbers were progressively reduced over the next 10 years by active shepherding, from 2.5 sheep to 0.18 sheep per hectare. By 1992, with lowering of the grazing pressure, Heather and Bilberry were spreading and most of the bare ground was revegetated (Table 11.5).

	1983	1992
Deschampsia flexuosa	41	83
Calluna vulgaris	0	25
Vaccinium myrtillus	1	16
Bare ground	51	6

Table 11.5. The plant cover (% cover) along 12 permanent transects on the west slopes of Kinder Scout at 450–530m altitude in 1983 and 1992 (data of Anderson 1997).

Sheep numbers on the moorlands of the Peak District are known to have risen steadily over the period from 1950 to 1989, with a fourfold increase in some areas (Figure 11.6). Before that, numbers had remained fairly stable back to at least 1914 (Yalden 1981b) – and indeed may not have been very different in 1690 (Shimwell 1974). The high stocking rates over the last 45 years have

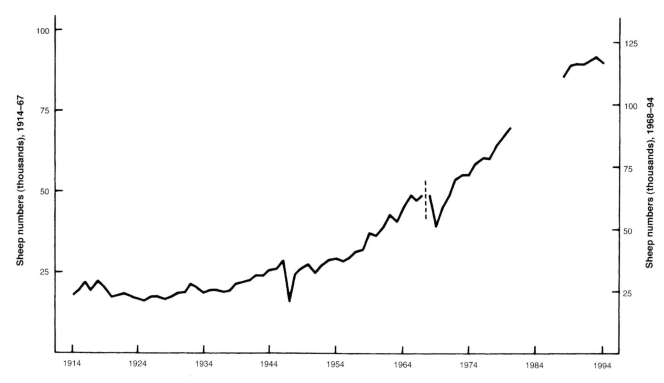

Figure 11.6. Trends in sheep numbers on the moorlands of the Peak District, 1914–1994. Because of differences in the way MAFF sheep statistics have been collected, the graph utilises two overlapping sets of data (with different vertical axes): 1914–1977: combined parishes of Hope Woodlands, Derwent, Edale, Charlesworth, Hayfield and Chinley/Buxworth/Brownside; 1968–1994: above parishes, plus Aston, Bamford, Brough and Shatton, Castleton, Hope, Thornhill, Chisworth, Glossop, Tintwistle, Whaley Bridge and New Mills. For problems in the collation of the data, see Yalden (1981b) and Anderson, Tallis & Yalden (1997).

undoubtedly been a major factor in the decline of Heather and the spread of grasses – though recent work in the Netherlands suggests that in the future the increased nitrogen concentrations in acid rain (see below) could further favour the spread of Wavy Hair-grass and Purple Moor-grass at the expense of Heather (de Smidt 1995).

The blighted landscape

A record of the rampant industrial growth (and concomi-tant population increase) that has occurred in the Mersey Basin lowlands over the last 250 years is preserved in the soot-blackened uppermost layers of the upland peats, and their associated higher concentrations of lead, zinc, copper and nickel (Figure 11.7). The upland peats and soils have also been watered over this time-period by acid rain, so that their pH may be as much as one unit lower than that of soils and peats in other, cleaner, upland areas. The upland plants are inevitably also affected, with higher internal concentrations of poten-tially toxic heavy metals (Livett, Lee & Tallis 1979) and of nitrogen (Press, Woodin & Lee 1986).

It is tempting to ascribe the scarcity of certain plant groups in the Mersey Basin uplands to this pollution. There is certainly direct experimental evidence that some *Sphagnum* spp. are deleteriously affected by acid rain (Lee 1981; Ferguson & Lee 1983; Lee, Tallis & Woodin

1988). The peat stratigraphy shows that over large areas of the higher moorlands of the Mersey Basin *Sphagnum* was a major component of the bog vegetation from early Medieval times through to about AD 1750 (Tallis 1965, 1987). Its remains disappear from the peat at the level when soot contamination first becomes apparent and heavy metal concentrations rise. The implication is clear: that *Sphagnum* was killed off by the products of the Industrial Revolution.

The growth of at least two other common moorland plants – Heather and Common Cotton-grass (*Eriophorum angustifolium*) has also been shown to be adversely affected at pH values below about 3.3 (Richards 1990; Caporn 1997). Some southern Pennine peats are as acid as pH 2.8 (Anderson, Tallis & Yalden 1997). There is thus good reason to believe that the Mersey Basin uplands are indeed a blighted landscape, caused by pollution from the surrounding lowlands.

It would be a mistake, however, to ascribe all the pecu-liarities of the upland plant cover to pollution. Thus the *species-poverty* of the vegetation is the result of at least three processes acting in combination:

1. the intrinsic infertility of many of the upland soils (especially those overlying Millstone Grit);

2. the selective action of several centuries of grazing and burning (allowing only the more resistant plant species to survive); and

3. air pollution (damaging the bryophytes and lichens in particular).

Even though large areas of *Sphagnum* were undoubtedly killed off by pollution in the 18th and 19th centuries, its abundance had already been reduced in many places long before that. The peat stratigraphy of deeply-gullied sites shows that *Sphagnum* had disappeared at these sites by the 16th century, if not earlier (Tallis 1985, 1987; Montgomery & Shimwell 1985). The *scarcity of dwarf-shrub heath* (and particularly Heather moorland), and the prevalence of grassland and sedge moorland, is again the product of several centuries of grazing and burning, with the effects already noticeable probably by the late 17th century. Thus, pollen analysis of sites currently in uneroded moorland (where the patterns of vegetation change have not been influenced by erosion) show that Heather pollen was predominant until about AD 1500, but that subsequently grass pollen rose steadily, most notably after *c*.AD 1650 (Figure 11.8). This is when hill farming became widely established in the Mersey Basin uplands (human impact episodes 4 and 5), with the plant cover increasingly exposed to the debilitating and selective action of sheep grazing.

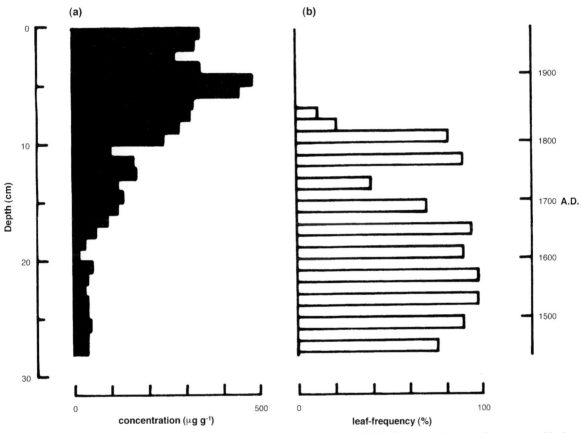

Figure 11.7. Concentration of lead (a) and frequency of *Sphagnum* leaves (b) in the upper layers of a peat profile from Featherbed Moss, Snake Pass (data of Livett (1982) and unpublished).

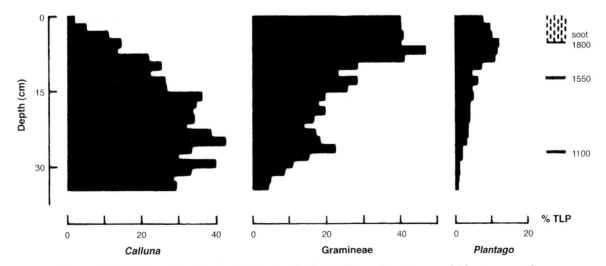

Figure 11.8. Pollen values (as % total land pollen) of *Calluna*, Gramineae and *Plantago* over the last 1,000 years at Featherbed Moss, Southern Pennines (Site 2 of Tallis 1965).

The scars of battle

The prevalent erosion of the plant cover at the present day in the Mersey Basin uplands is a more complex issue. The landscape is stressed and scarred as a result of a succession of human impacts (Figure 11.2), with erosion of different types, and resulting from different processes, often superimposed one upon another. The erosion is very much a feature of the higher ground, so that it is far more prevalent in the southern Pennines than in Rossendale. Three different sets of *processes* contribute to it:

1. disruption or destruction of the plant cover and exposure of bare peat or soil;

2. prevention of recolonisation of these bare areas, once formed, by plants; and

3. physical and chemical removal of the unstabilised bare peat and soil.

Physical and chemical removal (by water, wind and biochemical oxidation) affects only the surface peat/soil layers that have been loosened by frost and drought (Burt & Gardiner 1984; Francis 1990; Labadz, Burt & Potter 1991), and operates in all upland regions. It is a consequence of the harsh climatic conditions there. The initial disruption of the vegetation cover, and subsequent prevention of recolonisation, on the other hand, results from the action of largely extrinsic *agents* of erosion. Those that can be identified for the Mersey Basin uplands include pollution, grazing, fire, trampling, peat cutting and cloudbursts.

In non-polluted upland areas of Britain, mosses and liverworts 'fill in' the spaces between the individual higher plants in the vegetation, and cover over any unprotected soil and peat; *Sphagnum* is particularly characteristic of the wetter places. In the Mersey Basin uplands this group of plants has been killed off by pollution (and, to a lesser extent, by burning), exposing unprotected soil and peat. The ability of the higher plants subsequently to grow together, and minimise the damage caused by the death of the bryophytes, has been hampered by sheep grazing and intermittent burning. Heavy sheep grazing, particularly on steeper slopes, may even exacerbate the vegetation break-up. The recovery of the plant cover when sheep grazing is removed has been demonstrated in a variety of experimental trials in the Peak District National Park (Tallis & Yalden 1983; Pigott 1983; Anderson, Tallis & Yalden 1997).

Another widespread cause of erosion is damage from accidentally-started moorland fires ('wildfires') during dry weather. More than 300 wildfires were recorded on the moorlands of the Peak District National Park in the period from 1970 to 1995, mostly in years with prolonged dry weather in spring or summer (notably 1976). In total, about 42km² of the moorland was burnt (some 8% of the total moorland area – Anderson, Tallis & Yalden 1997). Some 45% of the fires are known to have started next to roads and close to the Pennine Way and other footpaths, so that much fire damage is clearly a product of the increasing accessibility of the uplands over the last 30 years. The larger and more severe fires have contributed substantially to the bare and eroding ground visible today, because recolonisation by plants has often been slow. Extensive bare areas still persist from major fires in 1976, 1959 and 1947 (Radley 1965; Tallis 1981). Up to 15–20% of the bare ground present in the Peak District National Park could have originated from wildfires (Tallis 1982), some of it from fires that occurred more than 200 years ago (Tallis 1987).

The periodic high concentrations of carbonised plant material down the peat profile at many sites suggest that wildfires have always been a hazard on the moorlands. At Alport Moor there are at least 15 such 'burning peaks' within the peat formed over the last 2,500 years (Tallis & Livett 1994). In addition, frequent, if rather haphazard, rotational burning of the vegetation has been carried out deliberately over the last few centuries, to encourage regrowth of either the coarse moorland grasses (at burning intervals of 5–7 years) or of Heather (at intervals of 10–15 years; Hobbs & Gimingham 1987).

Trampling, peat cutting and cloudbursts result in more localised erosion. The southern section of the Pennine Way was described in 1989 as 'a man-made quagmire in wet conditions' (Porter 1989). Nevertheless, the total contribution of trampling is small in the context of the whole landscape: no more than 56ha of eroded ground along moorland footpaths in the Peak District National Park (total area 52,000ha – Yalden 1981c). Nevertheless, more than £200,000 has already been spent in restoration work (Anderson, Tallis & Yalden, 1997). Peat cutting on the moorlands may have been more extensive in the past than is generally realised (P. Ardron, personal communication), but much of it occurred in the Middle Ages and the cut-over areas are now stabilised. Two sites of severe erosion in the Mersey Basin uplands have been attributed to storm damage: Cabin Clough, near Glossop, is known to have been devastated by a cloudburst on 30 July 1834 (Montgomery & Shimwell 1985), whilst the products of a peatslide, perhaps around AD 1770, are still visible on the northeastern side of Holme Moss (Tallis 1987). Other sites may well exist that have not yet been documented.

Sheep play a further role in the erosion process. Heavy grazing limits the supply of native seed available to recolonise bare ground, and damages the new plant cover as it attempts to get established. Sheep thus reinforce the inherent harshness of the upland environment, which restricts seedling establishment by promoting surface instability of the bare peat and soil. A further constraint derives from the additional acidity produced by acid rain. Even native moorland plants now find it difficult to germinate and establish on the most acid peats (Richards 1990; Richards, Wheeler & Willis 1995).

The Badlands of Britain

The oldest, and also the most widespread, erosion features on the moorlands are the gully systems, which are found particularly in the deeper peats above 450m altitude. The most intense erosion, by close-set reticulate gullying, normally occurs only above about 550m altitude (Bower 1960; Anderson & Tallis 1981). The development of the gullies appears to have coincided with periods of increased environmental stress on the Mersey Basin uplands in the last 450 years. Thus three phases of gullying have been distinguished. The deepest gullies, incised through 2m depth of peat or more into the underlying mineral substrate, probably started to form 400–500 years ago (Tallis 1995b). The shallower gullies, still contained entirely within the peat mass, represent a second phase of extension beginning 200–250 years ago (Tallis 1965). Very recent gully systems can also be observed, forming on peat areas bared by wildfires (Turtle 1984).

These three phases of gullying certainly match in time with episodes of perturbation of the bog surface (by post-disafforestation sheep farming, air pollution and wildfires, respectively). However, it is possible that the initial damage that led to the gullying occurred naturally, by extreme desiccation of the peat mass during the drier conditions of the so-called Early Medieval Warm Period of c.AD 1050–1200 (Tallis 1995b). The colder wetter conditions of the Little Ice Age, from c.AD 1550 to 1800, also led to an increased abundance of *Sphagnum* on as yet uneroded areas of the moorlands (Tallis 1987), so that the subsequent effects of pollution could have been enhanced by this natural climatic change.

Postscript

Gully erosion in the southern Pennines is exceptionally severe as compared with other upland areas of Britain. Gullying is most marked in the deeper peats, and deep peat is unusually widespread in the southern Pennines. This could be a consequence of the long time-period over which peat has accumulated there (see above). The early onset of peat formation could conceivably be linked to human perturbation also (burning of the upland scrub by Mesolithic hunters – Tallis 1991). If so, then the upland landscapes of the Mersey Basin are a product of human-induced environmental stress extending back not hundreds but thousands of years.

Acknowledgements

The following colleagues contributed substantially to the ideas in this chapter, through discussions and the provision of data: Mrs Penny Anderson, Dr Malcolm Bain, Ms Daryl Garton, Dr David Shimwell and Dr Derek Yalden.

Appendix

Sources of data used in the compilation of Figure 11.3.

Site	Altitude (m)	Date (yrs BP)	Code	Source*
A. Initial rise in Birch pollen, decline in Juniper pollen				
Red Moss	120	9798	Q-924	1
Middle Seal Clough	490	9230	SRR-336	2
Robinson's Moss	500	8950	Q-2320	3
B. Increase in Pine pollen, initial rise in Oak pollen				
Red Moss	120	8196	Q-918	1
Soyland Moor	385	8110	Q-2931	4
Robinson's Moss	500	7675	Q-2273	3
C. Initial rise in Alder pollen				
Red Moss	120	7107	Q-916	1
Soyland Moor	385	7640	Q-2390	4
D. Basal peat, radiocarbon date				
Robinson's Moss	500	8950	Q-2320	3
Soyland Moor	385	8650	Q-2392	4
Holme Moss	535	5370	GU-5376	5
Arnfield Moor	445	3610	GU-5378	5
Ogden Clough	370	3560	GU-5380	5
Black Brook	290	4740	HAR-6210	6
Round Loaf	305	3880	BIRM-1161	6
Winter Hill	427	3750	BIRM-1162	6
Pikestones	275	1710	HAR-6209	6
Rishworth Moor	410	5490	GAK-2822	7
Extwistle Moor	360	2260	BIRM-689	8
E. Basal peat, pollen-analytical estimated date				
Alport Moor	540	8100		9
Robinson's Moss B	495	7800		9
Bleaklow	622	7200		9
Ringinglow C	410	7000		9
Kinder	625	7000		9
Featherbed Moss	510	6800		9
Featherbed Moss	490	6500		9
Salvin Ridge	525	6300		9
Featherbed Top	540	5500		9
Tintwistle High Moor	467	6000		9
F. Tree remains at peat base				
Lady Clough Moor	470	5410	Q-1349	10
Coldharbour Moor	450	4670	Q-1407	10
Featherbed Moss	475	4570	Q-1346	10
Tintwistle Knarr	470	4475	Q-2314	10
G. Tree remains within peat				
Deep Clough	340	3540	BIRM-147	11
Over Wood Moss	540	7350	Q-1404	10
Laund Clough	455	5110	Q-1402	10
Rawkin's Brook	465	4620	Q-1408	10
Lady Clough Moor	480	4340	Q-1350	10
Lady Clough Moor	480	4495	Q-1348	10
Featherbed Moss	475	4320	Q-1347	10
Laund Clough	455	4250	Q-1401	10
Tintwistle Knarr	470	4210	Q-1405	10

Continued overleaf

* 1 = Hibbert, Switsur & West 1973
2 = Johnson, Tallis & Wilson 1990
3 = Tallis & Switsur 1990
4 = Williams 1985
5 = Unpublished data
6 = Bain 1991
7 = Bartley 1975
8 = Bartley & Chambers 1992
9 = Tallis 1991
10 = Tallis & Switsur 1983
11 = Tallis & McGuire 1972
12 = Conway 1954
13 = Tallis & Switsur 1973
14 = Tallis & Livett 1994.

| Far Back Clough | 400 | 3995 | Q-1406 | 10 |
| Tintwistle Knarr | 470 | 4000 | Q-2315 | 10 |

H. Rise in Ericaceae pollen

Bleaklow	622	5500		12
Robinson's Moss	500	5470	Q-2434	3
Alport Moor	540	>4900	Q-2431	3
Rishworth Moor	410	4010		7
Round Loaf	305	3200		6

I. Rise in Gramineae pollen

Featherbed Moss	500	491	Q-849	13
Alport Moor	515	430	SRR-4783	14
Extwistle Moor	360	1460AD		8

References

Allaby, M. (1983). *The Changing Uplands*. Countryside Commission, Cheltenham.

Anderson, P. (1997). Changes following the reduction in grazing pressure on the west face of Kinder Scout. *Restoring Moorland. Peak District Moorland Management Project, Phase 3 Report* (eds P. Anderson, J.H. Tallis & D.W. Yalden). Peak Park Joint Planning Board, Bakewell, Derbyshire.

Anderson P. & Tallis, J. (1981). The nature and extent of soil and peat erosion in the Peak District – field survey. *Moorland Erosion Study. Phase 1 Report* (eds J. Phillips, D. Yalden & J. Tallis), pp. 52–64. Peak Park Joint Planning Board, Bakewell, Derbyshire.

Anderson, P., Tallis, J.H. & Yalden, D.W. (eds) (1997). *Restoring Moorland. Peak District Moorland Management Project, Phase 3 Report*. Peak Park Joint Planning Board, Bakewell, Derbyshire.

Anderson, P. & Yalden, D.W. (1981). Increased sheep numbers and the loss of heather moorland in the Peak District, England. *Biological Conservation*, **20**, 195–213.

Bain, M.G. (1991). *Palaeoecological Studies in the Rivington Anglezarke Uplands, Lancashire*. PhD thesis, University of Salford.

Barnes, B. (1982). *Man and the Changing Landscape*. Merseyside County Museums, Liverpool.

Bartley, D.D. (1975). Pollen analytical evidence for prehistoric forest clearance in the upland area west of Rishworth, West Yorkshire. *New Phytologist*, **74**, 375–81.

Bartley, D.D. & Chambers, C. (1992). A pollen diagram, radiocarbon ages and evidence of agriculture on Extwistle Moor, Lancashire. *New Phytologist*, **121**, 311–20.

Birks, H.J.B. (1988). Long-term ecological change in the British uplands. *Ecological Change in the Uplands* (eds M.B. Usher & D.B.A. Thompson), pp. 37–56. Blackwell, Oxford.

Bower, M.M. (1960). Peat erosion in the Pennines. *Advancement of Science, London*, **64**, 323–31.

Burt, T.P. & Gardiner, A.T. (1984). Runoff and sediment production in a small peat-covered catchment: some preliminary results. *Catchment Experiments in Fluvial Geomorphology* (eds T.P. Burt & D.E. Walling), pp. 133–51. Geo Books, Norwich.

Caporn, S.M. (1997). Air pollution and its effects on vegetation. *Restoring Moorlands. Peak District Moorland Management Project, Phase 3 Report* (eds P. Anderson, J.H. Tallis & D.W. Yalden), pp. 28–35. Peak Park Joint Planning Board, Bakewell, Derbyshire.

Conway, V.M. (1954). The stratigraphy and pollen analysis of southern Pennine blanket peats. *Journal of Ecology*, **42**, 117–47.

de Smidt, J.T. (1995). The imminent destruction of north-west European heaths due to atmospheric nitrogen deposition. *Heaths and Moorland: Cultural Landscapes* (eds D.B.A. Thompson, A.J. Hester & M.B. Usher), pp. 206–17. Scottish Natural Heritage, Edinburgh.

Ferguson, P. & Lee, J.A. (1983). Past and present sulphur pollution in the Southern Pennines. *Atmospheric Environment*, **17**, 1131–37.

Francis, I.S. (1990). Blanket peat erosion in a mid-Wales catchment during two drought years. *Earth Surface Processes and Landforms*, **15**, 445–56.

Hibbert, F.A., Switsur, V.R. & West, R.G. (1973). Radiocarbon dating of Flandrian pollen zones at Red Moss, Lancashire. *Proceedings of the Royal Society, Series B*, **177**, 161–76.

Hobbs, R.J. & Gimingham, C.H. (1987). Vegetation, fire and herbivore interactions in heathland. *Advances in Ecological Research*, **16**, 87–173.

Jacobi, R.M., Tallis, J.H. & Mellars, P.A. (1976). The Southern Pennine Mesolithic and the ecological record. *Journal of Archaeological Science*, **3**, 307–20.

Johnson, R.H., Tallis, J.H. & Wilson, P. (1990). The Seal Edge Coombes, Derbyshire – a study of their erosional and depositional history. *Journal of Quaternary Science*, **5**, 83–94.

Labadz, J.C., Burt, T.P. & Potter, A.W.R. (1991). Sediment yield and delivery in the blanket peat moorlands of the Southern Pennines. *Earth Surface Processes and Landforms*, **16**, 255–71.

Lee, J. (1981). Atmospheric pollution and the Peak District blanket bogs. *Moorland Erosion Study. Phase 1 Report* (eds J. Phillips, D. Yalden & J. Tallis), pp. 104–08. Peak Park Joint Planning Board, Bakewell, Derbyshire.

Lee, J.A., Tallis, J.H. & Woodin, S.J. (1988). Acidic deposition and British upland vegetation. *Ecological Change in the Uplands* (eds M.B. Usher & D.B.A. Thompson), pp. 151–62. Blackwell, Oxford.

Livett, E.A. (1982). *The Interaction of Heavy Metals with the Peat and Vegetation of Blanket Bogs in Britain*. PhD thesis, University of Manchester.

Livett, E.A., Lee, J.A. & Tallis, J.H. (1979). Lead, zinc and copper analyses of British blanket peats. *Journal of Ecology*, **67**, 865–91.

Livett, E. & Tallis, J. (1994). *North Peak ESA: Pollen and Charcoal Analyses from Tintwistle Moor and Holme Moss*. Internal Report to the Peak Park Joint Planning Board, Bakewell, Derbyshire.

McEwen, M. & Sinclair, G. (1983). *New Life for the Hills*. Council for National Parks, London.

Montgomery, T. & Shimwell, D. (1985). *Changes in the environment and vegetation of the Kinder–Bleaklow SSSI, 1750–1840: historical perspectives and future conservation policies*. Internal Report to the Peak Park Joint Planning Board, Bakewell, Derbyshire.

Moore, P.D. (1975). Origin of blanket mires. *Nature, London*, **256**, 267–69.

Moore, P.D. (1993). The origin of blanket mire, revisited. *Climate Change and Human Impact on the Landscape* (ed. F. Chambers), pp. 217–24. Chapman & Hall, London.

Moss, C.E. (1913). *The Vegetation of the Peak District*. Cambridge University Press, Cambridge.

Phillips, J., Yalden, D. & Tallis, J. (1981). *Moorland Erosion Project. Phase 1 Report*. Peak Park Joint Planning Board, Bakewell, Derbyshire.

Pigott, C.D. (1983). Regeneration of oak–birch woodland following exclusion of sheep. *Journal of Ecology*, **71**, 629–46.

Porter, M. (1989). *Pennine Way Management Project. Second Annual Report 1988–89*. Countryside Commission, Cheltenham.

Press, M.C., Woodin, S.J. & Lee, J.A. (1986). The potential importance of an increased atmospheric nitrogen supply to the growth of ombrotrophic *Sphagnum* species. *New Phytologist*, **103**, 45–55.

Radley, J. (1965). Significance of major moorland fires. *Nature, London*, **205**, 1254–59.

Richards, J.R.A. (1990). *The Potential Use of Eriophorum angustifolium in the Revegetation of Blanket Peat*. PhD thesis, University of Sheffield.

Richards, J.R.A., Wheeler, B.D. & Willis, A.J. (1995). The growth and value of *Eriophorum angustifolium* Honck. in relation to the revegetation of eroding blanket peat. *Restoration of Temperate Wetlands* (eds B.D. Wheeler, S.C. Shaw, W.J. Fojt & R.A. Robertson), pp. 509–21. Wiley, Chichester.

Shimwell, D. (1974). Sheep grazing intensity in Edale, 1692–1747, and its effect on blanket peat erosion. *Derbyshire Archaeological Journal*, **94**, 35–40.

Shimwell, D. (1981). People pressure. *Moorland Erosion Study. Phase 1 Report* (eds J. Phillips, D. Yalden & J. Tallis), pp. 148–59. Peak Park Joint Planning Board, Bakewell, Derbyshire.

Smith, W.G. & Moss, C.E. (1903). Geographical distribution of vegetation in Yorkshire. *Geographical Journal*, **21**, 375–401.

Spratt, D.A. (1981). Prehistoric boundaries on the North Yorkshire Moors. *Prehistoric Communities in Northern England: Essays in Economic and Social Reconstruction* (ed. G. Barker), pp. 87–104. Dept. of Prehistory and Archaeology, University of Sheffield.

Tallis, J.H. (1965). Studies on southern Pennine peats. IV. Evidence of recent erosion. *Journal of Ecology*, **53**, 509–20.

Tallis, J. (1981). Uncontrolled fires. *Moorland Erosion Study. Phase 1 Report* (eds J. Phillips, D. Yalden & J. Tallis), pp. 176–82. Peak Park Joint Planning Board, Bakewell, Derbyshire.

Tallis, J.H. (1982). The Moorland Erosion Project in the Peak Park. *Recreation Ecology Research Group Report*, **8**, 27–36.

Tallis, J.H. (1985). Mass movement and erosion of a southern Pennine blanket peat. *Journal of Ecology*, **73**, 283–315.

Tallis, J.H. (1987). Fire and flood at Holme Moss: erosion processes in an upland blanket mire. *Journal of Ecology*, **75**, 1099–129.

Tallis, J.H. (1991). Forest and moorland in the South Pennine uplands in the mid-Flandrian period. III. The spread of moorland – local, regional and national. *Journal of Ecology*, **79**, 401–15.

Tallis, J.H. (1995a). Blanket mires in the upland landscape. *Restoration of Temperate Wetlands* (eds B.D. Wheeler, S.C. Shaw, W.J. Fojt & R.A. Robertson), pp. 495–508. Wiley, Chichester.

Tallis, J.H. (1995b). Climate and erosion signals in British blanket peats: the significance of *Racomitrium lanuginosum* remains. *Journal of Ecology*, **83**, 1021–30.

Tallis, J.H. & Livett, E.A. (1994). Pool-and-hummock patterning in a southern Pennine blanket mire I. Stratigraphic profiles for the last 2800 years. *Journal of Ecology*, **82**, 775–88.

Tallis, J.H. & McGuire, J. (1972). Central Rossendale: the evolution of an upland vegetation. I. The clearance of woodland. *Journal of Ecology*, **60**, 721–37.

Tallis, J.H. & Switsur, V.R. (1973). Studies on southern Pennine peats. VI. A radiocarbon-dated pollen diagram from Featherbed Moss, Derbyshire. *Journal of Ecology*, **61**, 743–51.

Tallis, J.H. & Switsur, V.R. (1983). Forest and moorland in the South Pennine uplands in the mid-Flandrian period. I. Macrofossil evidence of the former forest cover. *Journal of Ecology*, **71**, 585–600.

Tallis, J.H. & Switsur, V.R. (1990). Forest and moorland in the South Pennine uplands in the mid-Flandrian period. II. The hillslope forests. *Journal of Ecology*, **78**, 857–83.

Tallis, J.H. & Yalden, D.W. (1983). *Peak District Moorland Restoration Project, Phase 2 Report*. Peak Park Joint Planning Board, Bakewell, Derbyshire.

Turtle, C.E. (1984). *Peat Erosion and Reclamation in the Southern Pennines*. PhD thesis, University of Manchester.

Vine, P.M. (1982). The Neolithic and Bronze Age Cultures of the Middle and Upper Trent Basin. *British Archaeological Reports, British Series*, **105**, 1–410.

Williams, C.T. (1985). Mesolithic exploitation patterns in the Central Pennines. A palynological study of Soyland Moor. *British Archaeological Reports, British Series*, **139**, 1–179.

Wymer, J.J. (1977). *Gazetteer of Mesolithic Sites in England and Wales*. Council for British Archaeology, London.

Yalden, D. (1981a). Sheep and moorland vegetation – a literature review. *Moorland Erosion Study. Phase 1 Report* (eds J. Phillips, D. Yalden & J. Tallis), pp. 132–41. Peak Park Joint Planning Board, Bakewell, Derbyshire.

Yalden, D. (1981b). Sheep numbers in the Peak District. *Moorland Erosion Study. Phase 1 Report* (eds J. Phillips, D. Yalden & J. Tallis), pp. 116–24. Peak Park Joint Planning Board, Bakewell, Derbyshire.

Yalden, D. (1981c). Loss of grouse moors. *Moorland Erosion Study. Phase 1 Report* (eds J. Phillips, D. Yalden & J. Tallis), pp. 200–03. Peak Park Joint Planning Board, Bakewell, Derbyshire.

CHAPTER TWELVE

Urban wastelands – new niches and primary succession

A.D. BRADSHAW

Introduction

In most of Great Britain plant and animal communities are old – old in relation to our own life span at least. Indeed, except in special industrial habitats, such as those discussed by Ash (chapter eight, this volume), they have been in existence for several millennia, usually since the last glacial period. They may have changed under the influence of climate or human activity, but some sort of integrated plant and animal community has been present for a very long time.

This means that there has been a cover of plants which has accumulated nutrients and generated organic matter. The organic matter has become incorporated into the soil and then decomposed in the soil material by the combined effects of different soil organisms. As a result, relatively benign environments with fertile soils have been produced, which because of the time available have become colonised by many different species. There have been time and opportunity for full development of the features of structure and function that make up a complete and viable ecosystem.

The environments within the middle of cities are very different. Firstly, they are very recent; they have only been in existence since disturbance last occurred – due to road work, housing clearance, building construction, or waste disposal. Secondly, they are often ephemeral; they may exist for two or three years between the clearance of one building and the start of another. Thirdly, the soils are mostly inorganic, best described as skeletal; there has been no time for organic matter development and incorporation, only sand or gravel or mortar and broken concrete. Fourthly, they are likely to be physically and chemically stressful to plants; the surface of brick wastes can be hot and dry, and high in alkaline substances such as lime.

The plants which grow under these conditions might be considered to be of little interest. The situations themselves could be considered of little interest too, because they are so unlike normal habitats. But what occurs in them is what has occurred in the early development of the mature ecosystems to which we usually pay attention. The development is called primary succession, because of the succession of species which appear as time progresses and the environment changes.

It is an important process because it is what is involved in the origin of all ecosystems. It is obvious that early succession species are mostly annuals and biennials and that late succession species are perennials and woody species. There are, however, three different models of the processes involved (Connell & Slatyer 1977). In the first, species arrive as they can and their appearance is related solely to their life history characteristics; those remaining at the end are those which can tolerate the changing environmental conditions that develop – the *tolerance* model. In the second, the early species modify the environment so it is more suitable for the later species – the *facilitation* model. In the third model, once the early species have developed, they prevent the later species from developing until they have died and vacated the space they occupy – the *inhibition* model.

It is usually difficult to see succession in progress, because it took place a long time ago. Because of their history of change and clearance, however, the inner urban areas of the Mersey Basin provide excellent examples of primary succession in progress. The areas are scattered throughout the region and are easy to find; some are very spectacular, and there are many different types of site and environment. What is very clear is that there are a large number of species to be found in all of them and that these species change with time. So what processes are involved? I will take just two, urban clearance areas and disused railway lines, of which there are plenty in the Mersey Basin, to show what there is to discover – how new niches for plants, and for whole ecosystems, develop. For what happens more generally the reader is recommended to consult the excellent account by Gilbert (1989).

Site characteristics

Urban clearance areas and railway lands are both physically extreme. The soils are skeletal; when buildings have been cleared very stony substrates are left, pieces of brick and concrete with rather sandy material (commonly > 50% sand) in the cracks between. Sometimes the surfaces are hard and compacted. Railway lands are similar; coarse stony ballast was brought in to make the foundations for the sleepers. Both can be very dry at the surface, although there may be moisture underneath.

Chemically, both are bound to be deficient in nitrogen, since this element, which is the nutrient required in the greatest amount by plants (Bradshaw 1983), is accumulated almost entirely by biological processes and held in organic matter. This will not have had time to occur in skeletal soils. Other nutrients can vary. Calcium is always high in brick wastes, from mortar and cement; potassium and phosphate are present in the clays from which the bricks were made and become available as the bricks weather (Dutton & Bradshaw 1982) (Table 12.1). The railway ballast was commonly just crushed limestone rock from Derbyshire and North Wales, containing little but calcium; but in some places it was cinders and boiler ash, low in calcium and most other important elements. Underneath the ballast however there may be a quite fertile subsoil.

There are a number of places in the Mersey Basin where there are special toxicities due to past industry, especially chromates from tanning, and cyanides and phenols from gasworks wastes. But this sort of toxicity is unusual in brick wastes and railway land, and at the same time these areas are mostly nearly neutral in soil pH. However where ash has been used as ballast for rail tracks, there can be considerable acidity, giving pH 4–4.5, as low as in heathlands.

The sites are therefore not necessarily impossible for plant growth, and this applies even to very small areas such as in the spaces between bricks and lumps of concrete, which may be filled with sandy-loam material. So what actually happens?

Colonisation and chance

When these areas are first formed it is unusual for them to be quickly covered by plants, although a wide range of species may be represented. If the site characteristics suggest that conditions are not hopeless, what is restricting an immediate outburst of plants? The first problem is the availability of suitable colonists. This is partly a matter of chance, but partly a matter of biological logic. There are several reasons why a species does not arrive at a particular site.

Distribution characteristics
Species with light, easily dispersed seeds such as Oxford Ragwort (*Senecio squalidus*), Mugwort (*Artemisia*

vulgaris) and Annual Meadow-grass (*Poa annua*) occur on every urban clearance site in Liverpool, but Fat-hen (*Chenopodium album*) and clovers (*Trifolium* spp.), with much heavier seeds do not. Among trees and shrubs, it is conspicuous that Willow species, especially Goat Willow (*Salix caprea*) and Sallow (*S.cinerea*), will turn up everywhere; their minute seeds with a bundle of delicate hairs, the pappus, can be carried many kilometres. By contrast, despite its visible local successes, Sycamore (*Acer pseudoplatanus*) only appears in the vicinity of a suitable parent. Ash (*Fraxinus exelsior*) is similar. Both may have wings, but the seeds are relatively heavy. Birch (*Betula* spp.) is intermediate; its light seeds with wings mean it can spread widely over an area, such as in the derelict railway land at Garston docks, but only because seed parents exist in the vicinity.

It is obvious that a species which can produce many seeds is likely to be more successful than one which produces few. But in colonising situations a species that produces seeds within the first year can multiply much more quickly than one producing seeds only after two years – 100 seeds produced by a plant of annual meadow grass can produce 100 x 100 seeds, i.e., 10,000, by the end of the second year. Little wonder that many of early colonists are annuals. There is even evidence from local evolution of this pressure to produce seed. The inner city populations of Annual Meadow-grass from Liverpool have been shown to flower earlier and more freely than populations from pastures (Figure 12.1).

A perennial species has, however, the advantage that once it has arrived it has the potential to hold on to its living space. If it can then spread vegetatively, by rhizomes or stolons, it can expand its position without going through the risks of a seed and seedling stage. In this category are species which are very successful on many different sites, such as Yorkshire Fog (*Holcus lanatus*) and Creeping Bent (*Agrostis stolonifera*); they tend to become dominant after the annuals. False Oat-grass (*Arrhenatherum elatius*) and Bramble (*Rubus* spp.) are slower to establish and tend to be become common later. The long stolons of Creeping Bent allow it to spread over hard surfaces and root where conditions are more favourable; Colt's-foot (*Tussilago farfara*) spreads underground and Bramble, of course, above the ground.

Opportunity
Whatever its distribution characteristics, a species can only spread into a new area if it is already present some-

Site	Nitrogen[1]	Phosphorus[2]	Potassium[2]	Magnesium[2]
Good garden soil	1027(119)	30	83	66
Urban site 1	480 (3)	12	81	72
Urban site 2	480 (-9)	22	173	145
Urban site 3	405 (-6)	5	77	323
Urban site 4	870 (-4)	65	94	393

Table 12.1. Chemical characteristics of soils in urban clearance areas in Liverpool (from Dutton & Bradshaw 1982) .

[1] total (mineralisable in brackets) (ppm)   [2] available (ppm)

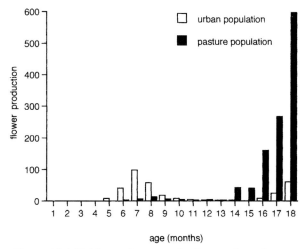

Figure 12.1. Evidence for natural selection for colonisation; much earlier flower production in an urban, Liverpool, population of *Poa annua* than in a pasture population from Cheshire (Law *et al.* 1977).

Calcareous wastes	Acid wastes
early stages	
Poa annua	*Agrostis capillaris*
Senecio squalidus	*Polytrichum* sp.
Holcus lanatus	*Aira praecox*
Agrostis stolonifera	*Hieracium* sp.
Trifolium repens (L)	*Lotus corniculatus* (L)
Sagina procumbens	*Rumex acetosella*
Matricaria recutita	*Vulpia bromoides*
Reseda luteola	*Campylopus inflexus*
Trifolium dubium (L)	*Aulacomium palustre*
	Cladonia sp.
middle stages	
Plantago lanceolata	*Lotus corniculatus* (L)
Dactylis glomerata	*Luzula campestris*
Buddleja davidii	*Deschampsia flexuosa*
Arrhenatherum elatius	*Carex ovalis*
Juncus effusus	*Festuca rubra*
Artemisia vulgaris	*Chamerion angustifolium*
Lotus corniculatus (L)	
Trifolium pratense (L)	
Rubus spp.	
Erysimum cheiranthoides	
Vicia angustifolia (L)	
late stages	
Salix cinerea	*Betula pubescens/pendula*
Salix caprea	*Calluna vulgaris*
Betula pubescens/pendula	*Hedera helix*
Quercus robur	*Lonicera periclymenum*
Alnus glutinosa	*Quercus robur*
Acer pseudoplatanus	*Salix caprea*
Fraxinus excelsior	*Salix cinerea*
Crataegus monogyna	

Table 12.2. Plant species particularly characteristic of primary successions on wasteland in the Mersey Basin, arranged in order of usual commonness, legumes indicated by (L); in many sites species indicated as belonging to early, middle or late stages may occur together, and because of the effects of chance there are likely to be considerable other deviations from this list.

where in the neighbourhood. Species such as Yellowwort (*Blackstonia perfoliata*) and Blue Fleabane (*Erigeron acer*) that occur 30 kilometres away clearly find it difficult to immigrate into sites in the Mersey Basin however suitable these might be (Ash, chapter eight, this volume). But it is almost as important whether the species is available in the immediate vicinity, within a few hundred metres. In cities such as Liverpool and Birkenhead there have been always a number of urban clearance areas in existence at any one time which can be colonised by a range of species (Table 12.2). But as colonisation of different sites takes place, quite startling differences in species occurrence can develop. This is particularly obvious with the Butterfly-bush or Buddleia (*Buddleja davidii*). It is an introduced shrub well adapted to the growth on brick waste. Yet it is only found on certain sites, related to its presence in a neighbouring area, usually upwind, where it may often have been planted for ornamental purposes. White and Red Clover (*Trifolium repens* and *T.pratense*), species which have an important role in soil development on brick wastes, will nearly always eventually be found on urban sites, but their initial appearance is strongly influenced by the occurrence of seeding plants in the immediate vicinity. The same applies to other more ubiquitous species such as mayweeds and Rosebay Willowherb (*Chamerion angustifolium*).

Establishment and ecological adaptation

It is not enough for propagules to arrive at a site. They have to be able to establish and grow. To do this the individual plants must firstly possess specific adaptations to overcome the substrate conditions. These adaptations are innumerable and their significances are well discussed elsewhere (Harper 1977), but evidence for their influence can readily be seen in the urban areas of the Mersey Basin.

Adaptation to soil physical characteristics

On urban brick wastes surfaces are usually hard and covered with fine textured material. On this only small seeds can grow because they can make good physical contact over a relatively large area of their seed coat and can therefore absorb water easily; large seeds can make contact only over a small part. It is therefore not surprising to find on brick wastes that it is small seeded species such as grasses that are initially successful. When, by contrast, a vegetation cover has begun to develop, the whole situation changes. Now the seedlings have to be able to compete with this cover, by having enough reserves to grow through it, and at the same time the surface of the ground is more moist. So species with bigger seeds are usually more successful. This includes such species as White Clover, Treacle Mustard (*Erysimum cheiranthoides*) and plantains (*Plantago* spp.).

Smooth surfaces, however, offer only rather exposed conditions, difficult for any seeds. A rough surface due to gravel or other coarse material provides protected microsites in which germinating seeds can prosper. This is well known in the colonisation of areas destroyed by volcanic eruptions such as Mount St Helens (del Moral 1993). The stony surfaces of old railway lines such as the Halewood Triangle of the Cheshire Lines, disused since 1964, have been spectacularly colonised by seedlings from the very small wind blown seeds of Sallow and Birch. This colonisation is helped by the initial absence of vegetation, which allows the seedlings to grow without interference. Stony brick wastes can be similar. By contrast, in the grass covered areas in the middle of the Halewood Triangle there is no sign of such colonisation. Instead the main woody colonists have been Hawthorn (*Crataegus monogyna*) and Oak (*Quercus* sp.), both species with large seeds.

Adaptation to soil chemical characteristics

There is a danger of assuming that the chemical conditions of urban areas are all similar or, if they are different, have no influence on primary succession. Since the urban site materials are mostly poor and calcareous there tends to be a common range of species. But the ballast of some railway lines was ash and cinders, not only poor in many nutrients but also very acid, with a pH of about 4. On such material a different range of species is to be found, since very few of the herbaceous species mentioned so far can survive. The early colonists are restricted to acid tolerant mosses such as *Polytrichum* spp. and annuals such as Early Hair-grass (*Aira praecox*). The main grass is Common Bent (*Agrostis capillaris*), normally to be found on heathland (Table 12.2). Then as a great surprise, in some places, such as the disused marshalling yards serving Garston Docks, Heather (*Calluna vulgaris*) appears, with Wavy Hair-grass (*Deschampia flexuosa*) and Oval Sedge (*Carex ovalis*). Where these have come from is unknown, but there is now so much Heather that the planners have given the area the name Cressington Heath, and it has been recognised as a potential local nature reserve. A number of other species typical of local heathland, however, such as Sheep's-fescue (*Festuca ovina*), Heath Bedstraw (*Galium saxatile*), Tormentil (*Potentilla erecta*) and Bell Heather (*Erica cinerea*), are missing, giving support to the idea that chance and opportunity are as important factors as ecological adaptation.

In damp or wet areas, in poorly drained railway tracks or where there have been excavations, such as in the old Garston gasworks site, moisture-loving species can occur. Typical are Hard and Soft Rush (*Juncus inflexus* and *J. effusus*). However, these are uncommon in most urban areas and therefore will not always be available for colonisation. Other examples of differentiation in the type of primary succession in relation to soil characteristics occur, but too few studies have been undertaken so far to distinguish them properly.

Growth and development

In a skeletal habitat growth may be able to start, but can it be sustained? This depends on adequate continuing supplies of the materials provided by the soil – water and nutrients. In primary successions failure or poor growth is easy to see; it is less easy to discover what are its critical causes.

Importance of water

A stony soil such as brick waste, containing no organic matter, can hold very little water; less than a third that of a normal soil. Although it may be moist over the winter, once drying conditions occur in the spring there can be little moisture left in the surface layers. For this reason many of the early colonists such as Annual Meadow-grass are annuals which germinate in the winter or early spring, and die, having set seed, by the early summer.

There is an alternative strategy, however. In all such situations there is always water lower in the soil profile. This will be available to any plant that can develop a large root system. This essentially requires the plant to be able to grow longer, even to be perennial. Woody plants and perennials with deep root systems, such as False Oat-grass and Mugwort, colonising urban areas, show no signs of water shortage. But this more substantial growth requires a supply of nutrients.

Importance of nutrients

A prime requisite for substantial growth is therefore an adequate supply of nutrients. From what was said earlier, the nutrient most likely to be deficient is nitrogen, without which little or no growth is possible. For good growth a continuous vegetation cover requires an annual supply of about 100kg N ha^{-1}. The over-riding importance of nitrogen has been demonstrated in fertiliser experiments exploring the problems facing vegetation artificially established on Liverpool brick wastes (Figure 12.2). It is significant that where nitrogen is applied once, growth falls off in the second year unless further nitrogen is applied. A continuing supply is needed. Similar experiments have not been carried out on a natural urban primary succession, but conspicuous green patches of better growth can be seen in naturally colonised urban areas where dogs have urinated or rabbits have left a patch of droppings.

How is this nitrogen to become available? It normally comes from the release of mineral nitrogen by the decay of organic matter. Some nitrogen is bound to be present in the small amount of organic matter in the skeletal soil, but it will only be enough to allow very limited growth. A second source is the nitrogen contained in rain. In country areas this contributes no more than 10kg N ha^{-1} yr^{-1}; in urban areas of the Mersey Basin this figure can be doubled or trebled by aerial pollution. A prime requisite of plants is therefore to have an extensive root system to scavenge for this and other nutrients. Woody plants such as Goat Willow and Buddleia particularly can achieve

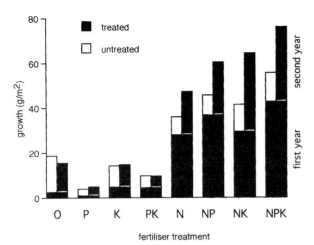

Figure 12.2. The effects of addition of fertilisers to grassland newly established on brick waste; the main effect is from nitrogen, but the need to repeat the addition in a subsequent year indicates the serious lack of nitrogen capital (Bloomfield *et al.* 1982).

Figure 12.3. Excellent growth of Buddleia where everything else is rather moribund; the extensive root system of the Buddleia allows it to scavenge successfully for nutrients (*Photo: A.D. Bradshaw*).

this, and can be very successful in nutrient poor sites when herbaceous species are doing badly (Figure 12.3). In consequence they are often termed scavengers.

The most successful strategy is, however, that adopted by legumes. They are able to make use of the limitless supplies of nitrogen in the air by the activities of nitrogen-fixing bacteria living in nodules on their roots. The plant provides protection and energy to the bacteria, and the bacteria provides fixed nitrogen which can be used immediately by the plant, making it independent of external supplies of nitrogen. If therefore a species such as White Clover arrives in an area of brick waste in Liverpool it immediately grows and spreads (Figure 12.4). It can fix over 100kg N ha^{-1} yr^{-1}, quite enough for its own growth and even for accompanying plants. After one or two years the latter begin to go green and grow vigorously. The whole succession takes on a new lease of life.

There are a number of legumes adapted to the poor calcareous conditions of brick wastes, notably White and Red Clovers and Common Bird's-foot-trefoil (Table 12.2). Which are to be found on a particular site seems to depend very much on chance immigration. The arrival of legumes in a succession means that there is an immediate improvement in plant cover due to their own growth, but because of their contribution to the overall nitrogen of the site there is a marked increase in the growth of all species.

Development of ecosystem processes

The effect of the developing vegetation is to contribute organic matter to the soil and visibly alter its texture and structure. This is visible on many sites. At the same time the nitrogen content of the soil is increased and the carbon/nitrogen ratio decreased. All this favours an increase in the biological activity of the soil, in particular

Figure 12.4. A single plant of White Clover; because it is a nitrogen-fixer it is itself growing excellently and will soon encourage the growth of associated species (*Photo: A.D. Bradshaw*).

| | reclaimed brick wastes | | | normal soil |
	raw brick waste	grass only	grass and clover	good lawn
nitrogen – total (ppm)	500	850	1600	2440
– mineralisable (ppm)	32	12	84	165
C/N ratio	38	46	29	21
yield of bioassay (g/m²)	510	450	940	950

Table 12.3. Changes occurring in nitrogen levels and availability in urban brick waste sites: a comparison of sites with raw brick waste with a) sites with grass only, b) sites with grass with good clover, and c) a long established ornamental lawn (means of at least nine sites in each category): bioassay is yield after 29 weeks of Rye-grass sown on soil samples in pots (from Marxen-Drewes 1983).

the organic matter turn over. This is best revealed by measurements of soil respiration – the amount of carbon dioxide being produced – but these are not available. However values for mineralisable nitrogen, the mineral nitrogen released by microbial processes when the soil is incubated, are perhaps the real key, since this fraction indicates what nitrogen the soil is able to offer growing plants on a continuous basis. The sort of changes that occur are illustrated in Table 12.3 by results for a set of soils under artificial grasslands established on brick waste in Liverpool compared with a garden soil as control.

A more direct measure of the improved fertility can be obtained by a bioassay in which plants are grown on the soils; this is included in Table 12.3. This confirms the poverty of the raw brick waste and the startling improvements that can occur once Clover has taken hold, improvements which make the Clover-rich site soil as good as soil from a long-established grass lawn. It is interesting that the site on which a poor grass sward was established behaves worse than the raw brick waste. This is probably because the grass has taken up all the available nitrogen and rendered it unavailable to other species. On this site the C/N ratio is higher, which would make the nitrogen less readily released by microbial activity.

The same processes of ecosystem development can be seen in the areas of disused railway lines. However where cinders were used as a ballast the process of nitrogen accumulation can be upset because the acid soil conditions do not permit the establishment of legumes. As a result nitrogen accumulation is slowed to what the vegetation can trap from the rain, and development is much slower.

Increasing competition

The soil studies show the way in which conditions progressively improve. This brings into importance a new factor, competition, which although it may have been present earlier, comes into play particularly once a vigorous vegetation has developed. This vegetation can interfere with establishment and growth of other species.

Its most obvious effects are in the disappearance of the annuals. The solid swards of perennial grasses such

as Yorkshire-fog and Cock's-foot (*Dactylis glomerata*) make it difficult for new generations of seedlings of annuals to establish. The succession may appear to reach a standstill.

But it is at this stage that the final major act of the succession occurs – woody species begin to become visible, although they have very often been present earlier. It is unfortunate for a student of primary succession that at this stage someone often notices the incursions of taller plants and orders a 'tidying up'. But if this does not occur, then the changes to scrub and woodland are most interesting.

Obviously once established, with their taller growth and more extensive root systems, woody plants should be able to win in the competition battle. But establishment may be a problem, depending on a chance gap, or safe site. As a result the establishment of woody vegetation may be delayed or only sporadic in well developed grassy areas. However, where conditions for establishment are favourable early on, and seed parents exist in the vicinity, woody vegetation can get a hold early and hasten the whole course of succession in a spectacular fashion, as on the rail tracks at Halewood.

The woody endpoint

If the areas remain untouched there is nothing to stop them developing into woodland. In areas where Buddleia gets a hold there can be a rather scruffy stage after about twenty years when it starts to die back, before more permanent species develop. On most sites in the Mersey Basin, however, because of their powers of immigration, dense Sallow or Birch woodland develops; for a decade or more this can be completely impassable, but then natural thinning processes start to occur and tree numbers drop from about 10m⁻² to less than 1m⁻².

In such woodland there is little scope for other species, and perhaps invasion by Oak and other species may have to wait until space appears. But in many of these dense woodlands, such as at Halewood, small numbers of Pedunculate Oak, Sycamore and sometimes Alder (*Alnus glutinosa*) are already present, which will be able to outlast the shorter lived species. Very few woodland herb species occur. This could be due to the density of the trees, but it is equally likely that

Figure 12.5. Nearly the end of the process – woodland over 100 years old on disused railway land; but a number of typical woodland species are missing – they have not yet arrived (*Photo: A.D. Bradshaw*).

reduce its density. The fine soil they cast onto the surface rapidly covers the brick waste, at a rate of about 4mm per year, so that after 10 years or so brick waste sites in Liverpool can be found to be covered with over 4cm of excellent material (Figure 12.6). This matches what over 100 years ago Darwin (1881) reported occurring in normal soils. At the same time the earthworms take down into their burrows large amounts of leaf material which becomes incorporated into the soil. Their contribution to the succession is therefore very important.

The animals in the Mersey Basin have not been extensively studied; however, good descriptions exist for other areas (Gilbert 1989). It must be remembered that there are many animal species waiting to take advantage of the improving conditions. If the vegetation of any well colonised grassy site is parted to ground level, it will reveal a network of trackways formed by small mammals such as voles and shrews, living off either the vegetation itself or the substantial developing insect populations. Overhead somewhere will be a kestrel, roosting in an old building but feeding off the small mammals.

The areas with shrubs begin to support excellent populations of Blackbirds (*Turdus merula*), Wrens

Figure 12.6. The contribution of earthworms; after 18 years this grassed brick waste is covered with over 5 cm of fine soil due to the earth brought to the surface by the well-developed earthworm population (*Photo: A.D. Bradshaw*).

suitable species are not available in the vicinity.

There is one wood on railway land of considerable age. This is between the converging tracks at the tip of the Halewood Triangle. It was allowed to start its succession when the railway was first built through farmland in 1879, and it is now a fine Oak–Birch wood (Figure 12.5). However a simple survey shows that the succession is not complete even after 100 years. There is a strange absence of a typical woodland ground flora, species such as Dog's Mercury (*Mercurialis perennis*) and even Bluebell (*Hyacinthoides non-scripta*) – although the latter is present at the entrance, obviously planted. A similar absence is found in the planted woodlands of the Mersey Basin. It is well know that the species of ancient woodland have very limited powers of dispersal (Peterken 1974) and there are no sources of these species in the vicinity. Perhaps under modern conditions the final stages of the succession can only be achieved with human assistance.

Animals

Animals depend on plants for their food and energy, so in many ways they play a secondary role in successions. They are important in the cycling processes, none more so than the earthworms which rapidly colonise brick wastes, finding the calcareous soil very much to their liking. In making burrows they loosen the soil and

(*Troglodytes troglodytes*), Robins (*Erithacus rubecula*) and tits (*Parus* spp.). Perhaps what is most outstanding, however, is to hear, every spring, the soft falling song of Willow Warblers (*Phylloscopus trochilus*) which have migrated from Africa to take up residence in the new Sallow woodlands in the Mersey Basin which they find much to their liking.

The succession telescoped

On some sites many of the late succession species can be found appearing early. This suggests that the constraints on growth which have been discussed may not always be operating. Species of relatively fertile soils, such as Perennial Rye-grass (*Lolium perenne*), Meadow Fescue (*Festuca pratensis*), Common Nettle (*Urtica dioica*), typically associated with human habitation and buttercups (*Rununculus* spp.), as well as many of the woody species, can be found on rather young inner city wasteland, growing vigorously.

There are two possible explanations. The first is that the soils in these sites are not actually as poor as they appear at first sight. In many situations the site may have been formed from the clearance of old housing and contain a considerable amount of garden soil, albeit mixed with bricks and other stony material. This sort of soil can be very fertile. In these circumstances the succession that is occurring is more like that found in situations where, although the vegetation has been destroyed, the soil has not. This type of succession is described as secondary and is typical where arable fields have gone out of cultivation. The main constraints in these situations are the availability and colonising power of the individual species. In urban areas there is always the possibility that several species may have come through from previously existing gardens, in soil and other materials, something which does not usually occur in normal secondary successions, from arable fields for instance.

The second possible explanation is that the idea that the occurrence and growth of species in primary successions are always limited by fertility does not always apply. Many species characteristic of the more fertile soils found in the middle stages of the succession can grow relatively well on infertile soils. In the list given in Table 12.2 this could certainly apply to Ribwort Plantain (*Plantago lanceolata*), Cock's-foot, Buddleia and Sallow, and perhaps to some of the other species too. Unfortunately there is little critical evidence about this. It is one more of the ecological and intellectual challenges of urban sites and of succession in general.

Conclusions

The new habitats of the Mersey Basin may look unprepossessing. But they provide good evidence not only of the way plants and animals colonise, and of the ways in which ecosystems are formed, but also of the different models, already mentioned, suggested to be operating in primary succession (Connell & Slatyer 1977). In all models the early species cannot grow once the site is occupied by the later species. But the long held view is that succession is controlled and driven by the processes of environmental improvement, particularly by the improvements arising from the growth of the early species allowing subsequent species to colonise and grow – the *facilitation* model. This is the way in which most of the evidence presented here would seem to argue. But since it is perfectly possible for late species to arrive and grow early on, it is possible to argue that the observed changes occurring in succession are driven more by accident of arrival combined with speed of growth – the *tolerance* model. There are also signs that later species may not be able to invade because the site is already occupied – the *inhibition* model.

If the evidence is taken at its face value it would appear that all these processes are operating in urban areas, and that no one process has overall control. It would be of great value to study what is going on in more detail, and with experimental interventions to test the significance of individual factors. The advantage of urban areas is that this can be done with little expense. There is also the advantage that changes can be watched and followed as they occur week by week without having to try to guess from an annual visit what has happened. All that occurs in an urban succession has a reason and a cause that ought to be definable, if one can only look closely enough.

At the same time it must not be forgotten that these same processes can produce plant communities of considerable attraction and landscape value, at little or no cost. It is too easy to sweep them away in their younger stages in the causes of tidiness or development. We should learn to cherish them.

References

Bloomfield, H.E., Handley, J.F. & Bradshaw, A.D. (1982). Nutrient deficiencies and the aftercare of reclaimed derelict land. *Journal of Applied Ecology*, **19**, 151–58.

Bradshaw, A.D. (1983). The reconstruction of ecosystems. *Journal of Applied Ecology*, **20**, 1–17.

Connell, J.H. & Slatyer, R.O. (1977). Mechanisms of succession in natural communities and their role in community stability and organisation. *American Naturalist*, **111**, 1119–44.

Darwin, C. (1881). *The Formation of Vegetable Mould through the Action of Earthworms*. John Murray, London.

Dutton, R.A. & Bradshaw, A.D. (1982). *Land Reclamation in Cities*. HMSO, London.

Gilbert, O.L. (1989). *The Ecology of Urban Habitats*. Chapman & Hall, London.

Harper, J. (1977). *Population Biology of Plants*. Academic Press, London.

Law, R., Putwain, P.D.P., & Bradshaw, A.D. (1977). Life history variation in *Poa annua*. *Evolution*, **6**, 233–46.

Marxen-Drewes, H. (1983). *Nitrogen in Urban Ecosystems*. MSc thesis, University of Liverpool.

del Moral, R. (1993). Mechanisms of primary succession on volcanoes: a view from Mount St Helens. *Primary Succession on Land* (eds J. Miles & D.W.H. Walton), pp. 79–100. Blackwell, Oxford.

Peterken, G.F. (1974). A method for assessing woodland flora for conservation using indicator species. *Biological Conservation*, **6**, 239–45.

Changing Habitats 2

S. WALKER

Most will be aware that the Environment Agency was formed on 1 April 1996 from the amalgamation of the National Rivers Authority, Her Majesty's Inspectorate of Pollution and staff from the Metropolitan and County Councils responsible for Waste Regulation. The Agency's vision is to provide a better environment in England and Wales for present and future generations. It aims to protect and improve the environment as a whole by effective regulation, by its own actions and by working in partnership with others.

This section continues the theme on changing habitats with the focus now shifting to the water and waterside environments.

In introducing the topic I would like to draw your attention to the increasingly 'good news' story of the Mersey Estuary and the improvements in quality that have been seen in recent years. These improvements are well described in the National Rivers Authority report of 1995 entitled *The Mersey Estuary : A Report on Environmental Quality* (NRA Water Quality Series No. 23). The document describes the physical and biological processes and the ecological and chemical quality of the estuary. It describes the potential for the future in the context of the regulatory framework. Though the quality of the Mersey Estuary has significantly improved in recent times, there is still much more to be achieved.

This section gives an insight into the potential, not just for the estuary but for the Mersey Basin as a whole. Chapters thirteen and fourteen describe the habitats and the changes that have taken place to the freshwaters of the Mersey Basin. These have been largely man induced and have been considerable with heavy pollution of many rivers. The authors consider that future prospects are mixed and depend upon long-term developments and management policies.

Chapters fifteen and sixteen continue with an assessment of estuarine and coastal habitats. Again the massive impact of human intervention and pollution is recorded even in the supposedly more natural sand dunes and saltmarshes. However, recent measures to remove pollution and restore habitats in the Mersey Estuary are being rewarded by an increase in biodiversity and a more optimistic future is foreseen. The coastal sand dunes and saltmarshes are dynamic habitats influenced by human activities but there are opportunities to influence the changes to provide more 'natural' and self sustaining systems than at present.

CHAPTER THIRTEEN

Biological change in the freshwaters of the Mersey Basin

J.W. EATON, D.G. HOLLAND, B. MOSS AND P. NOLAN

Introduction

The Mersey Basin offers a diversity of freshwater habitats. Fast-flowing stony and peaty upland headwaters of the River Mersey contrast with slower, silty channels in lowland areas. The Cheshire meres to the south of the river are a group of standing waters of various sizes and depths; some are connected to the river system, others are isolated. To the north as well as the south, former lake basins have succeeded to raised bogs, or mosses, many of them now drained and degraded. Man has added to this diversity by creating many artificial water-

bodies within the Mersey Basin. These include reservoirs, ornamental lakes, navigation canals, ditches and ponds.

The biology of most of these freshwaters only began to be recorded in any detail in the second half of the 20th century and then only to a limited extent. Before this a systematic search of the copious but scattered records left by 19th-century natural historians would almost certainly provide historic information. Nevertheless, knowing the enormous changes in land- and water-use which have occurred in areas such as the Mersey Basin since the start of the Industrial Revolution (Eaton 1989),

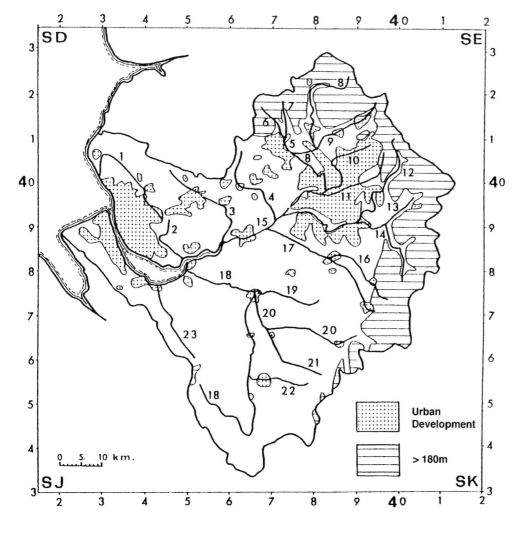

Figure 13.1. Rivers of the Mersey Basin, with topography and urban development.

1. River Alt
2. Ditton Brook
3. Sankey Brook
4. River Glaze
5. River Croal
6. Eagley Brook
7. Bradshaw Brook
8. River Irwell
9. River Roch
10. River Irk
11. River Medlock
12. River Tame
13. River Etherow
14. River Goyt
15. River Mersey
16. River Dean
17. River Bollin
18. River Weaver
19. Peover Eye
20. River Dane
21. River Wheelock
22. Valley Brook
23. River Gowy.

it is possible to deduce some of the corresponding biological changes which have probably occurred in the water-bodies involved.

This chapter outlines and interprets changes in freshwater habitats and their biota in the Mersey Basin, before giving a view of likely future trends, where these can be surmised.

Rivers

Habitats

Physical change

The rivers of the Mersey Basin (Figure 13.1) have been extensively modified during agricultural, industrial and residential developments in the catchment. This has been partly to exploit their value as water sources, effluent receptors and transport routes, but partly also to control their erosive energy and periodic flooding, so that their flood plains could be developed.

Intensive urban development in the Mersey Basin began at the start of the Industrial Revolution, when river management increasingly created walled channels, weirs, mill pools and leats as common features. Many of the rivers of the northern part of the Mersey Basin are now regulated by reservoirs, with resultant long-term changes in river flow, velocities and sediment transfers. Communities in the rivers have adjusted to these physical changes. Flow regimes vary according to the season and the drought which began in the 1994/95 winter resulted in compensation flows to some rivers being reduced in 1995 and 1996 (National Rivers Authority 1996a).

The lower sections of the River Weaver were extensively modified by the construction of the Weaver Navigation Channel. Remnants of the former meandering channel can be seen in the flood plain between Frodsham and Winsford, and provide habitats which are now valuable wildlife havens for wetland plants and insects such as dragonflies. The River Mersey itself was greatly altered by the construction of the Manchester Ship Canal, and impact on the river can still be seen today.

A systematic programme of 'river improvements' and land drainage continued here until the early 1980s. Especially during the 1960s and 1970s, considerable lengths of river were deepened, widened and straightened. Whilst this enabled intensified use of the flood plain, the natural storage capacity for flood waters was lost, requiring the construction of additional urban and agricultural flood defences to protect infrastructures such as roads and railways, industrial complexes and agricultural land.

Some stretches of rivers and streams in Manchester, Merseyside and urban areas elsewhere in the Mersey Basin are now hidden in pipes and tunnels. A culverted river is virtually uninhabitable to wildlife, and a canalised open concrete channel is little better. Uniform, featureless channels can support only the commonest and most tenacious river plants and animals, particu-

Figure 13.2. Bedford Brook managed for flood control 'improvements' (*Photo: D.G. Holland*).

larly when this habitat simplification is combined with poor water quality. Such uniformity, together with destruction of the natural, diverse bankside vegetation by the engineering construction works and subsequent intensive management, has helped spread aggressive alien species such as Indian or Himalayan Balsam (*Impatiens glandulifera*) and Japanese Knotweed (*Fallopia japonica*) which are widespread throughout the catchment.

By contrast, rivers such as the Dane and Dean in Cheshire, which have received little or no management in the past, retain physical diversity. Riffles and pools, together with varied banks, shoals, cliffs and wetland margins support diverse channel vegetation and a range of bankside trees and shrubs. Good examples of river cliffs, which provide valuable habitats for birds, e.g., Sand Martin (*Riparia riparia*) and Kingfisher (*Alcedo atthis*) and insects, e.g., solitary wasps and bees, survive on unconstrained stretches of the Rivers Mersey (between Ashton and Carrington), Bollin and Tame. On the flood plains, marshes, herb-rich pastures, wet woodlands, river terraces and ox-bow lakes can be found. Good examples of flood plain features can be seen on the River Dane and on the River Bollin upstream of Manchester Airport and Wilmslow.

In recent years river management policies have changed. The Environment Agency and its predecessor the National Rivers Authority, have been given conservation responsibilities in respect of wildlife, landscape and heritage. These statutory duties and obligations are currently detailed in Sections 6(1), 7(1), 8 and 9 of the Environment Act 1995. Under this Act, and earlier Acts relevant to the water environment, e.g., Wildlife and Countryside Act 1981, Water Resources Act 1991, management is increasingly focusing on techniques sympathetic to the protection and enhancement of wildlife. Conservationists and river engineers work together to ensure all capital construction schemes and routine maintenance programmes are designed to do minimal damage to the natural environment. In addition, the Agency is actively involved in recreating more natural conditions out of existing heavily engineered watercourses (Royal Society for the Protection of Birds, National Rivers Authority and Society for Nature Conservation 1994), e.g., restoration projects on Padgate and Whittle Brooks in Warrington and deculverting of sections of the River Alt (Nolan & Guthrie in press). Furthermore, the Agency comments on applications by third parties for new developments and has policies which seek to retain river corridors as features of such developments. To provide a basis for more sustainable management of the rivers, Local Environment Agency Plans have been prepared to identify issues which it is feasible to resolve, in a more balanced approach to new management (National Rivers Authority 1994, 1996c; Environment Agency 1996).

Water quality influences

The water quality in a river greatly influences which plant and animal species occur in it. Holland and Harding (1984) describe the appalling conditions created by the Industrial Revolution in the Mersey Basin and show how gradual recovery started in the 1950s. Figure 13.3, recording the oxygen conditions at three sites, illustrates the substantial improvements which have been made in the last 40 years in the more industrial part of the Mersey Basin. Aquatic plant and invertebrate communities are key indicators of water quality, and the following section shows how plants, invertebrates and fish have responded to changes during this period.

Life in the rivers

Plants

As is the case with many British rivers, the River Mersey and its tributaries have received little systematic survey from botanists, and historical documentation of plants is limited. River Corridor Surveys (National Rivers Authority 1992) and River Habitat Surveys (National Rivers Authority 1996b) provide useful information on the distribution of aquatic and marginal plants and their habitats, but they have not been collated into any systematic database. Most information on the Mersey system comes from data collected after 1978 by Harding (1981) and Holland & Harding (1984).

In prehistoric times many parts of the Mersey Basin were densely wooded and growth of riverine plants was restricted by dense shading. As woodland clearance and increased nutrient run-off from the land occurred, there was probably a marked increase in submerged plant growth. A few clues to the nature of these plant communities can be seen in present vegetation in, for example, the upper reaches of the River Etherow and River Goyt which have not been subject to the kind of pollution affecting the rivers downstream. Here the bed is rocky and the vegetation is dominated by mosses and liver-

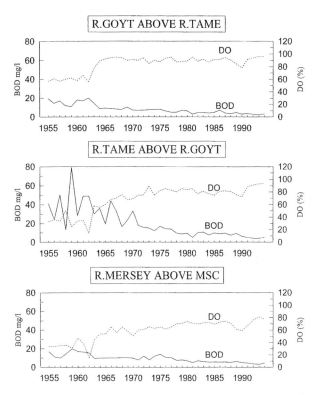

Figure 13.3. Dissolved oxygen (DO as % saturation) and biochemical oxygen demand (BOD as mg/l) in three rivers of the Mersey Basin. MSC – Manchester Ship Canal.

worts with four species, *Hygrohypnum ochraceum*, (Wils.) Loeske, *Fontinalis squamosa* Hedw., *Racomitriun aciculare* (Hedw.) Brid. and *Scapania undulata* (L.), being especially characteristic of these acidic upland rivers.

Although many of the rivers have probably always contained areas of sand, gravel and mud suitable for the establishment of rooted plants, none has remained unaffected by pollution or artificial channel management. Surveys of the distribution of River Water-crowfoot (*Ranunculus fluitans*) in the Mersey catchment (Harding 1981), indicated that although the species is unable to form large beds in rivers affected by severe organic pollution, occasional plants may survive for long periods in such situations. This can happen even in competition with dense growths of pollution tolerant plants such as Fennel Pondweed (*Potamogeton pectinatus*) which has become widespread in the Mersey Basin as organic pollution has decreased. Since Harding's survey in 1978/80, Fennel Pondweed, present then in the Rivers Bollin, Dane and Dean, has now been displaced there by River Water-crowfoot.

Improving water quality in the Mersey Basin generally has been associated with an increase in abundance and diversity of channel and marginal vegetation. The largely sandy/silty nature of the lowland rivers favours colonisation by plants such as Branched Bur-reed (*Sparganium erectum*), Reed Canary-grass (*Phalaris arundinacea*) both common throughout; Common Reed (*Phragmites australis*), e.g., River Mersey in Warrington; Reed Sweet-grass (*Glyceria maxima*), e.g., Rivers Gowy,

Figure 13.4. River Dane, Congleton showing natural river processes of erosion and deposition (*Photo: D.G. Holland*).

Figure 13.5. River Rehabilitation Scheme on Whittle Brook, Great Sankey. Low-level shelf for wet habit and bank slope reduced with grass mat to re-stablish flora (*Photo: D.G. Holland*).

Birket and Weaver; Yellow Iris (*Iris pseudacorus*), Water Forget-me-not (*Myosotis scorpioides*), Brooklime (*Veronica beccabunga*) and aquatics such as water-starworts (*Callitriche* spp.), Fool's Water-cress (*Apium nodiflorum*), Fennel Pondweed, Curled Pondweed (*Potamogeton crispus*) and Broad-leaved Pondweed (*P. natans*).

In Cheshire, the River Gowy and River Weaver receive nutrients from treated sewage effluents and agricultural run off, creating eutrophic conditions in which Yellow Water-lily (*Nuphar lutea*) and Flowering-rush (*Butomus umbellatus*) are common in slow-flowing sections, and Celery-leaved Buttercup (*Ranunculus sceleratus*) and the sweet-grasses *Glyceria fluitans*, *G. declinata* and *G. notata* are frequent on cattle-poached margins.

Ditch communities of regional conservation importance are found on drained wetlands of the Frodsham and Ince Marshes and Gowy Meadows, where the pond-like conditions have enabled uncommon plants such as Bladderwort (*Utricularia* spp.), Small Pondweed (*Potamogeton berchtoldii*) and Water-violet (*Hottonia palustris*) to survive, though they are now under threat from intensive agriculture.

In addition to the effects of changes in water and river maintenance practices, plant distribution has also been influenced by the introduction of alien species. These include Indian Balsam, Japanese Knotweed and Giant Hogweed (*Heracleum mantegazzianum*), which has increased dramatically in the last 10 years in the river systems of the Bollin and the Croal–Irwell. Aquatics such as New Zealand Pigmyweed (*Crassula helmsii*) and Water Fern (*Azolla filiculoides*) are currently invading some standing waters in the Mersey Basin. Canadian Waterweed (*Elodea canadensis*) became established widely during the 19th century, but in the last 30 years it has been displaced in many places by Nuttall's Waterweed (*E. nuttallii*).

Invertebrates

Holland (1976a) reported the distribution of *Asellus aquaticus* L. and *Gammarus pulex* L. in the rivers of the Mersey Basin during 1971 and related their presence to water quality. With the accumulation of a further 24 years of data collection (surveys of rivers two or three times each year), it is now possible to see how water quality improvements have altered the distribution of invertebrate animals. The 1971–72 results are compared with those from a much larger number of survey sites used in 1994–95 in Figures 13.6–13.9. The distributions of four groups are compared: *Asellus* Geoffrey, *Gammarus* Fabricius, *Baetis* Leach and Ephemeroptera (excluding *Baetis* spp.)/Plecoptera, which have been selected to cover the whole range of water quality from badly polluted to unpolluted. *Gammarus* includes the

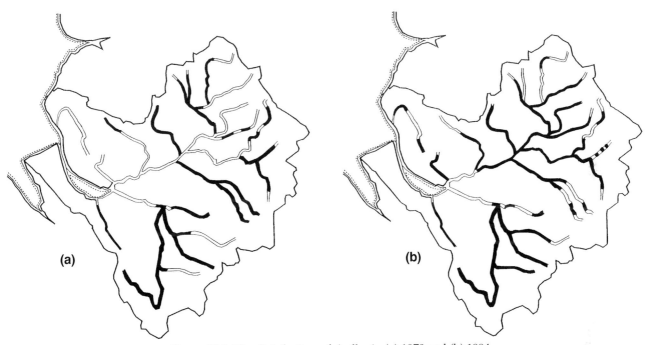

Figure 13.6. The distribution of *Asellus* in (a) 1970 and (b) 1994.

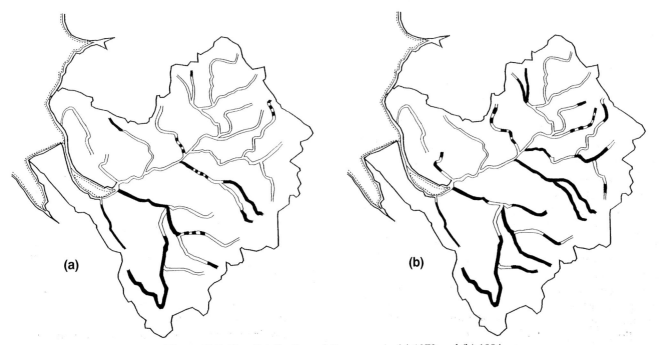

Figure 13.7. The distribution of *Gammarus* in (a) 1970 and (b) 1994.

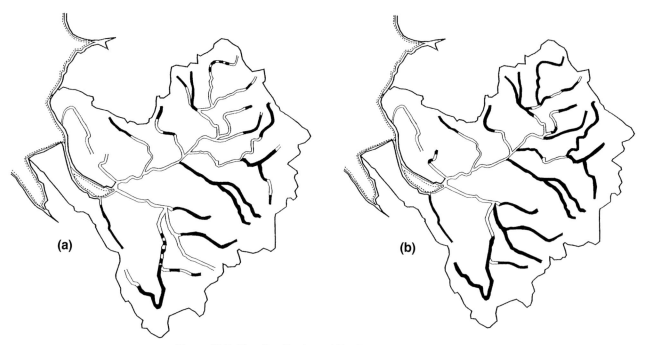

Figure 13.8. The distribution of *Baetis* in (a) 1970 and (b) 1994.

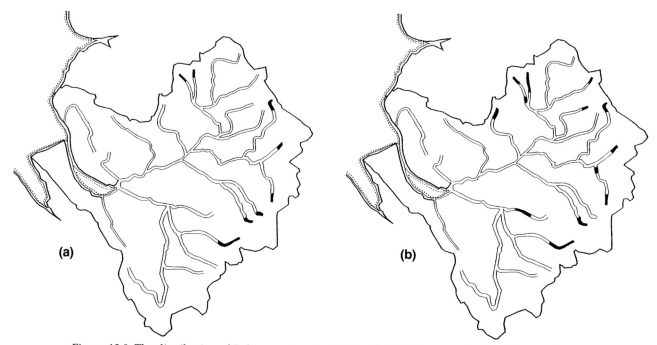

Figure 13.9. The distribution of Ephemeroptera (excluding *Baetis*)/Plecoptera in (a) 1970 and (b) 1994).

brackish-water species referred to below.

Comparisons show changes over the 24-year period which reflect the water quality improvements known to have taken place throughout the Mersey Basin. *Asellus* established itself in many stretches where previously it was absent, and it has disappeared in other lengths where it was once recorded. Appearances generally correspond with known water quality improvements from poor to moderate; disappearances are often displacements by clean water species where previously moderate quality water has been upgraded.

Freshwater species of *Gammarus* show a similar

advance into cleaner waters, mostly into the rivers of the upper Mersey, where in the early 1970s it was notably scarce. The spread of *Baetis* into many new river stretches provides a vivid biological statement of water quality improvements in the catchment. The poor distributions of clean-water Ephemeroptera and Plecoptera reflect pollution problems that still exist. Indeed this confirms what is widely acknowledged, that despite improvements, the River Mersey remains one of the most polluted rivers in Britain today

The Environment Agency has now adopted a more detailed form of invertebrate survey. Data from 1990 and

1995 are still being collated but should provide valuable information.

Holland (1976b) linked the distribution of the three species *G. duebeni, G. tigrinus* and *G. zaddachi* to the various inputs of saline water from direct discharges and the water-table. Conditions have altered markedly since then, but although detailed current records are not available, some notable changes can be described. The salt discharge to the Trent and Mersey Canal, which took the salinity to above 50% of sea water, ceased and in the fresh-water conditions now prevailing, the three species have probably been lost. However, *G. zaddachi* has firmly established itself in the River Weaver from Northwich to Frodsham, where a range of inorganic salts continues to be discharged to the river by chemical industries.

A notable 50-year series of papers reports the work of McMillan on the Mollusca of rivers and other freshwater habitats in the region. The accumulated record shows various changes in distribution of individual species, especially the spread of aliens via interconnected waterways. This literature is listed in the references under Fisher & Jackson (1936) and McMillan, with others, 1942 to 1991.

Fish

In the early 1970s, the River Mersey was widely regarded as an area devoid of river fisheries (Figure 13.10(a)). Futile attempts were made at various times in the 1970s to introduce fish into several stretches of river, but organic and toxic pollution were always too great for fish survival. Zinc in the River Etherow and other toxic metals in the River Tame provided particularly adverse conditions.

The successive agencies responsible for the management of the river system have always looked for the benefits of success in improving water quality in terms of fishery development. In 1976 the North West Water Authority established a fisheries department to serve the area, and since then, a policy of fish stocking has been pursued as environmental conditions have allowed. From time to time, the existence of unknown fish populations has emerged, either through the occurrence of fish mortalities in pollution events, or from local anglers pursuing their sport on stretches hitherto believed to be fishless.

The situation today is summed up in Figure 13.10(b). A significant reduction in the number of completely fishless rivers has taken place. Whilst the total of viable fisheries has not increased much, encouragement can be taken from the large number of rivers where minor fish such as Three-spined Stickleback (*Gasterosteus aculeatus*) and Stone Loach (*Noemacheilus barbatulus*) now exist, and where coarse fish populations are surviving.

Lakes

Natural standing waters in the Mersey Basin consist largely of a group of relatively deep lakes, known as the meres (Gorham 1957a & b; Reynolds & Sinker 1976; Savage & Pratt 1976; Reynolds 1979; Savage 1990; Savage, Bradburne & Macpherson 1992; Moss *et al.* 1992; Moss, McGowan & Carvalho 1994), situated in Cheshire and formed by the melting of icebergs buried in glacial drift, some ten thousand years ago. Also grouped as 'meres' are some small, shallow lakes formed by a variety of means. Although many were man-made in the last two centuries and are to be found on former great estates, some are probably natural. Of these there were originally many more, but they have naturally filled in with vegetation to form the raised bogs, or mosses (Sinker 1962;

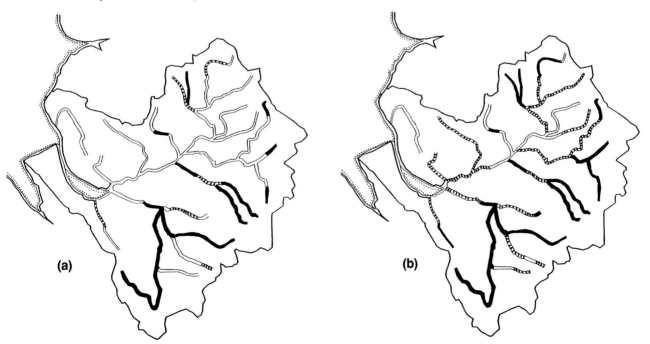

(a)

(b)

Figure 13.10. The distribution of fisheries in (a) 1972 and (b) 1995. Solid lines are established fisheries, hatched lengths are minor or emerging fisheries.

Figure 13.11. Mere Mere is a classic west midlands mere, with summer stratification, blue-green algal blooms and a rich littoral macrophyte community. Its use as a driving practice range for the local golf club poses no hazard to the lake, but makes sampling it occasionally exciting (*Photo: B. Moss*).

Tallis 1973), many of which have now been drained for agriculture or peat extraction.

The meres, with other similar lakes in adjoining river catchments, form a natural grouping from a limnological point of view. The deeper ones (operationally defined as >3m maximum depth and descending to 31m in Rostherne Mere) stratify in summer to form a distinct epilimnion and hypolimnion. The latter may often become anaerobic. The rate of water replacement is comparatively low because the meres are dominated by ground water supplies in many cases.

Following earlier suggestions by Reynolds (1979), recent surveys show that phytoplankton production is controlled by nitrogen supply (Moss *et al.* 1992; 1994) rather than the phosphorus supply that controls production in many, if not most lakes. The reason for this is not that the rate of input of nitrogen is necessarily very low, nor that that of phosphorus is particularly high. The lakes lie in agricultural catchments from which substantial amounts of nitrogen leach to the basins, and the nitrogen to phosphorus ratios of the inflowing or percolating waters are high, as is usual elsewhere.

However, denitrification in the wet meadows and reed fringes of the lakes may cause substantial loss of nitrogen, whilst other mechanisms, unusually, lead to retention of phosphorus in the water (Moss *et al.* 1997). In most lakes phosphorus is rapidly lost to the overflow or the sediments. In the deeper meres, the rate of flushing

is low, such that the main sink for phosphorus has been the sediments. Over many years, these sediments may have become saturated and can no longer absorb the incoming supplies of phosphorus, however small, from the catchment, so these accumulate in the water (Kilinc 1995).

The water itself thus has a low nitrogen to phosphorus ratio, a condition associated, together with thermal stratification and a base-rich water supply, with production of blue-green algae (properly called cyanoprokaryotes). These may sometimes float to the surface to form a bloom, a phenomenon known in the Shropshire meres as the 'breaking of the meres' (Griffiths 1925) by analogy with the scums of yeast that used to form in brewing vats. In recent years such blooms have come to be associated with eutrophication, the artificial enrichment of surface waters with nutrients from agricultural and domestic effluents.

Blooms in the meres are clearly an ancient phenomenon. They are referred to by Phillips (1884), and figure earlier in the folklore consciousness of the area (Webb 1924). Recent analyses of sediments from Whitemere and Colemere for pigments specific to particular algal groups, suggest that cyanoprokaryotes have been common in at least one mere for at least 6,000 years (McGowan 1996). There has nonetheless also been some recent eutrophication, superimposed on this ancient phenomenon and presumably driven by increased nitrogen run-off as agriculture has intensified in the post-war years (Brinkhurst & Walsh 1967; Grimshaw & Hudson 1970; Livingstone 1979; Nelms 1984; Reynolds & Bellinger 1992; Carvalho 1993; McGowan 1996).

Rostherne Mere, a National Nature Reserve, is one of the best-known of these lakes, with a lengthy record of scientific investigations (Banks 1970; Grimshaw & Hudson 1970; Goldspink 1978, 1981, 1983; Goldspink & Goodwin 1979; Carvalho 1993; Carvalho, Beklioglu & Moss 1995; Moss *et al.* 1997). It too is nitrogen limited and has extensive blue-green algal growths. The accumulations of blue-green algae, which may sometimes be highly toxic, have caused problems for local cattle. A combination of a desire to reduce these blooms, a conventional belief, based on a wealth of evidence from elsewhere, in the efficacy of phosphorus control in controlling eutrophication, and the overloading of a sewage treatment works designed for considerably fewer people than it now serves, led North West Water to divert the treated sewage effluent from the mere's inflow in 1991.

This discontinued the main (and considerable) source of phosphorus to the lake and there has been a modest decrease in the concentrations of phosphorus in the water (Carvalho 1993; Carvalho *et al.* 1995; Beklioglu 1995; Moss *et al.* 1997). There has been little change in the phytoplankton populations however, which, for the reasons adduced above, are nitrogen-controlled and are likely to remain so for the immediate future and probably much longer. Rostherne Mere has a greater flushing rate than the groundwater-fed meres and may, theoretically, reach a phase of phosphorus limitation eventually as the present stores are displaced and not replaced. For this reason it is a particularly interesting site. Deep, well-flushed lakes are usually readily restorable by phosphorus control.

Rostherne Mere is also of interest in having been among the first lakes to be designated 'guanotrophic' (Brinkhurst & Walsh 1967). Its nutrient supply was thought to have been dominated by the excreta of the birds roosting on it. There are genuinely guanotrophic lakes (Moss & Leah 1982), but quantitative studies have shown that Rostherne Mere is not one of them, and probably never was! Birds provide 1–2% of its phosphorus and nitrogen supplies (Carvalho *et al.* 1995).

Upstream of Rostherne Mere, and also originally a recipient of the sewage effluent, is Little Mere, an example of one of the group of shallow meres. Like many of the group, it was formed by the damming of a stream which emerged from the deeper Mere Mere (itself of interest, not least for its name), passed through Little Mere and then flowed down to Rostherne Mere. The shallow meres potentially contain ecosystems dominated by submerged and floating-leaved aquatic plants, but several have lost these as a result of changes consequent on eutrophication and other influences (Scheffer *et al.* 1993; Moss, Madgwick & Phillips 1996).

The plant communities are very resistant to increases in nutrients because of a suite of stabilising mechanisms that prevent phytoplankton from taking advantage of the nutrients and competitively displacing the plants. These include the provision of refuges against fish predation for invertebrates such as water fleas (particularly *Daphnia* spp.) which graze the algae growing suspended in the water, and snails which feed upon the algae coating submerged plants, thereby limiting the increase of the algae and keeping the water clear and leaf surfaces open to incoming light. The plants are therefore able to photosynthesise and grow without undue interference from the algae.

If the plants are damaged (for example by overcutting, boat propellers, grazing by carp, swans or geese), if the grazers are poisoned by pesticide run-off or other toxins, or if the balance of the fish community changes in favour of zooplankton-eating fish, the stabilising mechanisms break down and the algae can rapidly take over. The clear water plant-dominated state can then become a turbid phytoplankton-dominated one. Studies of the shallow meres have shown an inverse relationship between the crop of algae and the number of *Daphnia*, the main algal grazers, in the water (Moss *et al.* 1994). Where the *Daphnia* are few and the water is turbid with algae, the plants have been lost. In general the switch to turbid water is more likely to occur when the nutrient input is high, though one of the additional agents of change must also be operating.

Little Mere is of considerable theoretical interest among these shallow lakes because, before 1991, when the sewage effluent was diverted, it had huge concentrations of nutrients, yet clear water and a reasonable abundance of aquatic plants. The usual stabilising mechanisms of the plant community were fortified by a considerable population of bright cherry-red coloured *Daphnia magna* Sars (Carvalho 1994). This is an efficient grazer, but so large (up to 4mm) that it is a preferred prey of many fish, with which it cannot easily co-exist.

In Little Mere, however, the sewage effluent was not of high quality. It de-oxygenated the water, preventing sustainable populations of fish, but allowed the haemoglobin-rich *Daphnia magna* to survive and graze. Diversion of the effluent allowed fish to recolonise. The *Daphnia magna* was then replaced by less conspicuous *Daphnia* species, the stabilising mechanisms of the plant communities persisted and plant dominance continued and even extended in the mere (Beklioglu 1995; Beklioglu & Moss 1995, 1996).

Because there is considerable interest in the restoration (Moss *et al.* 1996) of the huge numbers of shallow lakes that have lost their plant communities in lowland Europe, particularly in Denmark, Sweden and the Netherlands, as well as in such well-known areas in England as the Norfolk Broads, the information still emerging from studies of Little Mere, no less than that from Rostherne and the other deep meres, is of considerable international interest.

Artificial standing waters

There are over 30 substantial reservoirs on upland tributaries in the Mersey catchment (Holland & Harding 1984). These are relatively acid for the most part, because of anthropogenic acidification of their gathering grounds. There is little published work on their biology, although unpublished data may well exist. Where they function with considerable drawdown of water level, their littoral communities are likely to be impoverished, although further north, e.g., at Grizedale Reservoir in Lancashire, a highly specialised flora, with Water-purslane (*Peplis portula*) and Shoreweed (*Littorella uniflora*) both rare in the region, may be found. More generally, they might be expected to be depauperate in their plankton and profundal benthos, because of acidity and low nutrient status, though the shaley catchments probably buffer the acidification to some extent.

In addition, numerous small reservoirs ('lodges') and holding lagoons were constructed for, and often within the premises of, factories and collieries. Where they received heated water from industrial processes, they sometimes supported exotic tropical waterplants (Fox 1963) presumably introduced by aquarists. De-industrialisation and the improved availability of public mains water supplies have been followed by the draining and infill of some of these waterbodies. Many do, however, survive. Some have been stocked with fish and are used for recreational angling. They represent a currently uninvestigated but potentially rich biological resource within the Mersey Basin.

The salt industry is responsible for having created a number of shallow, saline waterbodies or flashes caused partly by land subsidence over extraction workings, and these are used to accommodate saline wastes from the processes. The natural history of the flashes has been studied because of the importance of these waters to wading birds and also because of the unusualness of inland saline habitats and their associated faunas in a wet country like Britain (Holland 1976; Savage 1971, 1979, 1981, 1985).

Finally, there is a myriad of small agricultural ponds, usually groundwater fed, throughout the area, but especially in the lowlands south of the River Mersey. Their biology and future prospects are described by Boothby *et al.* (1995) and Boothby & Hull (chapter fourteen, this volume).

Canals

The Industrial Revolution created a demand for inland bulk transport which could only be satisfied to a small extent by navigation on the Rivers Mersey, Irwell and Weaver.

In response, the construction of artificial canals began with the Sankey Navigation, opened in 1757, followed by the Bridgewater Canal between the River Mersey at Runcorn and central Manchester in 1776. Most of the network shown in Figure 13.12 was completed over the next few decades, making the region second only to the Black Country in its density of navigations. Three trans-Pennine lines and links to the midlands and to the south were supplemented by internal routes, all constructed to small dimensions as compared with main rivers, to keep down excavation costs and subsequent water requirements for locks. Typically the channels are

Figure 13.13. Leeds & Liverpool Canal, Litherland. 1950s – coal barge approaching Liverpool (*Photo: British Waterways*).

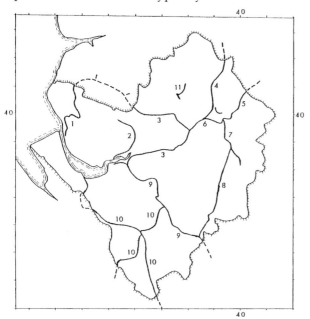

Figure 13.12. Canals of the Mersey Basin. 1. Leeds & Liverpool. 2. St Helens. 3. Bridgewater. 4. Rochdale. 5. Huddersfield Narrow. 6. Ashton. 7. Peak Forest. 8. Macclesfield. 9. Trent & Mersey. 10. Shropshire Union. 11. Manchester, Bolton and Bury.

10–15m wide and have maximum design depths in the range 1.0–1.7m.

Railway competition, beginning with the opening of the Liverpool & Manchester Railway in 1830, ended construction of small canals, but a final major addition came in 1894 with the opening of the Manchester Ship Canal. Designed for large, sea-going craft, this has minimum dimensions of 27m wide x 8.5m deep.

The early colonisation of these new freshwater habitats by flora and fauna is unrecorded, but it is clear from old illustrations and a few herbarium records that, in the cleanwater sections, pond-like species assemblages were present during the era of horse-drawn and sailed craft. Sedges (*Carex* spp.) were often planted to stabilise banks and mention of fisheries implied the presence of functioning food webs in the channels. Pollution affected some sections and probably intensified and spread as transport-requiring industries grew up in waterside locations and also used the canals for effluent disposal. River Board reports show that the Leeds & Liverpool Canal in Liverpool was still so grossly polluted by a range of city industries until the end of the 1960s that it was often black and anaerobic, with gas eruptions of hydrogen sulphide and methane.

The change from horse and sail to propeller-driven movement began on a large scale in the 1870s and led to greatly increased disturbance by each boat passage. Bank erosion became a problem and vegetation-based protection was replaced by stone, brick, concrete and metal hardening with associated loss of marginal plants and their fauna.

Rail and later road competition gradually reduced traffic on the canals. Whilst the main routes retained some freight until the middle of the 20th century, others declined into low activity. Some were filled in, but many were retained for their water supply and drainage functions. Reports of difficulties with weeds obstructing channels increased through the first half of the century (Murphy, Eaton & Hyde 1982), as boat traffic ceased to restrain their growth.

Nationally, the mid-20th century was the peak time for colonisation of the waterways by flora and fauna (Murphy, Eaton & Hyde 1982) and in the Mersey Basin network some diverse and conservationally important assemblages developed and, in some case, persisted. Shimwell (1984) and Bignall (1992) describe rich floras and faunas in some Manchester canals, but overall the second half of the century has been marked by a decline in species diversity and conservation interest. This has been caused by quite different circumstances in different canals.

Increasing use by recreational boat traffic is one factor, particularly in scenically attractive rural lengths, e.g., Shropshire Union and Macclesfield Canals (Murphy & Eaton 1983). Changing management is another factor. When gross pollution ceased in 1969 in the Liverpool section of the Leeds & Liverpool Canal submerged vegetation became abundant and by 1979 a total 24 species of aquatic angiosperms, bryophytes and charophytes was present, together with a rich invertebrate fauna (Eaton & Freeman 1982; Murphy & Eaton 1981; Murphy, Hanbury & Eaton 1981; Hanbury, Murphy & Eaton 1981). Further species colonised over the next 15 years, bringing the total to over 30 aquatic plant species (Eaton, personal observation). Then in 1994 a breach caused water loss and drying out. The length was refilled with brackish water from Liverpool Docks; the section nearest the terminus was then emptied again and deep-dredged before being re-watered in 1995. Six of the former species re-established during 1996.

A third factor has been the decrease in thermal pollution. Formerly, industrial discharges of hot water into the Stockport Branch of the Ashton Canal, combined presumably with aquarists' introductions, led to establishment of tropical vegetation, e.g., Large-flowered Waterweed (*Egeria densa*), Curly Waterweed (*Lagarosiphon major*) and *Najas graminea* (Bailey 1884; Weiss & Murray 1909; Kent 1955a, b; Shaw 1963). Another branch at Hollinwood sustained Tapegrass (*Vallisneria spiralis*) and Hampshire-purslane (*Ludwigia palustris*) in a heated section (Shaw 1963). The Pocket Nook Branch of the St Helens Canal received waste heat from the Ravenhead Glassworks and Large-flowered Waterweed and Curly Waterweed, one of the earliest northern populations of Water Fern and a range of tropical fish including large *Tilapia* spp., were present (Eaton, personal observation). All of these unusual species assemblages subsequently disappeared. The Stockport and Hollinwood sites are now filled in and at St Helens the main heating ceased in 1980, leading to the death of the tropical fish and their replacement by stocked cold water species. During the 1990s channel clearance as part of an environmental improvement project eliminated surviving unusual flora.

The final factor causing decline in species diversity

Figure 13.14. Rochdale Canal, Failsworth. Derelict: reeds and abundant Floating Water-plantain 1991
(*Photo: N.J. Willby*).

was natural vegetation succession and competitive exclusion of species, where no weed control took place on unnavigated canals (Murphy, Eaton and Hyde 1980). Murphy and Eaton (1981) showed how in the shallow water of the Huddersfield Narrow Canal succession to monocultures of Branched Bur-reed or Reed Sweet-grass could be reversed by clearance, but quickly resumed and returned to monoculture within a few years. The upper part of the remaining isolated length of the Hollinwood Branch developed into a Reed Sweet-grass swamp, but the lower part is kept open by conservationists using manual clearance, and 32 aquatic plant species and a further 19 associated wetland and riparian species were listed in the Nature Conservancy Council's 1981 survey which led to its designation as a Site of Special Scientific Interest. The Prestolee section of the Manchester, Bolton and Bury Canal, is cleared regularly for angling and is a locality for the rare introduced freshwater jellyfish *Craspedacusta sowerbyii* Lankester (Shimwell 1984).

The connections which canals have established with catchments outside the Mersey Basin seem to have been the means of species migrations across watersheds. Thus, Willby & Eaton (1993) provide evidence that Floating Water-plantain (*Luronium natans*) reached canals in the region via the Shropshire Union link from the Welsh catchment of the River Dee, whilst the eastern European amphipod *Corophium curvispinum* Sars. var. *devium* Wundsch spread northwards from its original site of introduction on the Avon Navigation at Tewkesbury via the Shropshire Union Canal to the Mersey Basin (Holland 1976; Pygott & Douglas 1989; the few Mersey Basin locations in their Figure 1 have since been supplemented and the species appears to be continuing its spread). The small freshwater snail, *Marstoninopsis scholtzi* Schmidt, is present in a few canals in the Manchester area. However, the origin of this stable population is unknown but its native population in East Anglia is declining (Bratton 1991).

The future

Man-induced change has been a major factor in all types of freshwaters of the Mersey Basin for at least the past 200 years and seems likely to continue in the future. In the case of rivers, whilst further local regulation and channelisation projects are inevitable adjuncts to catchment development, these are likely to be undertaken in a sensitive way to allow some return to richer floras and faunas, supported by further reductions in pollution.

For standing waters, control of the quantities and qualities of their water sources and the demands made upon them for water supply, fishing and other recreational uses will determine their future ecosystems. There is a strong argument for making conservation a high priority in the case of the meres, on account of their intrinsic scientific interest, and their proposed designation as an EU Special Area of Conservation under the Habitats Directive 1992.

Canal ecosystems have lost some features of interest in the past few decades and there are currently pressures which may continue this decline, notably complete clearances during urban regeneration projects and increasing use for recreational boating and fisheries.

Overall therefore, the freshwaters of the Mersey Basin have mixed prospects, the outcome of which depends upon long-term development and management policies within the catchments. Climatic change adds further uncertainty, especially since freshwaters seem peculiarly susceptible to colonisation by introduced species, and the range of those potentially able to establish could be greatly expanded at higher temperatures.

Acknowledgements

We are grateful to the Environment Agency and Mrs N. McMillan for information supplied for this paper and to Gill Haynes for technical assistance. The views expressed by the authors are their own and do not necessarily reflect those of the Environment Agency or any other organisation.

References

Bailey, C. (1884). Notes on the structure, the occurrence in Lancashire, and the source of origin, of *Naias graminea Delile*, var. *Delilei Magnus. Journal of Botany*, **22**, 305–33.

Banks, J.W. (1970). Observations on the fish population structure of Rostherne Mere, Cheshire. *Field Studies*, **3**, 357–79.

Beklioglu, M. (1995). *Whole lake and mesocosm studies on the roles of nutrients and grazing in determining phytoplankton crops in a system of shallow and deep lakes*. PhD thesis, University of Liverpool.

Beklioglu, M. & Moss, B. (1995). The impact of pH on interactions among phytoplankton algae, zooplankton and perch (*Perca fluviatilis*) in a shallow, fertile lake. *Freshwater Biology*, **33**, 497–509.

Beklioglu, M. & Moss, B. (1996). Mesocosm experiments on the interaction of sediment influence, fish predation, and aquatic plants with the structure of phytoplankton and zooplankton communities. *Freshwater Biology*, **36**, 315–25.

Bignall, M.R., (1992). *Rare Plants in the Canals of the Metropolitan Counties of NW England*. Report for English Nature, January 1992.

Boothby, J., Hull, A.P., Jeffreys, D.A. & Small, R.W. (1995). Wetland loss in North-West England: the conservation and management of ponds in Cheshire. *Hydrology and Hydrochemistry of British Wetlands* (eds J.M.R. Hughes & A.L. Heathwaite), pp.432–44. Wiley, Chichester.

Bratton, J.H. (1991). *British Red Data Books: 3. Invertebrates other than Insects*, pp.47–48. Joint Nature Conservation Committee, Peterborough.

Brinkhurst, R.O. & Walsh, B. (1967). Rostherne Mere, England: a further instance of guanotrophy. *Journal of the Fisheries Research Board of Canada*, **24**,1299–1309.

Carvalho, L.R. (1993). *Experimental limnology on four Cheshire meres*. PhD thesis, University of Liverpool.

Carvalho, L.R. (1994). Top-down control of phytoplankton in a shallow, hypertrophic lake: Little Mere, England. *Hydrobiologia*, **275/276**, 53–63.

Carvalho, L.R., Beklioglu, M. & Moss, B. (1995). Changes in a deep lake following sewage diversion – a challenge to the orthodoxy of external phosphorus control as a restoration strategy. *Freshwater Biology*, **34**, 399–410.

Eaton, J.W. (1989). Ecological Aspects of Water Management in Britain. *Journal of Applied Ecology*, **26**, 835–49.

Eaton, J.W. & Freeman, J. (1982). Ten Years' Experience of Weed Control in the Leeds & Liverpool Canal. *Proceedings EWRS 6th Symposium on Aquatic Weeds, 1982*, 96–104. Novi Sad, Jugoslavia.

Environment Agency (1996). *Alt Crossen Catchment Management Plan*. Environment Agency, N.W. Region, Preston.

Fisher, N. & Jackson, J.W. (1936). Early records of Lancashire and Cheshire non-marine Mollusca by James Wright Whitehead. *Journal of Conchology*, **20**, 275–81.

Fox, B.W. (1963). Plants of industrial tips and waste land. *Travis's Flora of South Lancashire* (eds J.P. Savidge, V.H. Heywood & V. Gordon), pp. 73–76. Liverpool Botanical Society, Liverpool.

Goldspink, C.R. (1978). Comparative observations on the growth rate and year class strength of roach *Rutilus rutilus* L. in two Cheshire lakes, England. *Journal of Fish Biology*, **12**, 421–33.

Goldspink, C.R. (1981). A note on the growth rate and year class strength of bream, *Abramis brama* L. in three eutrophic lakes, England. *Journal of Fish Biology*, **19** , 665–73.

Goldspink, C.R. (1983). Observations on the fish populations of the Shropshire–Cheshire meres with particular reference to angling. *Proceedings 3rd British Freshwater Fisheries Conference*, University of Liverpool, Liverpool.

Goldspink, C.R. & Goodwin, D.A. (1979). A note on the age composition, growth rate and food of perch *Perca fluviatilis* L., in four eutrophic lakes, England. *Journal of Fish Biology*, **14**, 489–505.

Gorham, E. (1957a). The chemical composition of some waters from lowland lakes in Shropshire, England. *Tellus*, **9**, 174–79.

Gorham, E. (1957b). The ionic composition of some lowland lakes from Cheshire, England. *Limnology and Oceanography*, **2**, 22–27.

Griffiths, B.M. (1925). Studies on the phytoplankton of the lowland waters of Great Britain. III. The phytoplankton of Shropshire, Cheshire and Staffordshire. *Botanical Journal of the Linnean Society of London*, **47**, 75–92.

Grimshaw, H.M. & Hudson, M.J. (1970). Some mineral nutrient studies of a lowland mere in Cheshire, England. *Hydrobiologia*, **36**, 329–41.

Hanbury, R.G., Murphy, K.J. & Eaton, J.W. (1981). The ecological effects of 2-methylthiotriazine herbicides used for aquatic weed control in navigable canals. II. Effects on macroinvertebrate fauna and general discussion. *Archiv für Hydrobiologie*, **91**, 408–26.

Harding, J.P.C. (1981). *Macrophytes as Monitors of River Water Quality in the Southern NWWA Area*. North West Water Authority Rivers Division Ref. No. TSBS-81-2. British Lending Library Loan Collection.

Holland, D.G. (1976a). The distribution of the freshwater Malacostraca in the area of the Mersey and Weaver River Authority. *Freshwater Biology*, **6**, 265–76.

Holland, D.G. (1976b). The inland distribution of brackish-water *Gammarus* species in the area of the Mersey and Weaver River Authority. *Freshwater Biology*, **6**, 277–85.

Holland, D.G. & Harding, J.P.C. (1984). The Mersey. *Ecology of European Rivers* (ed. B.A. Whitton), pp. 113–44. Blackwell Scientific Publications, Oxford.

Kent, D.H. (1955a). *Egeria densa* Planch. *Proceedings of the Botanical Society of the British Isles*, **1**, 322.

Kent, D.H. (1955b). *Lagarosiphon major* (Ridley) C.E. Moss. *Proceedings of the Botanical Society of the British Isles*, **1**, 322–23.

Kilinc, S. (1995). *Limnological studies on the North West Midland meres, with special reference to Whitemere*. PhD thesis, University of Liverpool.

Livingstone, D. (1979). *Algal remains in recent lake sediment*. PhD thesis, University of Leicester.

McGowan, S. (1996). *Ancient cyanophyte blooms. Studies on the palaeolimnology of White Mere and Colemere*. PhD thesis, University of Liverpool.

McMillan, N. (1942). Cheshire conchological notes. *North-Western Naturalist*, **16**, 328.

McMillan, N. (1944). *Planorbis corneus* (L.) in Wirral. *Journal of Conchology*, **22**, 103.

McMillan, N. (1947). Cheshire conchological notes no. 2. *North-Western Naturalist*, **21**, 103–04.

McMillan, N. (1947). Further notes on *Planorbis corneus* (L.) in Wirral. *Journal of Conchology*, **22**, 248.

McMillan, N. (1947). The land and freshwater Mollusca of the Wirral peninsula of Cheshire. *Report and Proceedings of the Chester Society of Natural Science, Literature and Art: Robert Newstead Memorial Volume*, pp. 83–93.

McMillan, N. (1948). *Bithynia tentaculata* (L.) in "closed" ponds. *Journal of Conchology*, **23**, 22.

McMillan, N. (1953). Cheshire conchological notes no. 3. *North-Western Naturalist*, New Series, **1**, 96.

McMillan, N. (1955). Notes on local non-marine Mollusca. *Proceedings of the Liverpool Naturalists' Field Club for 1954*, 19–20.

McMillan, N. (1955). The range of *Planorbarius* in the British Isles. *Journal of Conchology*, **24**, 63–65.

McMillan, N. (1956). Notes on local Mollusca. *Proceedings of the Liverpool Naturalists' Field Club for 1955*, 13–15.

McMillan, N. (1959). Notes on the land and freshwater Mollusca of Wirral, Cheshire 1948–1958. *Proceedings of the Liverpool Naturalists' Field Club for 1958*, 10–21.

McMillan, N. (1959). The Mollusca of some Cheshire marl-pits: a study in colonization. *Journal of Conchology*, **24**, 299–315.

McMillan, N. (1962). *Pisidium pseudosphaerium* Favre in Cheshire. *Journal of Conchology*, **25**, 63.

McMillan, N. (1963). Non-marine Mollusca. *Lancashire and Cheshire Fauna Committee* 33rd. Reports, pp. 48–50.

McMillan, N. (1963). The *Pisidium*-fauna of Bromborough, Cheshire. *Journal of Conchology*, **25**, 183–88.

McMillan, N. (1964). Report of Field Meeting held 1st May at Guide Bridge, Lancashire. *Conchologists' Newsletter* no. **14**, 93–94.

McMillan, N. (1967). A Cheshire locality for *Anodonta complanata* Rossmassler. *Conchologists' Newsletter* no. **20**, 142.

McMillan, N. (1967). Field Meeting to Leeds and Liverpool Canal, Lancashire. *Conchologists' Newsletter* no. **21**, 8–9 .

McMillan, N. (1967). Field Meeting to Plumley Nature Reserve, Cheshire, 30th April 1966. *Conchologists' Newsletter* no. **20**, 142–43.

McMillan, N. (1969). The effect of the exceptionally severe winter of 1962/63 on the Mollusca of a Cheshire pond. *Conchologists' Newsletter* no. **33**, 155–56.

McMillan, N. (1970). Report on the Mollusca, 1968 and 1969. *Lancashire and Cheshire Fauna Society Publication*, **57, 12–13.

McMillan, N. (1977). Records of Cheshire non-marine Mollusca, mainly from the Wirral peninsula. *Lancashire and Cheshire Fauna Society Publication*, **71**, 5–6.

McMillan, N. (1989). Observations on the freshwater Mollusca of some Cheshire marl-pits over forty-four years. *Conchologists' Newsletter* no. **108**, 157–65.

McMillan, N. (1991). The history of alien freshwater Mollusca in North-West England. *Naturalist*, **115**, 123–32. (Cover dated 1990.)

McMillan, N. & Ellison, N.F. (1944). Some habitats of *Hydrobia jenkinsi* (Smith) in Wirral, Cheshire. *North-Western Naturalist*, **18**, 320–22.

McMillan, N. & Fogan, M. (1967). Non-marine Mollusca: some noteworthy records and a note on 'garden' species. *Lancashire & Cheshire Fauna Committee 37th Report*, pp. 39–40.

McMillan, N. & Fogan, M. (1969). Field Meetings at Huyton and Knowsley Park, South Lancashire, October and November 1968. *Conchologists' Newsletter* no. **29**, 98–99.

McMillan, N. & Greenwood, E.F. (1969). The giant *Anodonta cygnea* of Claughton, West Lancashire. *Conchologists' Newsletter* no. **29**, 102–03.

McMillan, N. & Millott, J.O'N. (1950). Records of non-marine Mollusca from Cheshire, Flintshire and Denbighshire. Cheshire and North Wales Natural History vol. III. *Proceedings of the Chester Society of Natural Science, Literature and Art for 1949*, pp. 165–72.

McMillan, N. & Millott, J.O'N. (1954). Notes on the non-marine Mollusca of Cheshire and North Wales. Cheshire and North Wales Natural History vol. V. *Proceedings of the Chester Society of Natural Science, Literature and Art for 1951, 1952 & 1953*, pp. 109–13.

McMillan, N. & Wallace, I.D. (1979). Non-marine Mollusca in the Wirral peninsula, Cheshire: records and notes. *Lancashire and Cheshire Fauna Society Publication*, **75**, 5–8.

McMillan, N., Edwards, W.F., Fogan, M. & Millott, J.O'N. (1966). The Mollusca of canals in Lancashire and Cheshire. *Lancashire & Cheshire Fauna Committee 36th Report*, pp. 36–41.

Moss, B., Beklioglu, M., Carvalho, L.R., Kilinc, S., McGowan, S. & Stephen, D. (1997). Vertically-challenged limnology; contrasts between deep and shallow lakes. *Hydrobiologia*, **342/343**, 257–67.

Moss, B. & Leah, R.T. (1982). Changes in the ecosystem of a guanotrophic and brackish shallow lake in eastern England: potential problems in its restoration. *Internationale Revue der gesamten Hydrobiologie*, **67**, 635–59.

Moss, B., Madgwick, J. & Phillips, G. (1996). *A Guide to the Restoration of Nutrient-Enriched Shallow Lakes*. Broads Authority and Environment Agency, Norwich.

Moss, B., McGowan, S. & Carvalho, L.R. (1994). Determination of phytoplankton crops by top-down and bottom-up mechanisms in a group of English lakes, the West Midland Meres. *Limnology and Oceanography*, **39**, 1020–29.

Moss, B., McGowan, S., Kilinc, S. & Carvalho, L.R. (1992). *Current limnological condition of a group of the West Midland Meres that bear SSSI status*. Final Report, English Nature Contract F72-06-14.

Murphy, K.J. & Eaton, J.W. (1981). Ecological effects of four herbicides and two mechanical clearance methods used for aquatic weed control in canals. *Proceedings of a Conference on Aquatic Weeds and their Control, Association of Applied Biologists*, pp. 201–17. Christ Church, Oxford.

Murphy, K.J. & Eaton, J.W. (1983). Effects of pleasure-boat traffic on macrophyte growth in canals. *Journal of Applied Ecology*, **20**, 713–29.

Murphy, K.J., Eaton, J.W. & Hyde, T.M. (1980). A Survey of Aquatic Weed Growth and Control in the Canals and River Navigations of The British Waterways Board. *Proceedings 1980 British Crop Protection Conference – Weeds*, **2**, 707–14.

Murphy, K.J., Eaton, J.W. & Hyde, T.M. (1982). The Management of Aquatic Plants in a Navigable Canal System used for Amenity and Recreation. *Proceedings EWRS 6th Symposium on Aquatic Weeds*, **1982**, 141–51. Novi Sad, Jugoslavia.

Murphy, K.J., Hanbury, R.G. & Eaton, J.W. (1981). The ecological effects of 2-methylthiotriazine herbicides used for aquatic weed control in navigable canals. I. Effects on aquatic flora and water chemistry. *Archiv für Hydrobiologie*, **91**, 204–331.

National Rivers Authority (1992). *River Corridor Surveys, Methods and Procedures*. Conservation Technical Handbook No. 1., National Rivers Authority, Bristol.

National Rivers Authority (1994). *River Irwell Catchment Management Plan*. National Rivers Authority, N.W. Region, Warrington.

National Rivers Authority (1996a). *An Initial Review of the 1995 Drought*. Internal Report, National Rivers Authority, N.W. Region, Warrington.

National Rivers Authority (1996b). *River Habitat Survey*. National Rivers Authority, Bristol.

National Rivers Authority (1996c). *Upper Mersey Catchment Management Plan*. National Rivers Authority, N.W. Region, Warrington.

Nelms, R. (1984). *Palaeolimnological studies of Rostherne Mere (Cheshire) and Ellesmere (Shropshire)*. PhD thesis, Liverpool Polytechnic.

Nolan, P.A. & Guthrie, N. (in press). River rehabilitation in an urban environment: examples from the Mersey Basin, NW England. *Proceedings of an International Conference on River Restoration, 1996*. European Centre for River Restoration – National Environment Research Unit, Silkeborg, Denmark.

Phillips, W. (1884). The breaking of the Shropshire meres. *Transactions of the Shropshire Archaeological and Natural History Society*, **7**, 277–300.

Pygott, J.R. & Douglas, S. (1989). Current distribution of *Corophium curvispinum* Sars. var. *devium* Wundsch (Crustacea: Amphipoda) in Britain with notes on its Ecology in the Shropshire Union Canal. *Naturalist*, **114**, 15–17.

Reynolds, C.S. (1979). The limnology of the eutrophic meres of the Shropshire–Cheshire plain. *Field Studies*, **5**, 93–173.

Reynolds, C. S. & Bellinger, E.G. (1992). Patterns of abundance and dominance of the phytoplankton of Rostherne Mere, England: evidence from an 18-year data set. *Aquatic Sciences*, **54**, 10–36.

Reynolds, C.S. & Sinker, C.A. (1976). The meres: Britain's eutrophic lakes. *New Scientist*, **71**, 10–12.

Royal Society for the Protection of Birds, National Rivers Authority & Royal Society for Nature Conservation (1994). *The New Rivers and Wildlife Handbook*. RSPB, The Lodge, Sandy, Bedfordshire, England.

Savage, A.A. (1971). The Corixidae of some inland saline lakes in Cheshire, England. *Entomologist*, **104**, 331–44.

Savage, A.A. (1979). The Corixidae of an inland saline lake from 1970 to 1975. *Archiv für Hydrobiologie*, **86**, 355–70.

Savage, A.A. (1981). The Gammaridae and Corixidae of an inland saline lake from 1975 to 1978. *Hydrobiologia*, **76**, 33–44.

Savage, A.A. (1985). The biology and management of an inland saline lake. *Biological Conservation*, **31**, 107–23.

Savage, A.A. (1990). The distribution of Corixidae in lakes and the ecological status of the North West Midland meres. *Field Studies*, **7**, 516–30.

Savage, A.A. & Pratt, M.M. (1976). Corixidae (water boatmen) of the Northwest midland meres. *Field Studies*, **4**, 465–76.

Savage, A.A., Bradburne, S.J.A. & Macpherson, A.A. (1992). The morphometry and hydrology of Oak Mere, a lowland, kataglacial lake in the north-west midlands, England. *Freshwater Biology*, **28**, 369–82.

Scheffer, M., Hosper, S.H., Meijer, M-L., Moss, B. & Jeppesen, E. (1993). Alternative equilibria in shallow lakes. *Trends in Ecology and Evolution*, **8**, 275–79.

Shaw, C.E. (1963). Canals. *Travis's Flora of South Lancashire* (eds J.P. Savidge, V.H. Heywood & V. Gordon), pp. 71–73. Liverpool Botanical Society, Liverpool.

Shimwell, D. (1984). *The wildlife conservation potential of the canals of Greater Manchester County*. Countryside Commission and Groundwork North West, Manchester.

Sinker, C.A. (1962). The North Shropshire Meres and Mosses; a background for ecologists. *Field Studies*, **1**, 101–38.

Tallis, J.H. (1973). The terrestrialisation of lake basins in North Cheshire, with special reference to the development of a 'Schwingmoor' structure. *Journal of Ecology*, **61**, 537–67.

Webb, M. (1924). *Precious Bane*. Jonathan Cape, London

Weiss, F.E. & Murray, H. (1909). Alien Plants of the Reddish Canal. *Manchester Memoirs*, **53**, no. 14.

Willby, N.J. & Eaton, J.W. (1993). The Distribution, Ecology and Conservation of *Luronium natans* (L.) Raf. in Britain. *Journal of Aquatic Plant Management*, **31**, 70–76.

CHAPTER FOURTEEN

Ponds of the Mersey Basin: habitat, status and future

J. BOOTHBY AND A. HULL

Introduction – a brief history of ponds

Within the Mersey Basin there are several thousand small water bodies including lakes, meres, moats, flashes, pits and ponds. Though some of these may be natural features, remnants of glacial and periglacial conditions, a man-made origin for most small water bodies seems very likely. Ponds in north-west England (Cheshire, Greater Manchester, Lancashire and Merseyside), as elsewhere in lowland Britain, have been dug in considerable numbers for many centuries, but unlike elsewhere many remain visible in the rural landscape. Today, the Mersey Basin is at the heart of the greatest concentration of lowland ponds remaining in the British Isles, or, indeed, in Europe.

Though brick-clay extraction and sand-and-gravel working helped to create this resource, overwhelmingly the ponds were created by the ancient practice of marling, a practice which was certainly recorded during the 13th century, and which reached its peak in the mid- to late 18th century (Hewitt 1919; Hewitt 1929; Porteous 1933). Many pits were dug in close proximity to each other – sometimes as many as ten ponds or more in a cluster – in order to exploit a particularly rich lode of this mineral manure. Following excavation, the clayey-marls were often subjected to a number of treatments (e.g., frost-weathering, baking) and then spread on the land to improve fertility. Typically small-scale workings (say 30m in diameter) would use a horse and cart or horse-driven gin (Middleton 1949; Prince 1964), though there are later records of more substantial mechanised workings (Grantham 1864; Ferro & Middleton 1949). The application of marl was fairly heavy, changing the soil composition and playing a significant role in increased agricultural productivity.

There is evidence to suggest that the benefits of marling were not only substantial but also reasonably persistent, and many land leases of the late 17th and early 18th centuries permitted (and later required) marl to be exploited. In time, farmers came to use a variety of fertilisers and nitrogen-fixing crops – vetch, clover, trefoils – in their search for higher productivity. By the 1830s, marl was almost completely displaced by some of the new manures (e.g., guano and sodium nitrate) distributed by a growing transport infrastructure. On the dairying lands much use was made both of bone-dust fertiliser 're-exported' from the towns, and of stable and other manure transported by canal barge (Fussell 1954); Manchester alone produced 63,000 tons of night-soil annually in the 1840s (Davies 1960). The rapid regional growth of urban areas forced more land into agricultural production; many fields were levelled out, drained and ponds filled in, a process which has continued to the present.

The pond habitat is one of continuous change and this discussion considers the nature of the pond habitat in the region, identifies changes in its status, and reviews the future for this distinctive regional landscape.

The pond habitat

What constitutes the pond habitat?

The pond habitat can be variously interpreted as 1. the stock of the pond resource; 2. the individual ponds; and 3. the pondscape. Each is now explored in turn:

1. The stock of the resource

We estimate the number of pond sites in north-west England in 1996 at no more than 36,000, compared with perhaps 125,000 in the late 19th century. In the Mersey Basin area the total would be approximately 18,000. Based largely upon estimates made for Cheshire (Boothby, Hull & Jeffreys 1995a) the number of 'wet' sites is probably only 11,000; many other formerly wet sites are in advanced vegetational succession. Figure 14.1 shows the current pond sites, located mostly in the lowlands.

2. Individual ponds

The individual pond and its surrounding terrestrial buffer zone constitute a habitat for many animals and plants. For some, e.g., fish, aquatic plants and certain invertebrates, the habitat requirements depend upon such physiographical variables as hydrochemistry, water depth, degree of overshading, interspecific predation, pollution events, rainfall levels, temperature, and elevation. Though there is significant variation in these elements across the region, the bulk of ponds are in

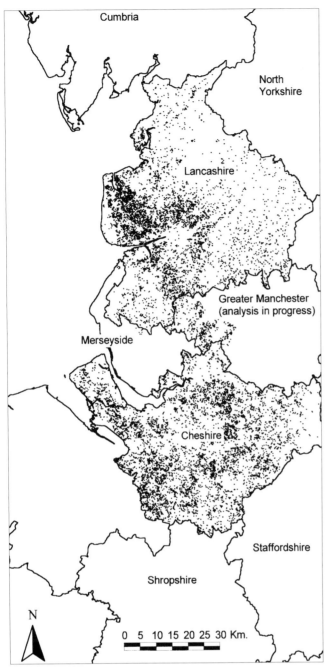

Figure 14.1. The ponds of north-west England.

lowland agricultural settings. Given their historical origins as marl pits, many have high pH values (Day, Greenwood & Greenwood 1982; Boothby *et al.* 1995b). Many pond species have specific requirements or tolerances, e.g., the necessity of emergent vegetation for dragonflies, gentle pond banks for amphibians, shaded ponds for certain invertebrates; there may be no such thing as an 'average pond', and all provide valuable habitats (Biggs *et al.* 1994).

Overall, the presence of some floral species seems to be fortuitous from pond to pond, and certainly the presence of a particular species cannot usually be inferred from a knowledge of its presence at a nearby pond; any individual pond could contain locally or regionally rare species.

3. From pond cluster to pondscape
For some species, a congregation of ponds is probably necessary for their persistence. Amphibians in particular exhibit marked mobility around the congregation in both foraging and colonisation behaviours. Indeed, many amphibians exist in metapopulations, in which 'subpopulations', with a varying home territory, interchange genetic material with nearby groups on a year-by-year basis, thus helping to ensure survival (Fahrig & Merriam 1993; Ebenhard 1991; Reading, Loman & Madsen 1991).

Seasonal mobility of amphibians could be 1,000m or more, but a more conservative estimate would be around 300m (Macgregor 1995). Individual ponds isolated at greater distances may have populations which are at risk from local extinction and which may not then be recouped by recolonisation (Sjögren 1991). It is thus safe to assume that a 'healthy' congregation of ponds contains not only suitable pond sites, but has a spatial arrangement of ponds which is also optimal. In this context, 'optimal' also includes the nature of the landscape matrix found between ponds (Ebenhard 1991). Grazed land is near to optimal in terms of its permeability to amphibians, ploughed land less optimal, and urban infrastructure may be an almost total impedance.

Ponds in clusters offer the 'safest' landscape for maintaining biodiversity. We may expect 'safe' landscapes to exhibit minimal spatial fragmentation, and that local clusters should be part of a wider 'pondscape'.

Biological status and change
Pond status
With some 18,000 individual pond sites within the Basin, it is perhaps unrealistic to suppose that we shall ever have complete, up-to-date biological records for each pond site, though this is already possible in some localities. However, sample information is becoming available on the overall status of ponds, and is forming a bench-mark for regional pond data (Pond *Life* Project 1996). This analysis shows only the broadest details of species composition, though inspection of the underlying statistical structure of the faunal, floral, and physiographical variables is expected to reveal more substantial patterns.

1. Plants
Of the 271 ponds surveyed throughout Lancashire and Cheshire in 1995, the maximum number of plant taxa recorded in any one pond was 45. Some 24% of ponds held 13 species or fewer, many of these sites being overshaded. Ponds with 29 or more plant species were, by this simple definition, in the top 10% of those surveyed (Table 14.1). Within this pattern, 26 ponds held between 5 and 10 taxa of submerged and floating species, and 39 ponds (14%) contained no aquatic plant species at all. Though aquatic vegetation was often to be found in small isolated patches, floating vegetation was often more extensively developed. Emergent vegetation was particularly prominent in the region's ponds; at 74 ponds,

Taxa recorded	<10	11–20	21–30	31–35	36–40	>40
Number of ponds in:						
Cheshire	21	81	43	4	1	2
Lancashire & Wigan	20	46	42	8	2	1
Full survey	41	127	85	12	3	3

Table 14.1. Distribution of abundance of aquatic plant taxa in surveyed ponds.
Adapted from Table 2.1, Pond *Life* Project (1996).

Bulrush (*Typha latifolia*) was often found as floating beds and is a rapid coloniser of ponds. Substantial numbers of ponds were found with stands of Branched Bur-reed (*Sparganium erectum*), Floating Sweet-grass (*Glyceria fluitans*) and Cyperus Sedge (*Carex pseudocyperus*) (Table 14.2); six 'rare' plant species were encountered in the survey: Soft Hornwort (*Ceratophyllum submersum*), Fringed Water-lily (*Nymphoides peltata*), Water Soldier (*Stratiodes aloides*), Tufted-sedge (*Carex elata*), Cowbane (*Cicuta virosa*) and Galingale (*Cyperus longus*). Of these Cowbane, Water Soldier, Fringed Water-lily and Galingale are nationally scarce (Stewart, Pearman & Preston 1993).

Acorus calamus	4
Bolboschoenus maritimus	1
Carex elata	1
Carex paniculata	13
Carex vesicaria	1
Carex acutiformis	9
Carex otrubae	17
Carex pseudocyperus	62
Carex rostrata	10
Eleocharis palustris	52
Equisetum fluviatile	56
Glyceria fluitans	106
Glyceria maxima	4
Phalaris arundinacea	87
Phragmites australis	12
Schoenoplectus lacustris	7
Sparganium erectum	142
Typha angustifolia	15
Typha latifolia	74

Table 14.2 . Potential dominants of swamp communities: total occurrences in the 271 ponds surveyed.
Note: from Table 2.10; Pond *Life* Project (1996).

2. Invertebrates

Across the 271 surveyed ponds, invertebrate species were found in all but 8 ponds; one pond held 59 species, but the majority held between 16 and 33 species. Whilst many species were common, several regionally or locally notable species were found, and 25 species with official

Joint Nature Conservation Committee (JNCC) scarcity status were found – these include one of the country's rarest beetles (Red Data Book (RDB) 1 Endangered), Lesser Silver Water Beetle (*Hydrochara caraboides*), until recently not encountered outside of the Somerset Levels. Specimens of scarce species were found in 102 ponds, some holding more than one such species. The JNCC's Invertebrate Sites Register Invertebrate Index (ISRII) allows any pond to be scored for its invertebrate interest. Of the 76 pond-clusters forming the focus of the survey, 41 score at least 20 points (significant) with a top-score of 350: this last is at a site in Wigan where opencast coal is to be extracted. Figure 14.2 shows the distribution of 'significant' invertebrate sites to be a region-wide phenomenon and not restricted solely to the remoter, more-rural environments.

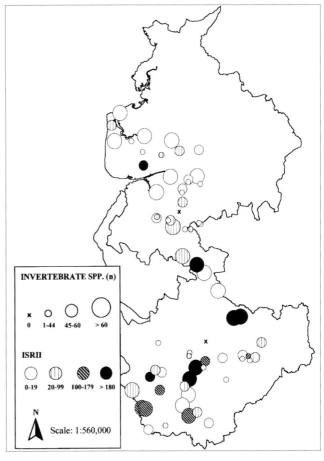

Figure 14.2. Aquatic invertebrate species at surveyed ponds/pond clusters. Symbol size denotes number of species; shading indicates ISRII value (see legend).

3. Amphibians

Amphibians are currently under considerable stress throughout the world (Wake 1991; Blaustein, Wake & Sousa 1994). The region contains all six of the recognised native British species, though the Natterjack (*Bufo calamita*) is found only on the coastal margins. In the survey, the number of species found at any one pond was very variable. One pond surveyed held 5 amphibian species and 3 other ponds held 4 species; about one third of ponds (31.2%) showed no evidence of amphibians at all, though this cannot be interpreted as permanent absence.

The Common Frog (*Rana temporaria*) is widely found in 41% of ponds, and the Common Toad (*Bufo bufo*) is found in over 17% of ponds. The protected species Warty or Great Crested Newt (*Triturus cristatus*) was found in 23% ponds, though it was much more common in Cheshire (35% of ponds); Palmate (*T. helveticus*) and/or Smooth Newts (*T. vulgaris*) were found in 18% of ponds. These data confirm the significance of the region and its ponds as substantial amphibian habitats, a picture which is not always accurately recorded (Hilton-Brown & Oldham 1991). Most importantly, the persistence of amphibians has widely been seen to be related to the integrity of the *pondscape* and not simply to the survival of *individual* ponds: it is to this wider context that we now turn.

The changing status of the pondscape

Since the cessation of agricultural pond excavation in the early 19th century:

1. many ponds have been lost; comparatively few new ponds have been added, perhaps fewer than 1000 across the region, principally for fishing;

2. vegetational succession has been marked;

3. pond-densities have declined (from $c.17 \text{km}^{-2}$ to $c.3 \text{km}^{-2}$);

4. the spatial extent of the pondscape has been much reduced; and

5. the fragmentation of the remaining pond landscape has greatly increased.

The fragmentation of the pondscape can be visualised using a procedure of distance-counting (Unwin 1979). In

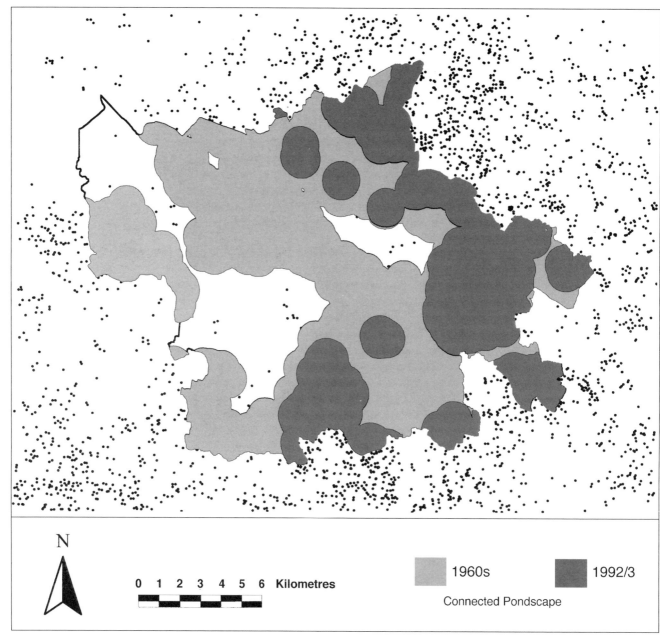

N

0 1 2 3 4 5 6 Kilometres

| 1960s | | 1992/3 |

Connected Pondscape

Figure 14.3. Pondscape of Vale Royal District, 1969 and 1992/93.

	1960s	1992–93
Total area of connected pondscape		
– (km²)	310	121
– as % District area	[81%]	[31.6%]
Number of fragments	3	10
Area of largest fragment (km²)	288	68
Area of second-largest fragment (km²)	23	23

Table 14.3. Pondscape change in Vale Royal District: summary statistics.

this, for each pond, the procedure counts the numbers of ponds within a specific search-radius, giving, for each pond, a value for connectedness. Mapping this connectedness shows the spatial extent of the pondscape and the increase of fragmentation. An example of this has been prepared for the District of Vale Royal in Cheshire.

In Figure 14.3, the pondscape is defined for the ponds of *c.*1969 (in light grey). The limit of the pondscape is defined as a line enclosing those ponds with the same value of connectedness (for Vale Royal – 19 ponds within 1,000m of any given pond). Other ponds lie outside this defined pondscape, but they are relatively isolated; there is almost uninterrupted coverage of the District by a connected pondscape. The configuration of the pondscape for 1992/93 is also shown (in darker grey) and the changes from the earlier map are clear to see: the pondscape now covers less than one quarter of the original area; it is fragmented into ten pieces, and the largest occupies only 68km² compared to 288km² (see Table 14.3). Figure 14.4 shows an attempt, using a similar approach, to map the pondscape for north-west England as a whole. Analysis for areas between Cheshire and Lancashire is incomplete at this time, but the connected pondscape certainly extends through Wigan, into Liverpool, Wirral and Knowsley / St Helens to the south and west, and into the fringes of Greater Manchester (see Figure 14.1).

Loss mechanisms

The era represented by the map of changing pondscape is a significant one both for the Mersey Basin and for elsewhere across England. Amongst the most significant features are: the intensification of agriculture, involving increased use of agri-chemicals; the growth of the transport infrastructure, including motorways; the spread of suburbs; and the ruralisation of manufacturing.

Though much of the farmland of the region has remained in pastoral use, helping ponds to survive, the search for increased profitability by farmers has meant that ponds, like hedgerows and trees, are often seen as archaic and therefore expendable landscape features. Perceived in this light, they are sometimes used for dumping farm waste, and some have been ploughed out or drained. As a generalisation, 'benign neglect' plays the most significant role in pond loss on agricultural land (Boothby *et al.* 1995b). With the loss of a purpose for most ponds, vegetational succession will often run

unchecked, aided by nitrate-rich run-off from agricultural fertilisers. This run-off would formerly have been intercepted by the soil and vegetation of the immediate pond buffer-zone, but many such areas have been incorporated into used agricultural land, allowing agri-chemicals to reach ponds. Given the origin of many of the ponds as sources of marl fertiliser, this is indeed ironic.

Across the region, urban spread and its associated infrastructural development has removed many pond sites. Only recently have building developers begun to incorporate existing ponds within new estates; only recently have new roads or other developments begun to incorporate substantial development mitigation packages within their proposals; only recently have golf-course architects realised that ponds can be

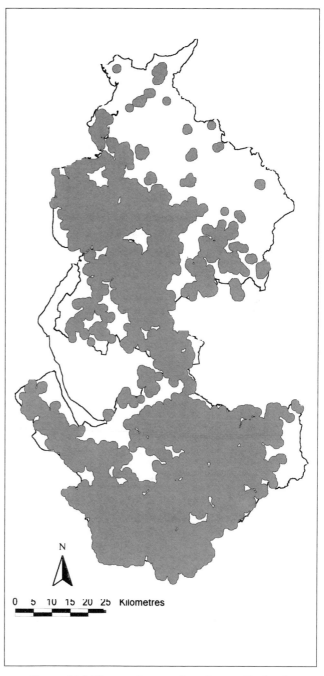

Figure 14.4. The pondscape of north-west England.

accommodated within their boundaries. Table 14.4 summarises the mechanisms of pond loss in Cheshire; only limited analysis is as yet available for Lancashire and other parts of the region (Boothby *et al.* 1995a), but research elsewhere (Heath & Whitehead 1992) suggests that similar forces will have been at work.

Looking to the future – threats and opportunities

We can evaluate the future for ponds using the concepts contained in Table 14.5. The *strengths* of the pondscape of the region lie in its diversity of animals and plants, which is considerable. For some substantial parts of the area, there is still sufficient congregation of ponds to afford habitats for those species which require 'connected' multi-patch habitats. Though evidence is sporadic, the waters of ponds are not known to be grossly or widely polluted (Boothby *et al.* 1995b). They are widely used for coarse fishing, they have provided generations of children with an educational and recreational resource, and, with their surrounding tree cover, they often provide prominent local landmarks. Where they are not over-managed, they provide local wilderness – or, at least, wild corners of nature and all this in the context of an historical landscape feature which represents once-widespread farming practices.

But there are *weaknesses* too. Earlier surveys notwithstanding (Day *et al.* 1982), our knowledge base for ponds is woefully thin; even counting the ponds accurately represents a major leap forward. Most ponds are on inaccessible farmland where, as a result of benign neglect, they are disappearing; their former economic functions have been lost and, with a few exceptions, new uses have not been found. The cost of surveying, monitoring, or actively managing the remaining stock could be prohibitive. Ponds stand relatively unprotected by legal designation other than for a relatively small and select(ed) number which have biological or archaeological interest.

The *threats* to ponds come, as we have seen, from a

	Ponds lost (%) to:			Ponds in advanced
	agriculture	*building and infrastructure*	*recreation*	*vegetational succession (%)*
Cheshire	19.1	7.5	2.2	40.0
Districts				
minimum	13.1	1.4	< 0.1	22.0
maximum	23.7	33.1	10.9	48.6

Table 14.4. Mechanisms of pond loss, *c.*1969–1992/93: Cheshire County.

variety of sources, but perhaps 'benign neglect' will continue to threaten; this is, in part, a perceptual problem, sometimes seen in the dumping of farm waste in a convenient 'damp hole in the ground'. This neglect will also be recognised in advanced vegetational succession, drying-out, and, ultimately, reversion to terrestrial habitat. As many ponds are in or are entering this phase, we may justifiably be concerned about the biodiversity of the stock of ponds in the region, especially as so few new ponds are being dug. At individual ponds, localised pollution events may render the site inhospitable, and those who walked the countryside in the mid-1990s will have had no difficulty in recognising the scale and criticality of the extended drought at that time. The implications of this 'catastrophic' event for species and habitat survival have yet to be quantified.

Lest all this sounds too negative, we must conclude by considering the *opportunities*. These arise from a popular concern to protect pond landscapes, and call into consideration a range of initiatives from new agricultural policies to the instigation of new planning policy potentialities.

We consider that, at the farm level, there may be some value in considering afresh the economic value of ponds, whether this be for the leasing of fishing rights, the exploration of new food sources, or the exploitation for scien-

Strengths	Weaknesses
high biodiversity	meagre knowledge-base
low incidence of pollution	often on inaccessible farmland
educational potential	loss of main economic function
recreational uses, especially angling	high costs of survey and management
significant landscape feature	general lack of legal protection
'wilderness' remnant	weak image amongst some farmers and land managers
historical/ archaeological continuity	
Opportunities	**Threats**
new economic value	benign neglect
changing agricultural policies	loss to other uses
strengthening of Countryside Stewardship	vegetational succession / drying-out
planning mitigation	lowering of biodiversity
pond-friendly planning policies	local pollution events
local enthusiasm and organisation	sustained drought
developing knowledge-base	fragmentation
	agricultural intensification

Table 14.5. Ponds: strengths, weaknesses, opportunities, threats.

tific and research purposes. We firmly believe that most farmers do not set out to delete ponds from their land: they simply fail to see an alternative which makes any economic sense. In this context, we now feel that the changing emphases of both agricultural policy and of nature conservation policies may be coinciding. The Countryside Stewardship initiative is now (1996) replacing the largely discredited Set Aside scheme as the appropriate vehicle for delivering nature conservation on farmland. For the first time this should allow some farmers to conserve ponds and small wetlands, and to be compensated for doing so. In this region, it may be possible to give the Countryside Stewardship a 'pondy' flavour in recognition of the distinctiveness of the regional landscape.

At other policy levels, there are opportunities to be seized. Though planning policy and practice is now required to consider explicitly the value of development, local planning authorities are also empowered to seek mitigation for any deleterious effects on landscape and habitat. That this can work to the benefit of ponds can be seen in the proposed mitigation package for the second runway at Manchester Airport: the developers are to provide 97 new or re-created ponds to re-establish habitat, both aquatic and terrestrial, for the important newt colony at that site. In all, 43 ponds will be destroyed, and the surrounding landscape changed significantly, if, in part, temporarily. This significant legal agreement is possible because of the principle of 'no net loss of environmental value' incorporated in the appropriate Structure Plan, and thence into Cheshire County Council policy (Cheshire County Council 1990). Whilst the proposed scale of environmental and landscape change at the airport creates concern in many people, the availability of mitigation should not be minimised in its importance.

Finally, opportunities also lie in local people and organisations. There is a massive will to undertake environmental conservation as witnessed by the size and scope of membership in voluntary bodies such as the Wildlife Trusts, BTCV, and others (Hull & Boothby 1996). In such a context, the Pond *Life* Project (Boothby *et al.* 1995a) plays a role in helping to co-ordinate volunteer attempts to survey, analyse and manage the ponds of the region. Operating through a 'pond warden' system we are attempting to focus enthusiasm, research capability, and practical experience towards sustaining and, where possible, increasing the stock of ponds, to reducing fragmentation of the pondscape, and to building a sustainable network of individuals, voluntary organisations, and planning authorities for the future.

In all of these opportunities, we are convinced of the necessity to focus not just on the individual pond, or even on several hundred individual ponds, but to stress the importance of the cohesion of the landscape, and to inform management actions accordingly.

Acknowledgements

The Pond *Life* Project is financially supported by the *Life* Programme of the European Union, by Liverpool John Moores University, and by co-operating partners in north-west England, the Netherlands, Belgium, and Denmark.

Surveys for the Project summarised here have been carried out by Jonathan Guest Ecological Survey; maps have been prepared by Jon M. Bloor.

References

Biggs, J., Corfield, A., Walker, D., Whitfield, M. & Williams, T. (1994). New approaches to the management of ponds. *British Wildlife*, **5(5)**, 273–87.

Blaustein, A.R., Wake, D.B. & Sousa, W.P. (1994). Amphibian declines: judging stability, persistence and susceptibility of populations to local and global extinctions. *Conservation Biology*, **8(1)**, 60–71.

Boothby, J., Hull, A.P. & Jeffreys, D.A. (1995a). Sustaining a threatened landscape: farmland ponds in Cheshire. *Journal of Environmental Planning and Management*, **38(4)**, 561–68.

Boothby, J., Hull, A.P., Jeffreys, D.A. & Small, R.W. (1995b). Wetland loss in North-West England: the conservation and management of ponds in Cheshire. *Hydrology and hydrochemistry of wetlands* (eds J. Hughes, & A.L. Heathwaite), Ch. 27. Wiley, London.

Cheshire County Council (1990). *County Structure Plan Review*. CCC, Chester.

Davies, C.S. (1960). *The Agricultural History of Cheshire, 1750–1850*. The Chetham Society, 3rd Series, 10, Manchester.

Day, P., Deadman, A.J., Greenwood, B.D. & Greenwood, E.F. (1982). A floristic appraisal of marl pits in parts of north-western England and northern Wales. *Watsonia*, **14**, 153–65.

Ebenhard, T. (1991). Colonisation in metapopulations: a review of theory and observations. *Biological Journal of the Linnean Society*, **42**, 105–21.

Fahrig, L. & Merriam, G. (1993). Conservation of fragmented populations. *Conservation Biology*, **8(1)**, 50–59.

Ferro, R.B. & Middleton, A.C. (1949). Clay marling: mechanised methods. *Agriculture*, **56**, 123–28.

Fussell, G.E. (1954). Four centuries of Cheshire farming systems, 1500–1900. *Transactions of the Historic Society of Lancashire and Cheshire*, **106**, 57–79.

Grantham, R.B. (1864). A description of the works for reclaiming and marling parts of the late Delamere Forest. *Journal of the Royal Agricultural Society of England, I*, **25**, 369–80.

Heath, D.J. & Whitehead, A. (1992). A survey of pond-loss in Essex, South-east England. *Aquatic Conservation: Marine and Freshwater Ecosystems*, **2**, 267–73.

Hewitt, E. (1929). *Medieval Cheshire*. Manchester University Press, Manchester.

Hewitt, W. (1919). Marl and marling in Cheshire. *Proceedings of Liverpool Geological Society*, **13**, 24–28.

Hilton-Brown, D. & Oldham, R.S. (1991). *The status of widespread amphibians and reptiles in Britain 1990, and changes during the 1980s*. Nature Conservancy Council, Peterborough.

Hull, A.P. & Boothby, J. (1996). Networking, partnership and community conservation in North West England. *Nature Conservation 4: The role of networks* (eds D.A. Saunders, J.L. Craig & E.M. Mattiske), pp. 341–55. Surrey-Beatty, Chipping Norton, NSW, Australia.

Macgregor, H. (1995). Crested newts – ancient survivors. *British Wildlife*, **7(1)**, 1–8.

Middleton, A.C. (1949). Clay marling: some historical notes. *Agriculture*, **56**, 80–84.

Pond *Life* Project (1996). *Critical pond biodiversity survey, 1995.* Pond *Life* Project, John Moores University, Liverpool.

Porteus, T.C. (1933). *Calendar of the Standish Deeds 1230–1575.* Public Library Committee, Wigan.

Prince, H.C. (1964). The origins of pits and depressions in Norfolk. *Geography,* **40,** 15–32.

Reading, C.J., Loman, J. & Madsen, T. (1991). Breeding pond fidelity in the common toad *Bufo bufo. Journal of the Zoological Society of London,* **225,** 201–11.

Sjögren, P. (1991). Extinction and isolation gradients in metapopulations: the case of the pool frog (*Rana lessonae*). *Biological Journal of the Linnean Society,* **42,** 135–47.

Stewart, A., Pearman, D.A. & Preston, C.D. (1993). *Scarce Plants in Britain.* JNCC, Peterborough.

Unwin, D.J. (1981). *Introductory spatial analysis.* Methuen, London.

Wake, D.B. (1991). Declining amphibian populations. *Science,* **253,** 860.

Map copyright

Boundaries for the Figures in this chapter are derived from OS Strategi data sets provided by Ordnance Survey under the terms of the CHEST Licensing Agreement.

CHAPTER FIFTEEN

Liverpool Bay and the estuaries: human impact, recent recovery and restoration

S.J. HAWKINS, J.R. ALLEN, N.J. FIELDING, S.B. WILKINSON
AND I.D. WALLACE

Introduction

Liverpool Bay and the estuaries that empty into it can be viewed as one interconnected ecosystem. The ecology of the estuaries and the Bay cannot be separated: considerable freshwater run-off enters the Bay via the estuaries from the notoriously wet north-western catchment; the large tidal range (10m) means high salinity water penetrates far up the estuaries on every tide. In order to be manageable, however, this chapter restricts attention to the Rivers Dee, Mersey and Ribble and the adjacent portion of Liverpool Bay (see Figure 15.1), focusing largely on the Mersey Estuary.

It is also impossible to separate the natural and anthropogenic factors acting on Liverpool Bay and its estuaries. Human activity has influenced the catchment and estuaries since earliest times (e.g., Chester was a Roman port). Human impacts have also been a major factor since industrialisation began in the 18th century. Indus-

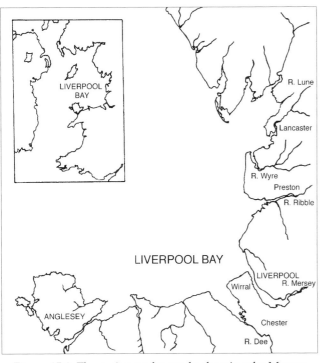

Figure 15.1. The region under study showing the Mersey Estuary.

trialisation and the development of the Port of Liverpool have led to considerable modification of the lower estuary, including an extensive series of walls first built in the last century to train the shifting sandbanks of the Bay to maintain navigation channels into the Narrows. The construction of the Manchester Ship Canal (opened 1894) changed the morphology of the upper Mersey Estuary and the normal cyclical channel migration (Bennett, Curtis & Fairhurst 1995). The River Ribble was also trained within levees when the docks were built in late 19th century. The upper Dee Estuary was subject to similar canalisation coupled with extensive land reclamation (see chapter sixteen, this volume).

As well as physical deformation there has been considerable input of waste via the rivers, directly to the estuaries and by dumping into the Bay itself. A variety of industrial and a huge quantity of domestic wastes, as well as dredging spoil have been discharged or dumped over the years. The River Mersey with its highly industrialised catchment has particularly suffered, earning the reputation of being the most polluted estuary in Europe (Clark 1989). In contrast the River Dee and River Ribble have been impacted to a much lesser extent, although point sources of pollution have had and continue to have some local effects.

We wish to highlight the lack of published work on the River Mersey; much of our review is derived from the 'grey' literature of reports, workshop proceedings and student theses. Early work on the region included floristic (e.g., Harvey-Gibson, 1890, 1891) and faunistic catalogues (e.g., Byerley, 1854; the work of Liverpool Marine Biology Committee edited by Herdman, 1886–92 summarised in Herdman, 1920). The story of the River Mersey up to the early 1970s is well told by Porter (1973). A wealth of anecdotal information and firsthand experience by local naturalists as well as professional investigations of the Mersey Estuary was summarised at a meeting of the Mersey Estuary Conservation Group in November 1988 (Curtis, Norman & Wallace 1995). An excellent recent summary on the water quality, pollution and ecology of the Mersey is the 'Mersey Estuary', a report on environmental quality (National Rivers Authority 1995).

Recent hydrocarbon exploitation has prompted

summaries of the environment and ecology of Liverpool Bay (Rice & Putwain 1987; Taylor & Parker 1993; Darby *et al.* 1994). There is also much relevant information in the various volumes of The Irish Sea Study Group (1990). Perhaps the most comprehensive recent study in the region was made to provide a baseline for the proposed construction of the Mersey Barrage which unfortunately is not in the public domain (Environmental Resources Limited report to Mersey Barrage Company 1993, copy in Zoology Department, Liverpool Museum). This work covered all aspects of the ecology of the River Mersey including a much needed recent survey of the benthos (Jemmett pers. comm.). These recent studies aside there has been surprisingly little published on the ecology of the various estuaries since the classic studies either side of the Second World War (Fraser 1931, 1932, 1935, 1938; Bassindale 1938; Pierce 1941; Corlett 1948; Popham 1966).

In the rest of this chapter background ecological information on Liverpool Bay, River Ribble and River Dee is briefly summarised. We then focus on the River Mersey, summarising what is known about its ecology before major impact and its subsequent recovery as a consequence of de-industrialisation and depopulation, tighter environmental controls and amelioration initiatives such as the Mersey Basin Campaign. Finally we describe pioneering work on restoration ecology of the former Merseyside docks. This is work that started in Liverpool and has been applied successfully elsewhere in the UK. It has shown what can be achieved in terms of water quality improvements as part of inner city renewal schemes. We finish with an appraisal of the current status of the region and its international importance for conservation.

Background information: Liverpool Bay, River Dee and River Ribble

Research on Liverpool Bay intensified in response to the dumping of sewage sludge, dredge aggregate and industrial waste in the 1970s (e.g., Department of the Environment 1972; Rees 1975). Elevations of organic matter, inorganic nutrients (nitrogen and phosphorus), heavy metals and resistant organics were detected in the areas near the sludge dumping grounds. Impacts on the benthos were less clear-cut due to the spatial and temporal variability of the highly unstable sediments (Rees *et al.* 1992). In particular there are highly localised pockets of muddy sediment which can support very dense communities of benthic animals* (Rees, Nicholaidou & Laskaridou 1977). The area is also notorious for wrecks of shellfish (Common Cockle, *Cerastoderma edule* and Rayed Trough Shell, *Mactra corallina*) in which large numbers of bivalves are washed up on the coast; these occur frequently and are natural

events caused by storms and very cold weather (McMillan 1975).

The inshore and intertidal communities are dominated by sandy or muddy sediments flanked at their terrestrial margins by extensive saltmarshes. The only significant area of natural rocky shore is at New Brighton and at Hilbre Island (reviewed in Craggs 1982) which supports an interesting rocky shore community truncated at its seaweed border by sand (Russell 1972, 1973, 1982) . Limpets (*Patella vulgata*), the dominant grazer in north-eastern Atlantic rocky shore (Hawkins *et al.* 1992) communities, are scarce. Their numbers are reduced by cold winters, e.g., 1947 (Burd 1947) and 1962–63 (B. Bailey pers. comm.). They are also vulnerable to siltation, and their numbers are probably low even where suitable habitat exists presumably due to poor larval supply from the nearest extensive breeding population on the Great Orme. Common Winkle (*Littorina littorea*), however, are plentiful and assume the role of major grazer. Silt is probably also the reason for the absence from the rocky littoral of chitons, top shells, sea-squirts and the scarcity of sponges. Rising sediment levels have greatly reduced the sub-littoral and as early as 1854 were blamed for the disappearance of Lobsters (*Homanus gammarus*) at Hilbre (Byerley 1854).

Artificial hard substrata abound in Liverpool Bay and the estuaries. The docks are considered in detail later. Pilings, sea-walls and sea defences, (e.g., the fish tail groynes on Wirral) all support impoverished rocky shore communities dominated by wrack seaweeds (*Fucus* spp.), barnacles (*Semibalanus balanoides* and *Elminius modestus*), Common Mussel (*Mytilus edulus*) and Common Winkle, plus plenty of ephemeral algae such as *Enteromorpha*. Limpets have appeared recently on the sea-wall at New Brighton and on the new defences along with Beadlet Anemone (*Actinia equina*). Offshore, various wrecks act as artificial reefs and are colonised by diverse assemblages of subtidal fouling organisms (see Hartnoll 1993 for review).

Coastal sediment communities are productive but of generally low diversity in part due to the very homogeneous nature of their habitats. Sedimentary substrates dominate in both the intertidal and immediate subtidal zones in the Liverpool Bay area (Mills 1991).

The relatively few studies made over the years (Bassindale 1938: Gardiner 1950; Stopford 1951; Perkins 1956; Popham 1966; Gilham 1978; Bamber 1988; Al-Masnad 1991; Davies 1991; Garwood & Foster-Smith 1991; and see Mills 1991; Davies 1992; Hawkins 1993; Darby *et al.* 1994 for reviews) have revealed typical communities whose composition reflects particle size and depth. Lowshore and shallow areas of muddy sand have a rich fauna with bivalves, e.g., Common Cockle, Banded Wedge Shell (*Donax vittatus*) and echinoderms, Brittlestar (*Ophiura ophiura*) and Heart Urchin (*Echinocardium cordatum*), being prominent. Coarser, more mobile areas have a more impoverished fauna dominated by Striped Venus Shell (*Chameiea gallina*) and Sand Mason Worm (*Lanice conchilega*). Coarse intertidal

---

* Names of invertebrate animals follow as far as possible Hayward & Ryland (1995).

sediments support amphipods (*Bathyporeia* spp., *Haustorius arenarius*), polychaetes (*Nepthys* spp, *Nerine cirratulus*) and low down the Thin Tellin bivalve (*Angulus tenuis*). In clean sand at the mouth of the River Dee, the worm, *Spio filicornis,* dominates along with Lug Worm (*Arenicola marina*) and Sand Mason Worm; in more sheltered and hence slightly muddier areas Baltic Tellin (*Macoma balthica*), the polychaete (*Pygospio elegans*) and higher densities of Lug Worm occur. At the seaward end of the Great Burbo Bank a patch of coarser sand, gravel and cobble occurs and here Dahlia and Cave Anemones (*Urticina felina* and *Sagartia trogloddytes*) and the King Rag Worm (*Neanthes virens*) have all been recorded (Bassindale 1938).

The large expanses of intertidal flats from Crosby to Fleetwood support a typical low biomass crustacean and small polychaete dominated community over much of their extent, giving way to a richer community supporting bivalves in the stabler sand of the lower shore of the Sefton coast. The finer, muddier sediments of the River Ribble (Popham 1966; Davies 1991) support typical estuary species, e.g., Baltic Tellin, Common Cockle, the crustacean *Corophium volutator* and Common Ragworm (*Hediste diversicolor*).

Changes are occurring in the River Ribble as the channel to the former port of Preston is no longer dredged and maintained (H.D. Jones pers. comm.). At Preston Docks, Conlan *et al.* (1988, 1992) showed the benthic communities to be very impoverished, an observation typical of a low salinity basin.

The effects of the extensive colonisation by Common Cord-grass (*Spartina anglica*), in both the Ribble and Mersey estuaries were reviewed by Doody (1984); Common Cord-grass has also covered large areas of the Dee Estuary. The spread of this species is of concern because it can result in the loss of intertidal feeding areas for wildfowl, and accelerates sediment accretion.

The Mersey Estuary

Historically, the Mersey Estuary contained sufficient Salmon (*Salmo salar*), Thick-lipped Grey Mullet (*Chelon labrosus*), Sturgeon (*Acipenser sturio*), Eels (*Anguilla anguilla*) and Smelt *(Osmerus eparlanus)* to support local fisheries for these species. Shrimp (*Crangon crangon*) and Flounder (*Pleuronectes flesus*) were also caught in the upper estuary (Cunningham 1898; Dunlop 1927). Thus it is safe to assume that human impact on the estuary was probably negligible until the Industrial Revolution. Although over 100 years of biological and chemical data are available for the River Mersey, differences in objectives, site selection, collection techniques, taxonomic expertise and analytical techniques between surveys mean that comparisons are exceptionally difficult to make. Furthermore, by the time surveys started the River Mersey was by no means pristine! Nevertheless,

Figure 15.2. Victoria Promenade, Widnes and the Mersey Estuary 1986 (*Photo: John Davies*).

Bassindale's 100 sampling site survey in 1933 (Bassindale 1938) provides a useful reference point for the River Mersey at the onset of the worst water quality conditions which probably occurred during the period 1940 to 1970 (Porter 1973). Another approach is the analysis of core samples coupled with radiometric dating. This work shows that the estuary was probably badly affected by pollution by the early years of the 20th century (National Rivers Authority 1995). Highest levels occurred for most heavy metals from the 1920s onwards, with decline starting in some cases as early as the 1930s (zinc, copper), but in others (e.g. mercury) only in recent years (see Figures 4.11 and 4.12 in National Rivers Authority 1995, based on Leah and co-workers, unpublished).

The major problem for the River Mersey has always been organic input rather than toxic chemicals. With the growing population and industry on the banks of the River Mersey, increased quantities of mainly domestic sewage and some industrial organic effluent introduced to the river reduced water quality considerably, so that anoxic conditions persisted for prolonged periods from the 1930s onwards (Bassindale 1938; Fraser 1938; Porter 1973 for review). In the upper estuary, a pronounced oxygen sag occurred regularly during summer low tide periods, causing anaerobic conditions and an associated offensive smell (Mersey & Weaver River Authority 1972; Irish Sea Study Group 1990). From the 1930s onwards, the general trend was one of reducing species abundance and diversity (Bassindale 1938; Holland & Harding 1984; Readman, Preston & Montoura 1986; Wilson, D'arcy & Taylor 1988; Environmental Advisory Unit of Liverpool University Ltd 1991). Salmon were effectively eradicated and the fisheries in the upper estuary severely limited (Liverpool Bay Study Group 1975; Wilson, D'arcy & Taylor 1988; Dempsey 1989). The loss of invertebrates due to anoxic bottom conditions, notably during the 1960s was considered a major factor in the decline of resident fisheries (Burt 1989). Around 1960 all but a few transient pelagic fish were absent from the inner and middle reaches of the estuary, and these did not penetrate much beyond the narrows (Porter 1973).

A number of heavy metals such as lead, mercury and arsenic and other persistent organic contaminants were, and still are, found in the water, sediments and marine organisms within the region at elevated concentrations, relative to other UK waters (Bull *et al.* 1983; Russell *et al.* 1983; Riley & Towner 1984; Norton, Rowlatt & Nunny 1984; Dickson & Boelens 1988; Jones & Head 1991; Law *et al.* 1992; Leah *et al.* 1992; Thompson *et al.* 1996). These concentrations tend to be highest close to the mouth of the River Mersey and to decrease with increasing distance offshore. There have also been acute incidents of lead poisoning of waders and wildfowl (Head, D'arcy & Osbaldeston 1980; Bull *et al.* 1983; Osborn, Every & Bull 1983). A study of the fauna and heavy metals present in the mud in Collingwood Dock found very high levels of lead and zinc with the polychaete, *Capitella capitata*, the dominant infaunal species (James & Gibson 1980).

In addition, the predominance of petrochemical and chemical industries led to chronic and acute pollution incidents, probably at their worst during the Second World War when pollution controls were relaxed (Hardy 1995). The most significant pollution incident in recent years involved a breakage in the pipeline, running from the Tranmere oil jetty to the refinery in 1989 (Hall-Spencer 1989; Taylor 1991). In addition to the 1989 pipeline spill, a series of smaller incidents involving the Stanlow refinery and other industrial plants in the Mersey Basin have continued to cause problems, with a number of prosecutions following.

The morphology of the Mersey Basin and estuary worsens the effects of pollution. The narrow mouth of the estuary and its rather long flushing time (1–3 weeks) hinders dispersal of pollutants. The freshwater reaches of the River Mersey and the Manchester Ship Canal have been efficient conveyers of the accumulated wastes of much of industrial Lancashire and Cheshire, which are then discharged into the upper estuary (see National Rivers Authority 1995 for review).

Pollution problems in the Mersey Estuary have decreased since the 1960s (Wilson, D'arcy & Taylor 1988; Irish Sea Study Group 1990). In the 1970s Liverpool Corporation (see Porter 1973) initiated water quality improvements, and in 1985 the Department of the Environment launched the Mersey Basin Campaign. In addition to maintaining dissolved oxygen levels in the estuary and preventing the fouling of beaches by gross solids, the Mersey Basin Campaign aims to ensure that all rivers in the region meet class 2 classification requirements by 2010 (Codling, Nixon & Platt 1991; Wood-Griffiths 1993). The completion of the major interceptor pipeline leading to the Sandon Dock sewage treatment system will have a major impact on water quality in the Mersey Estuary (Taylor & Parker 1993) as will secondary treatment at Sandon, Wirral and Widnes works.

A comparison between the findings of early studies (e.g., Herdman 1920 for summary; Dunlop 1927; Fraser 1935, 1938; Bassindale 1938; Hardy 1995) and more recent surveys (Carter 1985; Wilson, D'arcy & Taylor 1988; Environmental Advisory Unit 1991; Environmental Resources Limited 1993) provides an indication of biological recovery in the River Mersey. Recently there has been an increase in species diversity recorded within the Mersey Estuary including both marine invertebrates and fish. Carter (1985) recorded a more diverse fauna than in previous studies, and suggested that this apparent improvement had occurred since the clean up campaign was initiated. That fish are being caught by anglers in the upper estuary (Wilson, D'arcy & Taylor 1988), is encouraging, as is the indication of a return of salmonids to the estuary (Zheng 1995; Fielding 1997). The increasing abundance and diversity of fish observed by anglers probably reflects similar changes in the state of an important estuarine food resource, the benthic fauna (Environmental Advisory Unit 1991). In 1987, a total of 40 fish species were recorded, ten of which were freshwater. The species dominating the fish communities were Sprat (*Sprattus sprattus*), Herring (*Clupea*

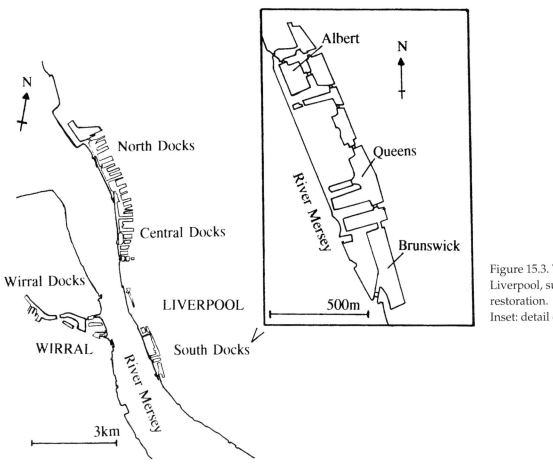

Figure 15.3. The docks in Liverpool, subject to restoration.
Inset: detail of South Docks.

harengus) and Sand Goby (*Pomatoschistus minutus*), while flatfish, Whiting (*Merlangius merlangus*) and Eels were also abundant (Henderson 1988). Increases in fish eating birds have also occurred since the early 1970s, with Cormorant (*Phalacrocorax carbe*), Grebe and Grey Heron (*Ardea cincerea*) all rising in numbers as the abundance of their food increased (Thomason & Norman 1995). Unfortunately, heavy metals and other contaminants present risks for the human consumption of fish, particularly bottom dwelling species such as Plaice (*Pleuronectes platessa*) and Flounder, which have greater contact with the contaminated sediments (see Henderson 1988; Collings, Johnson & Leah 1996; Leah *et al.* 1997).

The maintenance of a dense population of Common Cockles and Baltic Tellin recorded in 1985 at Dingle after a long absence, and the sporadic appearance of mussels at four middle zone sites point to long-term water quality improvement in recent years. A highly significant improvement in the freshwater fauna noted at Warrington in 1987 also provides biological evidence for an overall improvement in water quality in the inner zone in the last few years (Holland 1989). There are still some puzzles; Laver Spire Shell (*Hydrobia ulvae*) and *Corophium volutator* (which should be found on suitable muddy shores) were apparently absent or scarce (Ghose 1979) but have been found again in recent years (Environment Resources Ltd 1993) with *Corophium* occurring in large numbers. At the time of writing a fairly typical estuarine fauna dominated by the Baltic Tellin, Common Cockle and Common Ragworm was present.

Restoration of the Docks on Merseyside

History of the Docks

Liverpool was well situated for the growing trade with Africa and the colonies of the Americas at the end of the 17th century. However, the waters of the Mersey Estuary presented a major restriction to the growth of shipping trade, with strong tides, large tidal range and shifting shoals. In response to this pressure, construction of the first commercial maritime dock in the world began in 1710 (Ritchie-Noakes 1984) within the confines of a shallow creek (known as 'The Pool'), which had previously provided some shelter for ships. This dock (Old Dock) retained water at all states of the tide and allowed unrestricted loading and unloading of ships directly into warehouses on the quayside. The construction of this dock sparked off a spurt of trade centred on Liverpool which was sustained by the continual construction of new docks and industrialisation of the hinterland of the Mersey Basin. The docks expanded in a narrow ribbon along the north-eastern bank of the River Mersey, finally forming two main connected series of docks, known respectively as the North Docks and the South Docks (Figure 15.3). A third major area of dock basins was constructed on the south-western bank of the river, stretching inland from Birkenhead. At its peak Liverpool boasted over 100 docks stretching 10km upstream from the river mouth. The construction of docks and other structures along the banks of the River Mersey constituted a major modification of the shoreline habitats. Intertidal areas of soft sediment and a natural seawater pool were replaced by the hard substrate of retaining walls, permanently submerged sediment and extensive areas of standing water.

With the gradual change to larger ships, containerised shipping and road haulage, the older commercial docks throughout the UK suffered a serious decline in trade in the second half of this century, many falling into complete disuse and decay. These docks were often situated on prime land close to the centre of major cities and were targeted for redevelopment as part of the urban regeneration schemes of the early 1980s. In the majority of cases the water-space of the docks was retained as a central feature. Such development schemes have been carried out for the whole of the South Docks and, more recently, smaller areas of Wirral and the North Docks.

The Mersey Estuary provides the source of water for all docks on Merseyside and consequently water quality has often been poor. This is not a problem for docks used for shipping, but redevelopment schemes rely on the appeal of a waterside location. The problems encountered have proved similar throughout the redeveloped docks on Merseyside. Essentially, they are caused by the silty, nutrient rich and often sewage contaminated water from the River Mersey, and are of four main types: (1) phytoplankton blooms, resulting in turbid or brightly coloured water; (2) periods of low oxygen concentration in deeper water, with associated release of foul smelling hydrogen sulphide gas and mortality of fauna; (3) contamination of water with faecal indicator organisms and hence possibly pathogens; (4) fine silt forming a thick layer on the dock bottom containing high organic matter and pollutants such as heavy metals and persistent organics. These water quality problems are typical of many redeveloped docks throughout the UK (Hendry *et al.* 1988; Conlan, White & Hawkins 1992; Hawkins *et al.* 1992).

Redevelopment, water quality and ecology of the South Docks – a case study

The decline of the South Docks began after the Second World War, and by 1972 the docks were closed to commercial shipping. Subsequently the gates of the docks were left open, and silt deposited by the tides quickly built up to 10m deep in places. In 1981 the Merseyside Development Corporation was formed with a responsibility to develop the docks as a commercial project with funding from both public and private sectors. Dredging of the docks began in 1981, water was gradually replaced as dredging progressed, finishing in 1985 with replacement of the double lock gates at Brunswick river entrance. Thus, the South Docks became an interconnected chain of docks varying in depth from

3.5m to 6m (Figure 15.2). The water is brackish with salinity between 24‰ and 28‰; about three-quarters the salinity of the open Irish Sea. The South Dock area was redeveloped for a variety of uses. The historic warehouses hold offices, residential accommodation, the Tate Gallery North, the Merseyside Maritime Museum and shops. The Albert Dock is a major tourist attraction; there is a watersports centre in Queen's Dock and a marina in Coburg Dock. A new Customs and Excise building spans the Graving Dock at Queen's Dock. Developments are continuing with further recreational and housing projects planned in the vicinity of Princes Dock.

Severe water quality problems were recorded in the South Docks in the years following redevelopment, despite limited exchange of water with the Mersey Estuary. A series of projects supervised from the Universities of Liverpool and Manchester began in 1988 (Mincher 1988; Lonsdale 1990; Allen 1992; Wilkinson 1995; Zheng 1995; Fielding 1997). Throughout the first summer of monitoring severe problems were evident, similar to those recorded in previous years by the Merseyside Development Corporation (Allen 1992). Persistent dinoflagellate blooms coloured the water orange-brown and caused poor water clarity. The dominant species was *Prorocentrum minimum* which has been linked to paralytic shellfish poisoning outbreaks in other areas. Water clarity was found to be a good indicator of phytoplankton density as suspended sediment loads were very low due to conditions of little water movement (Figure 15.3). Dissolved oxygen levels above the sediment were often very low in deeper docks. The associated foul odours and occasional dying fish detracted from the appeal of the area. Oxygen depletion was attributed to decay of phytoplankton cells and a high demand from the organically rich sediments, coupled with a tendency for thermal stratification (Hawkins *et al.* 1992; Allen *et al.* 1992).

Remedial measures to improve water quality were attempted, based on previous experience at Sandon Dock, a north Liverpool dock used for a while as a fish and shellfish farm established by the Mersey Dock & Harbour Company (with advice from the Department of Zoology of the University of Liverpool). Here, between 1978 and 1983, artificial mixing was used successfully to eliminate anoxic conditions, and improvements in water clarity were attributed to the filtering effect of the large Common Mussel population growing on ropes and walls in the dock (Russell *et al.* 1983). In the South Docks, the Queen's Graving Dock was selected as an experimental site due to its semi-isolated nature. A helical type air lift mixer was deployed in this dock by Parkman Engineering to which we added a large population of mussels in mesh tubing suspended from a buoyed long-line (Allen *et al.* 1992; Allen & Hawkins 1993). Preliminary investigations of the walls and sediment throughout the South Docks at this time indicated that benthic communities were very poorly developed. However, in late summer and Autumn 1988, four months after the start of our project, a large natural settle-

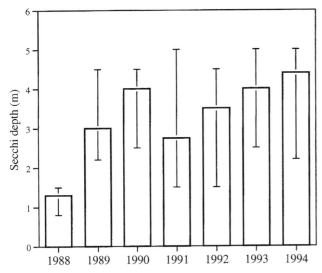

Figure 15.4. Improvements in water quality in the docks in Liverpool, as measured by Secchi disc extinction (i.e., depths at which a white disc lowered into the water disappears from view). Median ± ranges.

ment of mussels occurred in the South Docks. This may have been linked to much locking in and out of ships during the 'Tall Ships' race. Settlement was particularly dense in the Albert Dock, with mussels almost completely covering the dock wall. In the summers immediately following the introduction or settlement of mussels, marked improvements in water clarity were observed in both the Queen's Graving and Albert Docks (Figure 15.4). Bottom oxygen levels improved even in the Albert Dock, where no artificial mixing was initially used: with saturations of less than 20% being recorded for a maximum duration of almost three months in the summer of 1988, reducing to two to three weeks in 1989 and not occurring at all in 1990. Subsequently only very short periods of low bottom oxygen were recorded (e.g., in 1994) but oxygen levels were quickly raised using the simple mixing device installed in 1992 (Allen 1992; Zheng 1995; Wilkinson *et al.* 1996).

Water quality improvements were linked to a decline in the frequency and persistence of phytoplankton blooms, primarily of euglenoid and dinoflagellate species. The decline in phytoplankton blooms could not be explained by changes in nutrient availability, weather conditions or zooplankton grazing observed over the same period. Calculations of weight specific filtration rates for the mussel populations were made from published figures and from preliminary *in situ* measurements. It was estimated that the time taken for the Albert Dock mussels to filter a volume of water equal to that contained in the dock was in the order of one to three days (Allen & Hawkins 1993). It is thought likely, therefore, that the observed improvements in water quality were due, at least in part, to control of the phytoplankton populations by mussel filtration. Continued monitoring of water quality showed that improvements were sustained over a seven year period. During this time,

however, new recruitment of mussels on the walls was poor, and other filter feeding species became increasingly important, in particular a sea squirt *Ciona intestinalis*. In 1996 a new recruitment of mussels, which probably occurred in late 1995, became apparent (Fielding 1997). Therefore, while the relative abundance of species may fluctuate, the system appears to be sustainable with large numbers of filter feeders and good water quality.

The redevelopment of the South Docks and subsequent improvements in water quality have allowed colonisation by a relatively diverse estuarine/marine fauna. Once initial dredging of the South Docks was finished it is likely that the walls were affected by sediment, and the newly dredged bottom was devoid of benthic organisms. Since that time a gradual increase in the number of species recorded in the South Docks has been seen. By June 1988, one to five years after dredging was completed depending on the dock basin, encrusting bryozoans had become the major occupier of space on the walls with practically no other attached macrofauna present. The dense mussel settlement in Autumn 1988 provided a secondary substratum for a rich associated fauna (Allen, Wilkinson & Hawkins 1995). The Common Mussel very quickly became the dominant species and has remained so ever since. Recently, there has been an increase in the diversity and importance of other filter feeding species, particularly sea squirts (*Molgula manhattensis, Styela clava* and *Ciona intestinalis*) and the sponge *Halichondria bowerbanki*. The ephemeral nature of all the algal species so far recorded from the South Docks means that there has been no long-term dominance by this group (see chapter seventeen, this volume for details). Benthic fauna and flora are still largely confined to the wall. The majority of the species recorded from the sediment covered floor of the docks are short lived, and while longer lived species do colonise they apparently do not persist. Seventeen species of fish have been recorded from the South Docks to date, including a sea trout, *Salmo trutta* (Heaps 1988; Hawkins *et al.* 1993; Zheng 1995; Fielding 1997). The increased species diversity of the docks is reflected in the total species list which included over 90 species of macroflora and fauna by early 1994 (Allen, Wilkinson & Hawkins 1995).

New species are still being added to the list. Three of the species found in the South Docks may be regarded as lagoonal specialists (Barnes 1989). These are an amphipod *Corophium insidiosum*, the prawn *Palaemonetes varians* and a bryozoan *Conopeum seurati*. The potential of redeveloped docks as a resource in conservation of lagoonal and other species is an area warranting further investigation (Hawkins *et al.* 1993; Allen, Wilkinson & Hawkins 1995). The dockland areas on Merseyside represent a major habitat in the lower estuary, with the South Docks alone containing almost 30 hectares of water space.

Future prospects

Although Liverpool Bay and its estuaries are often considered as a highly polluted region, this chapter highlights the abundance of their marine life. Whilst not particularly diverse, in part due to a lack of habitat variety, the productivity of Liverpool Bay and its estuaries is high. They perform an important role as a nursery ground for Irish Sea juvenile fish and a winter feeding ground for birds from as far apart as Greenland and northern Russia. Various conservation designations testify to the recognition of the international importance of these feeding grounds (Taylor & Parker 1993; National Rivers Authority 1995; chapter seven, this volume).

In summary, following a concerted effort to improve the water quality there is a continuing recovery of the plant and animal populations which is reflected in an increasing biodiversity of the Mersey Estuary. Marine life *is* returning to the River Mersey. Broadscale and long term (20–30 years) improvements are, in part, an accidental by-product of de-industrialisation and depopulation of north-western England. They are also a response to environmental improvements throughout the region as part of the Mersey Basin Campaign. The neglect of the River Mersey since the Industrial Revolution was so extensive that a considerable time period seems likely before major recovery is evident. Complete recovery may never occur because of the amount of toxic material stored in waste tips and dumps lining the River Mersey and in the sediments themselves. The extensive reconfiguration of the estuary by engineering during development has inevitably led to irreversible change; but many of the newly created habitats provide opportunities for colonisation by a variety of species.

Locally, and on shorter timescales (<10 years), deliberate restoration of habitats is playing a role, coupled with natural advection of larvae and algal propagules from Liverpool Bay. Redevelopment of disused docks is ongoing in both Wirral and the North Docks. Experience from Sandon, the South Docks and Wirral waterfront has shown that water quality problems of severe phytoplankton blooms and anoxia are to be expected in newly redeveloped docks although remedial techniques are available (summarised in Allen & Hawkins 1993). Studies at Sandon and the South Docks have shown that a relatively diverse estuarine/marine flora and fauna can develop with time. Both the potential for biological diversity and good water quality could be increased if appropriate designs were considered at the planning stage. Such measures might include an increase in habitat diversity and the provision of suitable surfaces for colonisation by filter feeders, along with the use of water mixers where necessary.

Organisms can now live in the River Mersey and are not periodically killed by low oxygen. Ironically, organisms such as marine benthos and fish now survive long enough to accumulate pollutants: not only heavy metals (Leah *et al.* 1991; Leah, Evans & Johnson 1992) but also persistent organics (Thompson *et al.* 1996). Despite a possible risk to human health, the influence of these compounds on populations and communities is likely to be small. With tightening regulations, it is also unlikely that a major bird kill could occur again due to accumulation of pollutants, which pass unchanged up the food chain. Though by no means pristine, the River Mersey is recovering and beginning to match the biodiversity of the adjacent estuaries. The ecosystem is remarkably robust and has been enhanced by human provision of new habitats: coastal defences, forming artificial rocky shores and the 'lagoonoids' of the docks (Allen *et al.* 1995).

Acknowledgements

This work was mainly funded by the Merseyside Development Corporation with additional funding by the Department of Education Northern Ireland (SBW), NERC (NJF) and NMGM (IDW). Dr K.N. White, Dr H.D. Jones (University of Manchester) and Dr G. Russell (University of Liverpool) are thanked for their various inputs into the 'docks' project. Tony Tollitt, Jimmy McGill, Vanessa Wanstall, Wei Zhong Zheng and Gill Heyes are thanked for their help in the field.

References

Al-Masnad, F. (1991). *An assessment of the biological status of the Dee Estuary.* PhD thesis, University of Manchester.

Allen, J.R. (1992). *Hydrography, ecology and water quality management of the South Docks, Liverpool.* PhD thesis, University of Liverpool.

Allen, J.R., Hawkins, S.J., Russell, G.R. & White, K.N. (1992). Eutrophication and urban renewal: problems and perspectives for the management of disused docks. *Science of the Total Environment, Supplement 1992*, pp. 1283–95, Elsevier Science Publishers B.V., Amsterdam, Netherlands.

Allen J.R. & Hawkins S.J. (1993). Can biological filtration be used to improve water quality? *Urban Waterside Regeneration, problems and prospects* (eds K.N. White, E.G. Bellinger, A.J. Saul, M. Symer & K. Hendry), pp. 377–85. Ellis Horwood series in Environmental Management, Science and Technology, Ellis Horwood Press, Hemel Hempstead.

Allen, J.R., Wilkinson, S.B. & Hawkins, S.J. (1995). Redeveloped docks as artificial lagoons: the development of brackish-water communities and potential for conservation of lagoonal species. *Aquatic Conservation: Marine and Freshwater Ecosystems*, **168**, 299–309.

Bamber, R.N. (1988). *A survey of the intertidal soft-bottomed fauna of the Mersey Estuary.* March 1988, CEGB.

Barnes, R.S.K. (1989). The coastal lagoons of Britain: an overview and conservation appraisal. *Biological Conservation*, **49**, 295–313.

Bassindale, R. (1938). The intertidal fauna of the Mersey Estuary. *Journal of the Marine Biological Association of the U.K.*, **23**, 83–98.

Bennett, C., Curtis, M. & Fairhurst, C. (1995). Recent changes in the Mersey Estuary 1958–1988. *The Mersey Estuary – Naturally Ours* (eds M.S. Curtis, D. Norman & I.D. Wallace), pp. 12–23. Mersey Estuary Conservation Group, Warrington.

Bull, K.R., Every, W.J., Freestone, P., Hall, J.R. & Osborn, D. (1983). Alkyl lead pollution and bird mortalities on the Mersey Estuary 1979–1981. *Environmental Pollution*, **31 (Series A)**, 239–59.

Burd, A.C. (1947). Faunal notes on Hilbre. *Proceedings of the Liverpool Naturalists' Field Club for 1946*, 29–30.

Burt, A.J. (1989). An ecosystem approach to the Mersey barrage, Liverpool. *The Proceedings of the Mersey Barrage Symposium* (eds B. Jones & B. Norgain), pp. 17–26. The North of England Zoological Society, Chester.

Byerley, I. (1854). The fauna of Liverpool. *Proceedings of the Literary and Philosophical Society of Liverpool*, 8, Appendix.

Carter, J.J. (1985). *The influence of environmental contamination on the fauna of the Mersey Estuary*. MSc thesis, University of Manchester.

Clark, R.B. (1989). *Marine Pollution*. (2nd edn). Clarenden Press, Oxford.

Codling, I.D., Nixon, S.C. & Platt, H.M. (1991). *The Mersey Estuary. An assessment of the biological status of intertidal sediments*. Report to the National Rivers Authority. WRC report No. NR 2757, April 1991, 54 pp. Water Research Centre, Medmenham, Buckinghamshire.

Conlan, K., Hendry, K., White, K.N. & Hawkins, S.J. (1988). Disused docks as habitats for estuarine fish: a case study of Preston dock. *Journal of Fish Biology*, 33 (Supplement A), 85–91.

Conlan, K., White, K.N. & Hawkins, S.J. (1992). The hydrography and ecology of a re-developed brackish-water dock. *Estuarine, Coastal and Shelf Science*, 35, 435–52.

Corlett, J. (1948). Rates of settlement and growth of the 'pile' fauna of the Mersey Estuary. *Proceedings and Transactions of the Liverpool Biological Society*, 56, 2–25.

Collings, S.E., Johnson, M.S. & Leah, R.T. (1996). Metal contamination of angler-caught fish from the Mersey Estuary. *Marine Environmental Research*, 41, 281–97.

Craggs, J.D. (ed.) (1982). *Hilbre the Cheshire Island its history and natural history*. Liverpool University Press, Liverpool.

Cunningham, J.T. (1898). *The natural history of the marketable marine fishes of the British Isles*. Macmillan and Co., London.

Curtis, M.S., Norman, D. & Wallace, I.D. (eds) (1995). *The Mersey Estuary – Naturally Ours*. Mersey Estuary Conservation Group, Warrington.

Darby, D., Lawrence, S., Wolff, G. & Hawkins, S. (1994). *Environmental Assessment for blocks 110/13 and 110/15, Liverpool Bay*. Report commissioned by Hamilton Oil Company, London.

Davies, J. (1992). *Littoral survey of the Ribble, Duddon and Ravenglass Estuary systems, east basin of the Irish Sea*. Joint Nature Conservation Committee report, No. 37. (Marine Nature Conservation Review Report, No. MNCR/SR/21.)

Davies, L.M. (1991). *Littoral survey of the coast from Crosby to Fleetwood*. Nature Conservancy Council, CSD report No. MNCR/SR/17.

Dempsey, C.H. (1989). Implications of a barrage for present and future fisheries in the Mersey Estuary. *The Mersey Barrage. The Proceedings of the Mersey Barrage Symposium* (eds B. Jones & B. Norgain), pp. 72–91 The North of England Zoological Society, Chester.

Dent, D. (1986). *A survey of the mussel beds on the Ribble Estuary at Lytham*. BSc thesis, Department of Zoology, University of Manchester. (Copy at Department of Zoology, Liverpool Museum.)

Department of the Environment (1972). *Out of Sight, Out of Mind. Report of the Department of the Environment Working Party on sludge disposal in Liverpool Bay*. HMSO, London.

Dickson, R.R. & Boelens, R.G.V. (1988). *The status of current knowledge on anthropogenic influences in the Irish Sea*. I.C.E.S. Co-op Rep. No. 155, International Council for Exploration of the Sea, Copenhagen, Denmark.

Doody, J.P. (ed.) (1984). Spartina anglica *in Great Britain. A report of a meeting held at Liverpool University on 10th November 1982*. NCC. (Focus on marine nature conservation, No. 5), Peterborough.

Dunlop, G.A. (1927). Early Warrington fisheries: an historical sketch. *Proceedings of the Warrington Literary and Philosophical Society*, 1927–1929, 7–20.

Environmental Advisory Unit of Liverpool University Ltd (1991). *The Mersey oil spill project 1989–90. A summary report of the studies undertaken into the long term environmental impacts of the August 1989 oil-spill into the Mersey Estuary*. Mersey Oil Spill Advisory Group, Liverpool.

Environmental Resources Limited (1993). *Hamilton and Hamilton oil north gas platforms, block 110/13 Liverpool Bay: environmental assessment*. ERL, London.

Fielding, N.J. (1997). *Fish and benthos communities in regenerated dock systems on Merseyside*. PhD thesis, University of Liverpool.

Fraser, J.H. (1931). *The fauna and flora of the Mersey Estuary with special reference to pollution and sedimentary deposits*. MSc thesis, University of Liverpool.

Fraser, J.H. (1932). Observations on the fauna and constituents of an estuarine mud in a polluted area. *Journal of the Marine Biological Association of the UK.*, 18, 69–85.

Fraser, J.H. (1935). The fauna of the Liverpool Bay shrimping grounds and the Morecambe Bay spawning grounds as revealed by the use of a beam-trawl. *Proceedings and Transactions of the Liverpool Biological Society*, 48, 65–78.

Fraser, J.H. (1938). The fauna of fixed and floating structures in the Mersey Estuary and Liverpool Bay. *Proceedings and Transactions of the Liverpool Biological Society*, 51, 1–21.

Gardiner, A.P. (1950). Recorder's report: marine Mollusca. *Journal of Conchology*, 23, 124–26.

Garwood, P. & Foster-Smith, R. (1991). *Intertidal from Rhos Point to New Brighton*. Report contracted by the University of Newcastle and Dove Marine Laboratory, Cullercoats. Nature Conservancy Council, CSD Report No. 1194.

Gillham, R.M. (1978). *An ecological investigation of the intertidal benthic invertebrates of the Dee Estuary*. PhD thesis, University of Salford.

Ghose, R.B. (1979). *An ecological investigation into the invertebrates of the Mersey Estuary*. PhD thesis, University of Salford.

Hall-Spencer, J. (1989). Pipeline leak into the Mersey. *Marine Pollution Bulletin*, 20, 480.

Hardy, E. (1995). An introduction to the natural history of the Mersey Estuary. *The Mersey Estuary – Naturally Ours* (eds M.S. Curtis, D. Norman & I.D. Wallace), pp. 24–32. Mersey Estuary Conservation Group, Warrington.

Hartnoll, R.G. (1993). Shallow subtidal hard substrata. *The coast of North Wales and North West England, an environmental appraisal* (eds P.M. Taylor & J.G. Parker), pp. 42–43. Hamilton Oil Company, London.

Harvey-Gibson, R.J. (1890). Report on the marine algae of the LMBC district. *Proceedings and Transactions of the Liverpool Biological Society*, 3, 128–54.

Harvey-Gibson, R.J. (1891). A revised list of the marine algae of the LMBC district. *Proceedings and Transactions of the Liverpool Biological Society*, 5, 83–142.

Hawkins, S.J. (1993). Coastal habitats, communities and species. *The coast of North Wales and North West England: an environmental appraisal* (eds P.M. Taylor & J.G. Parker), pp. 19–24, Hamilton Oil Company, London.

Hawkins, S.J., Allen, J.R., Russell, G., White, K.N., Conlan, K., Hendry, K. & Jones, H.D. (1992). Restoring and managing disused docks in inner city areas. *Restoring the Nation's Marine Environment* (ed. G.W. Thayer), pp. 473–542. National Oceanic and Atmospheric Administration, Maryland Sea Grant College publication, Maryland, USA.

Hawkins, S.J., Allen, J.R., Russell, G., Eaton, J.W., Wallace, I., Jones, H.D., White, K.N. & Hendry, K. (1993). Former commercial docks as a resource in urban conservation and education. *Urban Waterside Regeneration, problems and prospects* (eds K.N. White, E.G. Bellinger, A.J. Saul, M. Symer & K. Hendry), pp. 386–99. Ellis Horwood series in Environmental Management, Science and Technology, Ellis Horwood, Press, Hemel Hempstead.

Hayward, P.J. & Ryland, J.S. (1995). *Handbook of the Marine*

Fauna of North-West Europe. Oxford University Press, Oxford.

Head, P.C., D'arcy, B.J. & Osbaldeston, P.J. (1980). *The Mersey Estuary bird mortality.* Autumn-winter 1979, preliminary report. Scientific Report DSS-EST-80-1, North West Water.

Heaps, L. (1988). *The fish populations of the South Docks, Liverpool.* BSc. Honours project, Department of Environmental Biology, University of Manchester. (Copy at University of Manchester Library.)

Henderson, P.A. (1988). The structure of estuarine fish communities. *Journal of Fish Biology,* **33 (Supplement A)**, 223–25.

Herdman, W.A. (1920). Summary of the history and work of the Liverpool Marine Biology committee. *Proceedings and Transactions of the Liverpool Biological Society,* **34**, 23–74.

Hendry, K. (1993). Former commercial docks as a resource in urban conservation and education. *Urban Waterside Regeneration, problems and prospects* (eds K.N. White, E.G. Bellinger, A.J. Saul, M. Symer & K. Hendry), pp. 386–99. Ellis Horwood series in Environmental Management, Science and Technology, Ellis Horwood Press, Hemel Hempstead.

Hendry, K., White, K.N., Conlan, K., Jones, H.D., Bewsher, A.D., Proudlove, G.S., Porteous, G., Bellinger, E.G. & Hawkins, S.J. (1988). *Investigations into the ecology and potential use for nature conservation of disused docks.* (Contractor: Department of Environmental Biology, University of Manchester.) Nature Conservancy Council, CSD Report No. 848.

Holland, D. (1989). 'Alive and Kicking'. The fish and invertebrates of the Mersey Estuary. *Proceedings of the Mersey Barrage Symposium* (eds B. Jones & B. Norgain), pp. 42–63. North of England Zoological Society, Chester.

Holland, D.G. & Harding, J.P.C. (1984). (ed. B.A. Whitton), pp. 113–44, *Ecology of European Rivers.*

The Irish Sea Study Group (1990). *The Irish Sea. An Environmental Review. Introduction and Overview.* Liverpool University Press, Liverpool.

James, C.J. & Gibson, R. (1980). The distribution of the polychaete *Capitella capitata* (Fabricus) in dock sediments. *Estuarine and Coastal Marine Science,* **10**, 671–83.

Jones, P.G.W. & Head, P.C. (1991). Turning the tide on pollution – the Mersey Estuary. *Proceedings of the International conference on environmental pollution (1),* Lisbon, Portugal.

Law, R.J., Jones, B.R., Baker, J.R., Kennedy, S., Milne, R. & Morris, R.J. (1992). Trace metals in the livers of marine mammals from the Welsh coast and Irish Sea. *Marine Pollution Bulletin,* **24**, 296–304.

Leah, R.T., Evans, S.T., Johnson, M.S. & Collings, R.T. (1991). Spatial patterns in accumulation of Hg by fish from the north-east Irish Sea. *Marine Pollution Bulletin,* **22**, 172–75.

Leah, R.T., Evans, S.J. & Johnson, M.S. (1992). Arsenic in plaice (*Pleuronectes platessa*) and whiting (*Merlangius merlangus*) from the north east Irish Sea. *Marine Pollution Bulletin,* **24**, 544–49.

Leah, R.T., Johnson, M.S., Connor, L. & Levence, C. (1997). Polychlorinated biphenyls in fish and shellfish from the Mersey Estuary and Liverpool Bay. *Marine Environmental Research,* **43**, 345–58.

Liverpool Bay Study Group (1975). *Liverpool Bay. An assessment of present knowledge.* The Natural Environment Research Council publications series 'C' No. 14.

Lonsdale, K. (1990). *The hydrology and ecology of the Albert Dock complex, Liverpool with particular reference to fish populations.* MSc thesis, Department of Environmental Biology, University of Manchester.

McMillan, N.F. (1975). "Wrecks" of marine invertebrates in the Liverpool Area 1946–1975. *Lancashire and Cheshire Fauna Society Publication,* **67**, 21–22.

Meng, F.H. (1976). *A pollution study of River Mersey by using natural algal communities and laboratory bioassay.* MSc thesis, University of Liverpool.

Mersey & Weaver River Authority (1972). *Seventh Annual Report.*

Mills, D.J.L. (1991). *Marine Nature Conservation Review. Benthic marine ecosystems in Great Britain: a review of current knowledge. Cardigan Bay, North Wales, Liverpool Bay and the Solway (MNCR coastal sectors 10 and 11).* Nature Conservancy Council, Peterborough.

Mincher, P.T. (1988). *The hydrography and ecology of the Liverpool docks.* MSc thesis, University of Manchester.

Moore, D.M. (1978). Seasonal changes in distribution of intertidal macrofauna in the lower Mersey Estuary. *Estuarine and Coastal Marine Science,* **7**, 117–25.

National Rivers Authority (1995). *The Mersey Estuary. A report on environmental quality.* Water quality series No. 23. HMSO, London.

Norton, M.G., Rowlatt, S.M. & Nunny, R.S. (1984). Sewage sludge dumping and contamination of Liverpool Bay sediments. *Estuarine, Coastal and Shelf Science,* **19**, 69–87.

Osborn, D., Every, W.J. & Bull, K.R. (1983). The toxicity of trialkyl lead compounds to birds. *Environmental Pollution,* **31 (Series A)**, 261–75.

Perkins, E.J. (1956). The fauna of a sandbank in the mouth of the Dee Estuary. *Annals and Magazine of Natural History.* **9, Series 12**, 112–28.

Pierce, E.L. (1941). The occurrence and breeding of *Sagitta elegans* Verill and *Sagitta setosa* J.Muller in parts of the Irish Sea. *Journal of the Marine Biological Association of the U.K.,* **25**, 113–24.

Popham, E.J. (1966). The littoral fauna of the Ribble Estuary, Lancashire, England. *Oikos,* **17**, 19–32.

Porter, E. (1973). *Pollution in four industrial estuaries. Four case studies for the Royal Commission on Environmental Pollution.* HMSO, London.

Readman, J.W., Preston, M.R. & Montoura, R.F.C. (1986). An investigated technique to quantify sewage, oil and PAH pollution in estuaries and coastal environments. *Marine Pollution Bulletin,* **17(7)**, 298–308.

Rees, E.I.S. (1975). Benthic and littoral fauna of Liverpool Bay. *Liverpool Bay An assessment of present knowledge.* (Compiled by members of the Liverpool Bay study group.) The Natural Environment Research Council publications series 'C' No. 14.

Rees, E.I.S., Nicholaidou, A. & Laskaridou, P. (1977). The effects of storms on the dynamics of shallow water benthic associations. *Proceedings of the 11th European Marine Biology Symposium* (eds B.F. Keegan, P. O'Ceidigh & P.J.S. Boaden), pp. 265–474. Pergamon Press, London.

Rees H.L., Rowlatt, S.M., Limpenny, D.S., Rees, E.I.S. & Rolfe, M.S. (1992). *Benthic studies at dredged material sites in Liverpool Bay.* Ministry of Agriculture, Fisheries and Food Directorate of Fisheries Research. Aquatic Environment Monitoring report, Number 28.

Rice, K.A. & Putwain, P.D. (1987). *The Dee and Mersey Estuaries: environmental background.* Shell UK Limited, London.

Riley, J.P. & Towner, J.V. (1984). The distribution of alkyl lead species in the Mersey Estuary. *Marine Pollution Bulletin,* **15, (4)**, 153–58.

Ritchie-Noakes, N. (1984). *Liverpool's historic waterfront – The world's first mercantile dock system.* HMSO, London.

Russell, G. (1972). Phytosociological studies on a two-zone shore. I. Basic pattern. *Journal of Ecology,* **60**, 539–45.

Russell, G. (1973). Phytosociological studies on a two-zone shore. II. Community structure. *Journal of Ecology,* **61**, 525–36.

Russell, G. (1982). The Marine Algae. *Hilbre. The Cheshire Island: its history and natural history* (ed. J.D. Craggs), pp. 65–74. Liverpool University Press, Liverpool.

Russell, G., Hawkins, S.J., Evans, L.C., Jones, H.D. & Holmes, G.D. (1983). Restoration of a disused dock basin as a habitat for marine benthos and fish. *Journal of Applied Ecology* **20**, 43–58.

Stopford, S.C. (1951). An ecological survey of the Cheshire foreshore of the Dee Estuary. *Journal of Animal Ecology*, **20**, 103–22.

Taylor, P.M. (1991). A pipeline spill into the Mersey Estuary, England. *Proceedings 1991 International Oil Spill Conference*, pp. 399–405. American Petroleum Institute, Washington DC, USA.

Taylor, P.M. & Parker, J.G. (1993). *The Coast of North Wales & North West England. An Environmental Appraisal.* Hamilton Oil Company Ltd., London.

Thomason, G. & Norman, D. (1995). Wildfowl and waders of the Mersey Estuary. *The Mersey Estuary – Naturally Ours* (eds M.S. Curtis, D. Norman & I.D. Wallace), pp. 33–59. Mersey Estuary Conservation Group, Warrington.

Thompson, A., Allen, J.R., Dodoo, D., Hunter, J., Hawkins, S.J. & Wolff, G.A. (1996). Distributions of chlorinated biphenyls in mussels and sediments from Great Britain and the Irish Sea Coast. *Marine Pollution Bulletin*, **32(2)**, 232–37.

Wilkinson, S.B. (1995). *The ecology of benthos in Liverpool Docks.* PhD thesis, University of Liverpool.

Wilkinson, S.B., Zheng, W., Allen, J.R., Fielding, N.J., Wanstall, V.C., Russell, G. & Hawkins, S.J. (1996). Water quality improvements in Liverpool Docks: the role of filter feeders in algal and nutrient dynamics. *P.S.Z.N.I. Marine Ecology*, **17(1–3)**, 197–211.

Wilson, K.W., D'arcy, B.J. & Taylor, S. (1988). The return of fish to the Mersey Estuary. *Journal of Fish Biology*, **33 (Supplement A)**, 235–38.

Wood-Griffiths, L. (1993). *Mersey Estuary management plan. Report on nature conservation and water pollution.* Department of Civic Design, University of Liverpool, Liverpool.

Zheng, W. (1995). *Water quality problems of the Liverpool docks system in relation to the adjacent estuary.* PhD thesis, University of Liverpool.

CHAPTER SIXTEEN

Saltmarshes and sand dunes – natural or not

J.P. DOODY

Introduction

Saltmarshes and sand dunes are described in most ecological text books as exhibiting primary succession. Coastal habitats are also often described as being amongst the most natural, and saltmarshes and sand dunes are the most frequently quoted examples. The classic studies of the saltmarshes and sand dunes on the north Norfolk coast (for example Chapman 1938, 1941, 1959; Steers 1960) or the Dovey Estuary in western Wales (Yapp, Johns & Jones 1917) emphasised the natural status of the vegetation. Indeed these studies looked at systems where the ecological processes appeared to have been virtually free from direct human influence.

The results of these and other studies often describe the vegetation as a series of types progressing from the early pioneer stages to more complex forms, which are related to the physical parameters affecting their development. In the case of saltmarshes and sand dunes this includes tides, waves and winds, sediment and soil characteristics. This has led to the impression that the process of succession takes place in a sequence which is determined exclusively by natural forces. Selection of areas for conservation designation frequently uses the existence of recognised vegetation patterns based on the understanding of this 'ecological succession'. Subsequent management often seeks to maintain this pattern in the face of change. This chapter examines the evidence from the natural world for the existence of natural habitats and asks if the influence of man has reached such a point that natural habitats (and nature reserves) in north-western Europe can only be sustained with direct human intervention.

Saltmarshes and sand dunes – as ecological systems

That there are zonations which can be attributed to environmental gradients in coastal vegetation is not in dispute. There are numerous examples of primary succession where specialised plant species adapted to the rigors of their environment, provide the main agents in stabilisation. Anyone looking at the early stages in saltmarsh or sand dune growth will need little convincing that there is a natural pattern to the mosaic of vegetation, which can be related to the sequence of development.

Accretion of sediment on the tidal flat surface occurs around primary colonisers of mudflats such as Common Glasswort (*Salicornia europea*) or Annual Sea-blite (*Suaeda maritima*) in north-western Europe. In saltmarshes these pioneers, in addition to being tolerant of immersion in sea water, depend on periods during the early stages in plant establishment when they are free from tidal movement. As sediment height increases the marsh is subject to progressively fewer tidal inundations, less sediment is deposited and a richer complement of plants and animals replaces the specialist salt tolerant species (Ranwell 1972a). Thus from an ecological point of view saltmarshes exhibit primary succession and a parallel spatial zonation related to tidal inundation, 'largely without human interference' which makes them ideal for studying the processes associated with 'natural' vegetation development (Gray 1992).

In the early stages of dune growth there are similarly a few species which can tolerate inundation with sea water or spray, rapid changes in rates of sand deposition, desiccation and exposure. The sometimes spectacular development of foredunes is directly related to the ability of plants such as Sand Couch (*Elytrigia juncea*) and Marram (*Ammophila arenaria*) to withstand burial by sand, together with the other stresses present in this unstable and inhospitable environment (Ranwell 1972a). In both cases the resulting succession is often described as a zonation which reflects the physical parameters influencing its development and shown in text books as a linear profile through the saltmarsh or sand dune.

Saltmarshes and sand dunes – as complex systems

Saltmarshes are much more complex systems than the straightforward succession described above appears to suggest. Accretion does result in more diverse forms of vegetation and associated animal species, and transitions to terrestrial vegetation can be the most complex. However, this apparently simple picture hides a complex relationship. Sea-level change may introduce long-term patterns which are difficult to separate from sequences of erosion and regrowth as estuary channels move. A series of steps can form as new saltmarsh develops to seaward of the eroding cliffed saltmarsh edge, as occurred in the Mersey Estuary before the building of the Manchester Ship Canal (Bennett, Curtis & Fairhurst 1995) or on the Solway in south-western Scotland (Marshall 1962). Salt-pans represent another element in the complex mosaic, and deposits of seaweed on the tide-line may smother the surface vegetation creating further spatial variation as the strandline deposits rot.

Sand dunes also show forms of successional development in the early stages of growth by the accretion of sand, aided by specialist plants. However, once the main body of the dune is formed, other processes come into play and the change from mobile foredunes and yellow dunes to grassland, heath, scrub and woodland is rarely a straightforward progression. Blowouts occur with or without the intervention of man and can be the precursors of dune slacks (Ranwell 1972a). Similarly the reprofiling of dune ridges under the influence of changing wind patterns brings an infinitely variable topography, the origins of which may be difficult or impossible to unravel.

Against this background of change associated with the physical characteristics of the habitat, the most common use of both saltmarshes and dunes is by grazing animals. Herbivores such as ducks and geese, on saltmarsh, and Rabbits (*Oryctolagus cuniculus*) on sand dunes, probably grazed both habitats before their extensive use by domestic stock. In addition to the direct effect on the nature of the vegetation, they also introduce structural change in the habitat through physical impacts such as trampling and burrowing.

Grazing occurs extensively on saltmarshes in Great Britain (especially in the west and north) and throughout the rest of north-western Europe (Dijkema 1984). It has a major effect on the structure and species composition of the marsh. In general, as grazing intensity increases, there is a loss of structural diversity as the standing crop is removed. At the same time grazing-sensitive species are removed from the sward, reducing species diversity. The loss of structural diversity also reduces the range and diversity of invertebrates and breeding bird populations (Figure 16.1). However, as the sward becomes shorter and dominated by tillering grasses (*Puccinellia maritima* and *Festuca rubra*), it is favoured by grazing ducks and geese. These effects are more pronounced as other, human activities associated with 'improving' grazing may cause further change. On saltmarshes these may include drainage and erection of summer dykes or banks.

On calcareous sand dunes grazing has helped to create a rich flora and fauna similar to that of calcareous grasslands. Continued grazing is necessary for the survival of the rich plant communities and helps prevent the growth of course grasses and scrub. On acid dunes, where heath is dominated by Heather (*Calluna vulgaris*), Crowberry (*Empetrum nigrum*) or other heathers, lower levels of grazing may also be important for maintaining the status quo, and preventing scrub development at the expense of the heath. In the past overgrazing on the dunes by cattle and sheep, particularly when combined with burrowing Rabbits, can lead to unstable conditions and eventually, large-scale erosion. Although this has resulted in many dunes where erosion control is a major consideration (Ranwell & Boar 1986) it can introduce further patterns into the vegetation as recolonisation takes place following removal of the destabilising vector.

Other human influences – the history of intervention in UK and Europe

As has already been intimated above, coastal habitats are

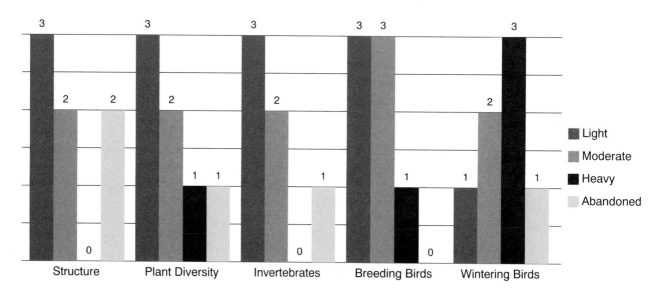

Figure 16.1. Summary of nature conservation interest in relation to grazing levels on saltmarshes in north-west Europe.

0–3 indicates level of nature conservation interest for each component of the system in relation to grazing pressure:
0 – little or no interest; 1 – minor interest; 2 – moderate interest; 3 – major and significant interest.

often considered amongst the most natural ecosystems. However, in addition to their extensive use by grazing animals, they have also been directly modified by human activities over many years. On the sedimentary shorelines of the Wash, artificial embankments to enable salt making, were present in some numbers in Roman times (Simmons 1980). Hay making, oyster cultivation, turf and reed cutting, and samphire gathering all take place or have taken place on upper marshes throughout Europe (Dijkema 1984). The deliberate planting of Common Cord-grass (*Spartina anglica*) – itself a hybrid fashioned from the interaction of a native and an introduced species – (Marchant 1967) has also been a major influence.

Occupation of sand dunes probably dates back several thousand years, and many sites have examples of middens with the remains of shellfish on them. An analysis of flint tools suggests that settlers of Torrs Warren, in south-western Scotland, may have appeared between 5,500–7,000 years ago (Coles 1964). It seems that cultivation has taken place since then, and in 1572 three farm houses were recorded on the Warren. As long as 4,500 years ago a small settlement existed on the shore of Skaill Bay in Orkney. The Neolithic people who inhabited the site, known today as 'Skara Brae', seem to have been agriculturists as well as hunter/gatherers (Ritchie & Ritchie 1978) until their village was overwhelmed by a sand storm. Since Medieval times dunes were used extensively as rabbit warrens (Thompson & Worden 1956), and are grazed by domestic stock to the present day.

More recent human activity has included major direct loss of habitat and the impact of this is considered in more detail in relation to the saltmarshes and sand dunes of the Mersey Basin.

The Dee, Mersey and Ribble Estuaries

Enclosure of the upper Dee Estuary began around 1730 and continued up to 1986 (Figure 16.2a). Much of the land gained was at the expense of the saltmarsh and was subsequently developed for industry, housing, roads and agriculture including grazing marsh. It is estimated that approximately 4,600ha were enclosed between 1737 and 1877 (Rice & Putwain 1987) for residential land, agriculture and industry. This gradual reduction of the size of the estuary was followed by an extension of saltmarsh. This is thought to be partly due to the reduction in tidal volume, which in turn has reduced scour and helped accelerate the natural sedimentation tendency within the estuary. This process was aided by the introduction of Common Cord-grass around 1930, and since then some 1500ha of intertidal mud was naturally 'reclaimed'. Marker (1962) estimated that at Parkgate between 1947 and 1963 vertical accretion of the *Spartina* marsh took place at a rate of approximately 25mm per year.

In the Mersey Estuary much of the tidal land around Frodsham was cut off from the main river by the building of the Manchester Ship Canal between 1887 and 1893. The land is now used for industry and agriculture. This not only had a profound effect on the shoreline of the south side of the estuary, but it also influenced the way in which the estuary channels behaved and their effect on the balance between erosion and accretion of saltmarsh (Bennett *et al.* 1995). When taken together with the major port and other infrastructure developments, the shoreline of the estuary has a high proportion of its waterfront composed of vertical sea-walls. Transitions to non-tidal habitats are virtually non-existent.

Land claim on the Ribble Estuary is less well documented, though the progress of the more recent loss of tidal land for agricultural use has been studied (Figure 16.3). Enclosure for industry and agriculture took place

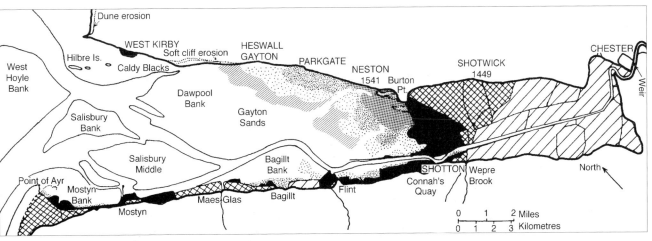

Figure 16.2. (a) Enclosure of the Dee Estuary 1730–1986;
(b) saltmarsh accretion 1910–1986.

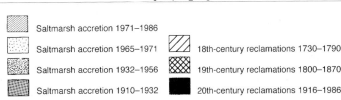

Saltmarsh accretion 1971–1986	
Saltmarsh accretion 1965–1971	18th-century reclamations 1730–1790
Saltmarsh accretion 1932–1956	19th-century reclamations 1800–1870
Saltmarsh accretion 1910–1932	20th-century reclamations 1916–1986

during the last century. An increase in siltation and salt-marsh growth is attributed to the development of industry and associated control of the river flows including the training of the Ribble Channel in the 1840s (Berry 1967; Robinson 1984). As in the Dee Estuary the growth of the saltmarsh appears to have been aided by the planting of Common Cord-grass, in this case in 1932. Not only did this fundamentally change the nature of the saltmarsh vegetation, but it also created concerns for amenity interests, as the grass expanded onto the Southport beaches (Truscott 1984). The extent of salt-marsh loss in the major estuaries of the north-west before 1911 is shown in Table 16.1.

Further enclosures have taken place since then including 320ha within the Ribble Estuary in the 1980s. Although not dealt with in detail here, other human uses of these estuaries have had a further effect on their status for both economic and cultural use, and nature conservation. Pollution, by a variety of heavy metals, sewage and radioactive discharges together with oxygen depletion, has affected fisheries and been linked to the death of bird populations in the past, particularly in the Mersey Estuary.

The Sefton coast

The original area of blown sand along the Sefton coast is amongst the most extensive in Great Britain. The original dune landscape may have extended for approxi-

Site name	Area (ha) enclosed	Source
Dee Estuary	3,160 (by 1857)	(Royal Commission on Coastal Erosion and Afforestation 1911)
	4,600 (1737–1877)	(Rice & Putwain 1987)
Mersey	492 (19th century)	(Royal Commission on Coastal Erosion and Afforestation 1911)
The Ribble	1,960 (19th century)	(Buck 1993)

Table 16.1. Loss of estuarine habitat.

mately 3,000ha, though today the area of vegetated dune is much smaller, probably about 2,113ha (all dunes in Merseyside). In Medieval times uncontrolled use of the dunes for grazing, use of Marram for thatch and cattle bedding lead to massive dune erosion. Greater control of the use of the dune, including prohibition of Marram cutting and the requirement for tenants to spend up to 6 days per year planting the species helped reinstate a more stable dune. By the early 1800s what was considered to be a more acceptable dune landscape had been created which, on the face of it, appeared largely unaffected by human activity. Things began to change again, and in the early 1900s further dune stabilisation, Asparagus (*Asparagus officinalis* ssp. *officinalis*) planting and afforestation all influenced the development of the

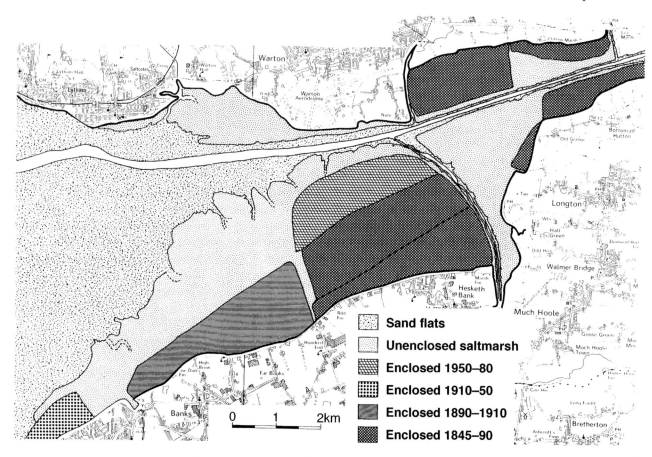

Sand flats
Unenclosed saltmarsh
Enclosed 1950–80
Enclosed 1910–50
Enclosed 1890–1910
Enclosed 1845–90

Figure 16.3. Saltmarsh enclosure in the Ribble Estuary, 1845–1980.

dune. It was during this period that Formby Point began to erode, following a period between 1845 and 1906 when it grew seawards under the influence of active management. The advent of tourism, promoted further stabilisation, including the introduction of Sea-buckthorn (*Hippophae rhamnoides*), and these and other influences such as the building of a golf course, rifle range and sand extraction reduced the area of open dune (Atkinson & Houston 1993).

The evidence, therefore, from the Mersey Basin and elsewhere shows that both saltmarshes and sand dunes have been influenced by human activity for centuries. This has ranged from total destruction of the habitat, to modification of natural forms and the creation of areas which appear natural but are largely the result of human intervention. Understanding the extent of human influence is a key element in assessing the value of habitats and sites for conservation purposes, and is crucial to management decisions.

Nature conservation and human activity

The identification of saltmarsh and sand dunes as important examples of particular types of habitats is based on the notion that there is a classification which can be recognised in the field, and that this can provide a basis for assessment. In Britain the early pioneers of the conservation movement, notably Tansley, described the vegetation of the British Isles in terms of zonations which for saltmarshes at least 'correspond, in a general way, with the saltmarsh succession' (Tansley 1953). His exposition of the conservation needs as expressed in 'Our heritage of wild nature' (Tansley 1945) continued the tradition of viewing the special habitat and species of importance to conservation to equate with those which the ecologist considered 'natural' or 'semi-natural'. In themselves these terms have come to mean something which suggests little or no human influence. Ratcliffe (1977) in his description of the assessment of the key conservation areas in Great Britain ascribes great importance to 'naturalness' and 'typicalness' as primary criteria for site selection. However this publication and the subsequent *Guidelines* for the selection of biological SSSIs (Nature Conservancy Council 1989) stress the fact that few if any habitats are really natural. Those which may be 'most natural' are confined to 'high mountains… and certain coastal features' and that up to 30% of the total is modified in some way and classed as '*semi-natural*'. However the selection of sites must 'satisfy a certain level of quality marked by lack of features which indicate gross or recent human modification'.

Thus, whilst recognising the influence of human activity on most habitats and species, the selection and subsequent management of many sites, particularly saltmarshes and sand dunes, has been concerned primarily with the conservation of the classic sequence of vegetation and the associated animals. Much activity has centred around the protection of the features for which

the sites, including nature reserves, were established. This is manifested in the importance attached to arresting succession particularly where the invasion of 'non-native' species such as Common Cord-grass (Doody 1984; Gray & Benham 1990) on mudflats and Sea-buckthorn (Ranwell 1972b) on dunes. In certain situations, notably on sand dunes, this approach may result in other potentially more damaging losses as, for example, when scrub invades open species rich dune vegetation (Doody 1989). This approach, which is based on the relatively narrow view of succession may also reduce the number of options considered by the conservation manager for the site (Doody 1993).

A key question from a nature conservation point of view therefore relates to whether saltmarshes and sand dunes are really 'natural' systems exhibiting inherent characteristics which can be used as a basis for determining management options. In particular does the spatial mosaic of vegetation used to assess the site for conservation purposes give an adequate representation of a successional sequence? In the situations where invasive, non-native plants and animals threaten the survival of existing habitats, such as the two cases mentioned above the answer may be obvious. However, most examples of saltmarsh and sand dune successions do not proceed in an orderly fashion from one stage to the next, and it is not possible to infer a successional relationship from the spatial zonation. Thus, in defining management options, much more complex patterns of change may need to be interpreted. At the same time, adhering to the notion that these systems have some inherent naturalness which can be considered in isolation from human influence, is clearly not sustainable.

Conclusion – are saltmarshes and sand dunes natural or man-made?

The view that the ecological or geomorphological development of coastal habitats is 'natural' must be tempered by the recognition that most, if not all examples, have been modified in some way by human activity. Not only are most coastal systems influenced by activities taking place in the hinterland which affect water quality and sediment movement, but also other habitats (salinas, in the Mediterranean and coastal grazing marsh, in south-eastern England, for example) have been created by past human use. Throughout Europe individual habitats have been modified by different levels of human activity. It is almost certain that the open areas of upper saltmarsh would have provided rich pasturage in the winter months, and the early settlers in America cut hay on saltmarshes, a practice borrowed from north-western Europe. Grazing and turf cutting continue to be practised today, especially in north-western England, and reed cutting for thatch occurs in the Tay Estuary in eastern Scotland. Almost all dunes have been used to graze domestic stock, and about 8,000 years BP sheep herding was prevalent in many areas of the Mediterranean. Sheep and goats continue to influence some of the remaining undeveloped dunes in the eastern Mediterranean, notably in Turkey. Historically there are also numerous examples of large-scale destabilisation due to overuse from other activities such as over-grazing and burning, and Marram grass cutting for bedding and thatch, which have resulted in major sand movement, e.g., in the Outer Hebrides (Angus & Elliot 1992), Denmark (Skarregaard 1989) and the National Park of Doñana (Garcia Nova 1979).

In considering, therefore, whether saltmarshes and sand dunes are really natural systems it is important to understand the extent of man's impact upon their development. In one sense, as has been described above, there are probably no saltmarshes or sand dunes in Europe which are entirely 'natural', i.e., have not in some way been influenced by human activity. Even the extensive unmodified lagoons and deltas with their apparently natural saltmarshes and sand dunes in Albania and eastern Turkey, grow rapidly today because of deforestation in the hinterland. Even here drainage and land 'improvement' for agriculture has destroyed large areas of transitional vegetation, and pollution is a major consideration. On closer inspection even the most 'natural' examples of succession may have been initiated by human activity. The growth of the 5km-long dune spit of Bull Island, Dublin Bay is directly attributable to the construction of a breakwater into the Bay between 1819 and 1823 (Jeffrey 1977). *Perhaps the best that can be said is that in the early stages of development saltmarshes and sand dunes exhibit characteristics of natural succession.*

This conclusion may have important consequences both for the development of conservation policy and the ecological principles upon which it has been based. Agenda 21, a programme for action agreed at the United Nations, Earth Summit in Rio in 1992 aims to help achieve the twin goals of sustainable human development and the maintenance of biodiversity. The statements clearly point the way towards recognition of the interrelationship between the so-called 'natural environment' and human economic and cultural activity. If policies are to be developed which fulfil the aspirations both of the politicians and the traditional conservationist, future studies of saltmarshes and sand dunes and their management for conservation must include human use as a key component of the 'natural' system.

The description of the coastal habitats of the Mersey Basin and human influence upon them suggests that the extent of modification, even of the apparently most natural areas, is such that continued human intervention is required for their conservation. This is almost certainly true if the aim of the conservation policy is to maintain the *status quo* for the habitats, identified as being of importance when the site was first assessed. However, the traditional approach which seeks to arrest succession may not be adequate, partly because of the resource implications. Too often other options may be overlooked because saltmarsh or sand dune erosion, especially when apparently caused by human use, are viewed as indicating the vulnerable and sensitive nature of the

habitats which as a consequence, require protection. It is not unusual to find rocks placed along the eroding edge of saltmarshes, e.g., in the Severn Estuary, and almost all major sand dune systems in north-western Europe have been 'protected' by chestnut paling fencing and/or Marram planting. If a longer view is taken, and it is accepted that past human use has played a significant part in the evolution of the current interest, we may decide on a different approach to management.

Change is an important component of coastal systems, and saltmarshes and sand dunes are particularly well adapted to accommodate environmental perturbations, whether due to natural processes or human intervention. Saltmarsh erosion and accretion are natural phenomena, taking place in response to changes, for example, in the location of the tidal channels or, over a longer time period, in sea-level. Mobile sand driven by wind is required for dune development, and erosion is essential for the development of dune slacks. Initiating change may provide an important opportunity for the rejuvenation of degraded habitats. In this context the alliance between the ecologist, geomorphologist and coastal engineer in recreating saltmarshes on the Essex coast as part of a new approach to sea defence, may yet instil a more pro-active approach to conservation than has hitherto been considered appropriate. Perhaps ultimately we will emulate our ancestors, and in some areas initiate major change and then step back and let nature take its course. They did it by accident; we can do it by design! In so doing we may not have the same nature conservation features as before, but if the area over which the processes are allowed to develop is large enough, the resulting system may be more 'natural' and self-sustaining than at present.

In this context the Mersey Basin and its coastal areas are no exception. The Dee Estuary will continue to accrete as the saltmarsh (mainly Common Cord-grass) expands onto the mudflats. This, in its turn, will reduce the available feeding habitat for the internationally important wintering populations of waterfowl. Whether control will become necessary, as has happened at other conservation sites in the UK and elsewhere around the world (Doody 1984; Gray & Benham 1990), to prevent significant habitat loss and threats to the wintering bird populations feeding on the mudflats, will need to be constantly assessed. What is clear is that the recent decision to allow the extension of the Mostyn dock on the Welsh shore must add to the pressures which enhance siltation. Both here and in the Mersey Estuary, where Common Cord-grass appears to be expanding, the spread of the species should be carefully monitored. In this context it will be important to understand the nature of the forces which may have helped stimulate its growth. Is there a fundamental change in the nature of the sediments? What is the impact of sand dredging in the outer Ribble Estuary? Has it shifted the balance of the sediment type from a sandy matrix to a smaller, muddy substrate more suitable for Common Cord-grass growth along the Sefton shore?

The situation in Albania, where new coastal habitats are formed from sediment released as a result of deforestation in the mountains, also reminds us that human activities can influence coastal developments far away from the activity itself. In looking, therefore, at the future of the coastal habitats of the Mersey Basin, the whole area, from the nearshore marine environment to the active intermediate coastal zone and the hinterland as far inland as the catchment of the three main rivers, must be considered. Integrating planning and management in this zone will help to ensure decisions are taken which help to prevent one sectoral activity damaging the interests of another.

References

Angus, S. & Elliot, M.M. (1992). Erosion in Scottish machair with particular reference to the Outer Hebrides. *Coastal dunes, geomorphology, ecology and management for conservation* (eds R.W.B. Carter, T.G.F. Curtis & M.J. Sheehy-Skeffington), pp. 93–112. A.A Balkema, Rotterdam, Netherlands.

Atkinson, D. & Houston, J. (eds) (1993). *The Sand Dunes of the Sefton Coast*. National Museums & Galleries on Merseyside in association with Sefton Metropolitan Borough Council, Liverpool.

Bennett, C., Curtis, S. & Fairhurst, C. (1995). Recent changes in the Mersey Estuary 1958–1988. *The Mersey Estuary – Naturally Ours* (eds M.S. Curtis, D. Norman & I.D. Wallace), pp. 12–23. Mersey Estuary Conservation Group, Warrington.

Berry, W.G. (1967). Saltmarsh development in the Ribble Estuary. *Liverpool essays in geography – A Jubilee Collection* (eds R.W. Steel & R. Lawton), pp. 121–35. Longmans, London.

Buck, A.L. (1993). *An inventory of UK estuaries, 3. North-west Britain*. Joint Nature Conservation Committee, Peterborough.

Chapman, V.J. (1938). Studies in saltmarsh ecology, I–III. *Journal of Ecology*, **26**. 144–79.

Chapman, V.J. (1941). Studies in saltmarsh ecology, IV. *Journal of Ecology*, **29**. 69–82.

Chapman, V.J. (1959). Studies in saltmarsh Ecology, IX. Changes in saltmarsh vegetation at Scolt Head Island. *Journal of Ecology*, **47**, 619–39.

Coles, J.M. (1964). New aspects of Mesolithic settlement of south-west Scotland. *Transactions of the Dumfries and Galloway Natural History and Antiquarian Society*, **XL**, 67–98.

Dijkema, K.S. (ed.) (1984). *Salt marshes in Europe*. Nature and environment series, No. 30. Council of Europe, Strasbourg, France.

Doody, J.P. (ed.) (1984). *Spartina anglica in Great Britain*. Focus on nature conservation, **No. 5**. Nature Conservancy Council, Attingham.

Doody, J.P. (1989). Management for nature conservation. *Coastal Sand Dunes* (eds C.H. Gimmingham, W. Ritchie, B.B. Willetts & A.J. Willis), pp. 247–65. *Proceedings of the Royal Society of Edinburgh*, **96B**.

Doody, J.P. (1993). Changing attitudes to coastal conservation. *ENact*, English Nature, Peterborough.

Garcia Nova, F. (1979). The ecology of vegetation of the dunes in Doñana National Park (south-west Spain). *Ecological*

Processes in Coastal Environments* (eds R.L. Jefferies, & A.J. Davy), pp. 571–92. (British Ecological Society Symposium) Blackwell Scientific Publications, Oxford.

Gray, A.J. (1992). Saltmarsh plant ecology: zonation and succession revisited. *Saltmarshes – Morphodynamics, conservation and engineering significance* (eds J.R.L. Allen & K. Pye), pp. 63–79. Cambridge University Press, Cambridge.

Gray, A.J. & Benham, P.E.M. (1990). *Spartina anglica – a research review*. ITE research publication, No 2. HMSO, London.

Jeffrey, D.W. (1977). *North Bull Island, Dublin Bay – a modern coastal natural history*. The Royal Dublin Society, Dublin.

Marchant, C.J. (1967). Evolution of *Spartina* (Gramineae). 1. The history and morphology of the genus in Britain today. *Journal of the Linnean Society, Botany*, **60**, 1–24.

Marker, M.E. (1962). The Dee Estuary: its progressive silting and saltmarsh development. *Transactions of the Institute of British Geographers*, **211**, 65.

Marshall, J.R. (1962). The physiographic development of Caerlaverock merse. *Transactions of the Dumfries and Galloway Natural History and Antiquarian Society*, **39**, 102.

Nature Conservancy Council (1989). *Guidelines for the selection of biological SSSIs*. Nature Conservancy Council, Peterborough.

Ranwell, D.S. & Boar, R. (1986). *The coast dune management guide*. Institute of Terrestrial Ecology, Monks Wood, Huntingdon.

Ranwell, D.S. (1972a). *Ecology of Saltmarshes and Sand Dunes*. Chapman & Hall, London.

Ranwell, D.S. (1972b). *The management of sea buckthorn Hippophae rhamnoides L. on selected sites in Great Britain*. Report of the Hippophae Study Group. Nature Conservancy (NERC).

Ratcliffe, D.A. (1977). *A nature conservation review*. Cambridge University Press, Cambridge.

Rice, K.A. & Putwain, P.D. (1987). *The Dee and Mersey Estuaries: environmental background*. Shell UK Limited, London.

Ritchie, A. & Ritchie, G. (1978). *The ancient monuments of Orkney*. HMSO, Edinburgh.

Robinson, N.A. (1984). The history of *Spartina* in the Ribble estuary. *Spartina anglica in Great Britain* (ed. P. Doody). Focus on nature conservation, No. 5, pp. 27–29. Nature Conservancy Council, Attingham Park.

Royal Commission on Coastal Erosion and Afforestation (1911). *The reclamation of Tidal lands*. Third (and final) Report Vol. III, Pt V, HMSO, London.

Simmons, I. (1980). Iron Age and Roman coast around the Wash. *Archaeology and coastal change* (ed. F.H. Thompson). Occasional Paper No. 1. The Society of Antiquities, London.

Skarregaard, P. (1989). Stabilisation of coastal dunes in Denmark. *Perspectives in coastal dune management* (eds F. van der Meulen, P.D. Jungerius and J.H. Visser), pp. 151–61. SPB Academic Publishing, The Hague, Netherlands.

Steers, J.A. (ed.) (1960). *Scolt Head Island*. Heffer, Cambridge.

Tansley, A.G. (1945). *Our heritage of wild nature*. Cambridge University Press, Cambridge.

Tansley, A.G. (1953). *The British Islands and their vegetation, volume II*. Cambridge University Press, Cambridge.

Thompson, H.V. & Worden, A.N. (1956). *The rabbit*. New Naturalist. Collins, London.

Truscott, A. (1984). Control of *Spartina anglica* on the amenity beaches of Southport. *Spartina anglica in Great Britain* (ed. P. Doody). Focus on nature conservation, No. 5, pp. 64–69. Nature Conservancy Council, Attingham Park.

Yapp, R.H., Johns, D. & Jones, O.T. (1917). The saltmarshes of the Dovey Estuary. II, The saltmarshes. *Journal of Ecology*, **5**, 65–103.

The Changing Flora

A.D. BRADSHAW

At first sight it might appear that the only lesson to be learnt from a complex urban area such as the Mersey Basin is that whenever conditions become difficult for them, plants and species disappear, and as a result the flora becomes more and more impoverished. This is indeed clear from the following chapters, and emphasises a serious message for all those caring for our environment.

Yet three notable principles are also apparent. The first is that if and when environmental damage abates there can be rapid recolonisation by the affected species. Yet this colonisation can be a fickle process, some species not showing the same ability to return as others, for reasons not yet always understood.

The second is that industrial activity creates new environments that may be suitable for species which did not find a suitable habitat previously. The third principle is that where there is so much human activity, there is a continuous in-flow of species which may be able to take advantage of the new habitats available.

From all this comes the not necessarily expected fact that the history of an industrialised region such as the Mersey Basin is not one of a downward spiral towards loss of diversity, but even an increase. There has been a remarkable process of change and adjustment as different species either lose or gain in relation to the complex environmental changes occurring, just as they have naturally over the past aeons of time.

CHAPTER SEVENTEEN

Marine algae: diversity and habitat exploitation

G. RUSSELL, A.W.L. JEMMETT AND S.B. WILKINSON

Introduction

The historical record of marine algae in the Mersey Basin starts in the second half of the 19th century. It therefore covers a period of considerable change both in the size of the human population and in the industrial activity of the region. Most of the information available has been obtained from studies of the estuaries of the Rivers Mersey and Dee, and it is on these that this paper will concentrate (Figure 17.1).

It is scarcely possible to investigate the marine vegetation of any major estuary in the UK without taking some account of the effects of industrial, agricultural or domestic pollution and, in that respect, the two estuaries offer an interesting contrast. The Dee catchment area includes the sparsely populated mountains of Snowdonia, and its water enters the tidal reaches in a condition that is generally Good or Very Good, although some of the smaller tributaries may be affected by agricultural and sewage pollution (NRA 1996). Industrial pollution (copper, zinc and organo-tins) may on occasion reach unacceptable levels in the Estuary. Sewage contamination of the tidal reaches has also reduced water quality and, at West Kirby, the water now (1996) fails to meet the EC Bathing Waters Directive. However, of the 65.5km of estuary, 63.0km is still classified as Good and the remaining 2.5km as Fair (NRA 1996).

The River Mersey and its tributaries, on the other hand, flow through large centres of population before becoming tidal, and water quality has been Poor or Bad over much of its length throughout this period (NRA 1995). Indeed, its pollution history can be traced back to the late 18th century, when the industrialisation of northwest England began, and quite recently the estuary was considered the worst polluted in Europe (Head & Jones 1991). This raises a rather obvious question: have the differences between the River Dee and River Mersey been expressed in the character of the marine algal vegetation? As with many simple questions, the answer has proved to be elusive and incomplete.

Macroalgae

Hilbre Island (Figure 17.1), at the mouth of the River Dee, has been visited for its seaweeds, certainly since the 1860s. F.P. Marrat (Marrat 1863–64), who worked in Liverpool City Museum, made several excursions at that time and the results of his observations were later incorporated into species lists by Gibson (1889, 1891). E.A.L. Batters, possibly the most eminent of late-Victorian phycologists in the UK, also came to the Island on at least one occasion (Batters 1902). His interest in Hilbre was probably occasioned by the fact that it is almost the only example of natural rocky shore in the long stretch of coastline between Cumbria and the Great Orme. It must have been seen therefore as important in the compilation of regional seaweed floras for the UK (Holmes & Batters 1890). However, these late-19th-century lists have since proved to be much more interesting as evidence of floristic change. Drawing from these and from personal records, Russell (1982) listed 151 species of macroalgae for Hilbre (Table 17.1) of which 38 had not been seen in recent years, while a further 33 species were considered to be definitely absent. Prominent among the latter were the kelps (*Alaria, Chorda, Laminaria* spp.) together with their common understorey species (*Delesseria, Membranoptera, Odonthalia, Palmaria, Phycodrys, Ptilota* spp.). Thus, an entire zone of vegetation has disappeared from an estuary that has been relatively unpolluted throughout its history. The build-up of sand and other sediments in the estuary seems likely to have buried some of the sublittoral reefs, but areas of lower-shore rock, potentially suitable for a kelp forest,

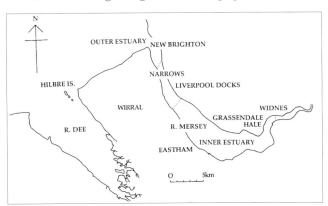

Figure 17.1. Outline map of estuaries of the Rivers Dee and Mersey showing locations of sites referred to in text.

NOSTOCOPHYCEAE
+ *Calothrix crustacea*
+ *Entophysalis conferta*
+ *E. deusta*
+ *Microcoleus lyngbyaceus*
+ *Oscillatoria lutea*
+ *Schizothrix calcicola*
+ *S. tenerrima*
+ *Spirulina subsalsa*

BANGIOPHYCEAE
? *Antithamnion cruciatum*
− *Ahnfeltia plicata*
+ *Audouinella daviesii*
? *A. floridula*
? *A. membranacea*
+ *A. purpurea*
+ *A. secundata*
+ *A. thuretii*
? *A. virgatula*
+ *Bangia atropurpurea*
? *Callithamnion corymbosum*
+ *C. hookeri*
+ *C. roseum*
+ *Catenella caespitosa*
? *Calliblepharis ciliata*
? *Ceramium circinnatum*
+ *C. deslongchampii*
? *C. fastigiatum*
+ *C. rubrum*
? *C. shuttleworthianum*
? *C. strictum*
? *C. tenuissimum*
+ *Chondrus crispus*
+ *Corallina officinalis*
− *Cryptopleura ramosa*
? *Cystoclonium purpureum*
− *Delesseria sanguinea*
+ *Dumontia contorta*
+ *Erythrotrichia carnea*
+ *Furcellaria lumbricalis*
+ *Gelidium pusillum*
? *G. sesquipedale*
+ *Gracilaria verrucosa*
− *Griffithsia flosculosa*
− *Gymnogongrus crenulatus*
− *Heterosiphonia plumosa*
+ *Hildenbrandia rubra*
− *Laurencia pinnatifida*
− *Lomentaria articulata*
− *Mastocarpus stellatus*
− *Membranoptera alata*
? *Monosporus pedicellatus*
− *Nemalion helminthoides*

− *Odonthalia dentata*
− *Palmaria palmata*
− *Petrocelis cruenta* (probably = *Mastocarpus stellatus*)
− *Phycodrys rubens*
− *Phyllophora crispa*
− *P. pseudoceranoides*
+ *Phymatolithon lenormandii*
? *Pleonosporium borreri*
− *Plocamium cartilagineum*
− *Plumaria elegans*
? *Polysiphonia fibrata*
? *P. fruticulosa*
d *P. lanosa*
+ *P. atlantica*
? *P. nigrescens*
+ *P. urceolata*
+ *Porphyra linearis*
+ *P. leucosticta*
+ *P. purpurea*
− *Pterosiphonia thuyoides*
− *Ptilota plumosa*
? *Rhodomela confervoides*
? *R. lycopodioides*
? *Spermothamnion repens*

FUCOPHYCEAE
− *Alaria esculenta*
+ *Ascophyllum nodosum*
? *Asperococcus fistulosus*
? *Chilionema ocellatum*
− *Chorda filum*
− *Chordaria flagelliformis*
+ *Cladostephus spongiosus*
− *Cutleria multifida*
d *Desmarestia aculeata*
? *Ectocarpus fasciculatus*
+ *E. siliculosus*
? *Dictyosiphon foeniculaceus*
+ *Elachista fucicola*
− *Fucus ceranoides*
+ *F. serratus*
+ *F. spiralis*
+ *F. vesiculosus*
+ *Giffordia granulosa*
+ *G. ovata*
? *Halopteris scoparia*
d *Halidrys siliquosa*
+ *Hecatonema maculans* (probably = *Punctaria sp.*)
+ *Isthmoplea sphaerophora*
− *Laminaria digitata*
− *L. hyperborea*
− *Leathesia difformis*

− *Litosiphon pusillus*
− *Mesogloia vermiculata*
+ *Myrionema strangulans*
? *Myriotrichia clavaeformis*
+ *Pelvetia canaliculata*
+ *Petalonia fascia*
? *P. zosterifolia*
+ *Petroderma maculiforme*
? *Punctaria latifolia*
+ *P. tenuissima*
+ *Ralfsia verrucosa*
+ *Scytosiphon lomentaria*
+ *Sphacelaria fusca*
? *S. cirrosa*
? *S. plumosa*
+ *S. radicans*
+ *Spongonema tomentosum*
+ *Stictyosiphon tortilis*
+ *Waerniella lucifuga*

CHLOROPHYCEAE
+ *Blidingia marginata*
+ *B. minima* (= *B. ramifera*)
? *Bryopsis plumosa*
? *B. hypnoides*
+ *Chaetomorpha linum*
+ *Chlorococcum submarinum*
? *Cladophora hutchinsiae*
+ *C. pellucida* (1 record since 1982)
+ *C. rupestris*
+ *C. sericea*
? *Epicladia flustrae*
+ *E. perforans*
? *Enteromorpha clathrata*
+ *E. compressa*
+ *E. intestinalis*
? *E. linza*
+ *E. prolifera*
+ *E. ralfsii*
? *E. ramulosa*
+ *Eugomontia sacculata*
+ *Percursaria percursa*
+ *Pringsheimiella scutata*
+ *Prasiola stipitata*
+ *Pseudococcomyxa adhaerens*
+ *Rhizoclonium riparium*
+ *Ulva lactuca*
+ *Ulothrix flacca*
+ *U. subflaccida*
+ *U. speciosa*
+ *Urospora penicilliformis*
− *Spongomorpha arcta*

Table 17.1. Benthic macroalgae of Hilbre Island (from Russell 1982). Species names, as far as possible, follow South & Tittley (1986), and readers are referred to that publication for the species authorities.

+ = species seen in recent years and probably still present
? = formerly recorded and possibly still present

− = formerly recorded but now considered definitely absent
d = present only as drift plants.

remain at the seaward end of Hilbre (Russell 1972). A possible explanation for the loss of these species is provided by Marrat (1863–64) who, in one of his reports, writes enthusiastically about the clear rock pools of Hilbre. Dee Estuary water is now very turbid as a result of suspended clay minerals, and light transmission is consequently poor. The similarly turbid water of the Bristol Channel was investigated by Dring (1987), who concluded that the lower 1–2m of the intertidal zone at Avonmouth receives insufficient light for the sustained growth of *Fucus serratus*. So, the absence of light necessary for the formation of a kelp forest may have been the crucial factor in its disappearance. The importance of light in the zonation of Hilbre seaweeds can also be inferred from the fact that the zone of maximum species diversity is the upper and not the lower eulittoral, as might be expected on most UK rocky shores (Russell 1973). Algae are adaptable plants, however, and the low-lying rock at Hilbre has been exploited, in summer especially, by ephemerals such as *Ulva, Enteromorpha* and *Pilayella* species (Russell 1973, 1982).

Clay minerals do not only reduce light transmission, however. During periods of calm weather, the sea conditions may permit sediments to settle out on rock surfaces, forming a thick blanket. This can be lethal to small algae and to barnacles and is, at the same time, an unsuitable surface for fresh algal colonisation.

Documentary evidence for floristic change in the River Mersey is less complete than that for Hilbre. Gibson (1891) chose, inexplicably, to combine algal records from the River Dee and River Mersey in a single list, while keeping those of Hilbre separate. While some of the records in his earlier list (Gibson 1889) clearly refer to locations in the River Mersey, the detail is insufficient to allow compilation of a flora comparable with that of Hilbre. However, Gibson (1889) quotes an observation by Marrat that many Mersey seaweeds had been more healthy in appearance in 1860, and it is certainly the case that numerous Mersey species recorded in this publication are now absent. According to Gibson, the deterioration noted by Marrat was due, without a shadow of doubt, to the impact of increased industrial effluent, and chemical pollution of the River Mersey has certainly been serious and continuous. Analyses of arsenic, chromium, copper, lead, mercury and zinc in mud cores at Widnes (Figure 17.1) show increases in all by 1890, but to levels that are very low in comparison with those reached in the mid-20th century (NRA 1995). Either the algae in 1889 were unexpectedly sensitive to industrial pollution or some other factor(s) had also been involved in their demise. The natural sandstone reefs of the Mersey Estuary at New Brighton and upstream at Grassendale and Eastham (Figure 17.1) remain poor in species.

Artificial substrates such as promenade walls and the more recently constructed coastal defences at New Brighton and along the north Wirral shore (Davies 1989) are also uninteresting floristically. Perhaps the most successful intertidal taxa are *Fucus vesiculosus* and several species of *Enteromorpha*. These are notoriously variable genetically, phenotypically plastic and with broad niche ranges. For example, it has been found that *F. vesiculosus* from strongly wave-exposed shores has a lower intrinsic growth rate than conspecific populations living in shelter (Bäck, Collins & Russell 1992a). Also, Baltic *F. vesiculosus* is more tolerant of greatly reduced water salinity, and less tolerant of high salinities, than populations from Atlantic coasts (Bäck *et al*, 1992a). The inability of Baltic plants to survive high salinities is now known to be due to their rather weak capacity for the biosynthesis of mannitol, an important osmolyte of brown algae (Bäck, Collins & Russell 1992b). In the case of the Dee and Mersey populations, the natural vari-

ability of *F. vesiculosus* has been further increased by introgression (Burrows & Lodge 1951; Russell 1995).

Enteromorpha intestinalis on Irish Sea coasts has been found to possess at least three different salinity ecotypes (Reed & Russell 1979). Estuarine populations are analogous to Baltic *Fucus* in being tolerant to low and intolerant to high salinities. Low-shore marine ecotypes lack tolerance to reduced salinities, but grow well in fully saline media. The third group comprises plants from high-shore marine rock pools. These pools experience major fluctuations in salinity arising from dilution (rainfall) and concentration (evaporation), and the *Enteromorpha* from such habitats has the broadest tolerance range of all (Reed & Russell 1979). Salinity has also been found to trigger plastic responses resulting in branch proliferation in this species (Reed & Russell 1978).

The success of these taxa in the River Mersey echoes an observation by Nevo *et al.* (1986) that narrow-niche gastropods with low genetic diversity are less able to survive treatment with marine pollutants than broad-niche species with greater variability.

Apart from *Fucus vesiculosus*, the only large brown seaweed in the River Mersey is *Ascophyllum nodosum*. At present, this species is confined to a small area of shore at Eastham (Figure 17.1) where it forms a narrow belt in the upper eulittoral zone. The fronds of these plants are of interest because they fail to branch until they reach an age of about 6 years and, consequently, they have a curiously linear appearance (Figure 17.2). It seems likely that the fronds have responded in an adaptive way to the poor light conditions in the Mersey water. Thus, the early linear development should enable fronds to maximise their rate of entry into the better-lit surface waters, where improved irradiance will permit branching, and hence increased tissue production, to occur.

It is the restored disused Liverpool Docks (Figure 17.1) which provide most information on the potential ability of seaweed species to colonise Mersey Estuary water. The restoration of Sandon Dock (Russell *et al.* 1983) and the South Docks (Allen, Wilkinson & Hawkins 1995) has created environments in which water clarity has been greatly improved, partly through settling of mud particles, and partly through filtration by mussels. In spite of the fact that the docks were filled, and are still topped up, with polluted Mersey water, their seaweed floras already include a number of species that are absent from the intertidal reefs of the estuaries (Allen *et al.* 1995; Wilkinson 1995); see also Table 17.2. Some are likely to have been recruited from relatively distant sites, such as Anglesey or possibly the Isle of Man (e.g., *Palmaria palmata, Giffordia sandriana, Sorocarpus micromorus, Desmarestia viridis, Cladophora vagabunda*).

A surprising feature of the dock floras is the continued absence of large perennial brown algae. An experimental introduction of *Laminaria saccharina* to Sandon Dock resulted in rapid losses of plants and, by the end of the experiment, the few survivors were overgrown by epiphytic ephemeral algae and sessile animals (Russell *et al.* 1983). A similar introduction of *Fucus vesiculosus* to

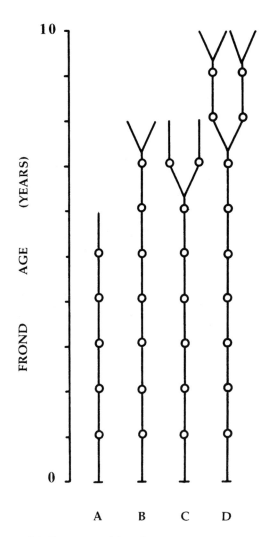

A   B   C   D

Figure 17.2. Diagrams of four fronds of Mersey *Ascophyllum nodosum* with small circles denoting gas bladders. After year 1, this species normally produces one bladder and one dichotomy from every tip each year. Frond A is therefore 6 years old, fronds B and C are 8 years old, while D is 10. Note absence of branching until approximately year 6.

NOSTOCACEAE
Oscillatoria margaritifera

BANGIOPHYCEAE
Audouinella secundata
Ceramium rubrum
C. strictum
Erythrotrichia carnea
Palmaria palmata
Polysiphonia urceolata
Pterothamnion plumula

FUCOPHYCEAE
Ectocarpus siliculosus
Giffordia granulosa
G. ovata
G. sandriana
Leptonematella fasciculata
Pilayella littoralis
Punctaria latifolia

Scytosiphon dotyi
Sorocarpus micromorus
Stictyosiphon soriferus

CHLOROPHYCEAE
Bryopsis hypnoides
B. plumosa
Cladophora vagabunda
Enteromorpha compressa
E. intestinalis
E. linza
Monostroma grevillei
Ulothrix speciosa
U. subflaccida
Urospora penicilliformis
Ulva lactuca

TRIBOPHYCEAE
Vaucheria litorea

Table 17.2. Benthic macroalgae found in Liverpool South Docks since their restoration in 1985. Princes Dock, which lies seaward of the South Docks, contains also *Desmarestia viridis* and occasional depauperate *Fucus* of uncertain species.

may be closed during periods of gross pollution in the estuary provides an additional element of protection. This proved to be the case during an episode of oil pollution in the River Mersey in August 1989 (Environmental Advisory Unit 1991) when the gates to the South Docks were kept shut.

Microalgae

The written history of microalgae in the estuaries is too fragmentary to give reliable evidence of historical change. The subfossil diatom record, used extensively in constructing lake histories, is unlikely to be reliable in most estuaries where water movement can disturb sediments. However, at Hale (Figure 17.1) on the Liverpool bank of the inner estuary, a small area of *Phragmites* (Common Reed) marsh has been cut into by a landward shift in the River channel. As a result, a face *c*.1.5m in height has been exposed, revealing alternating bands of leaf litter and estuarine mud (Figure 17.4). Radionuclide dating has confirmed that these are annual deposits, and that the marsh came into being at about 1912 (Jemmett 1991). The process of marsh development is essentially one of build-up of mud around the bases of Common Reed stems, which collapse during winter and are replaced in spring by a further crop of new shoots. Altogether, 88 species of diatom were identified in a monolith cut from the marsh face in November 1988, but, for the sake of simplicity, these have been reduced to three broad categories: *epiphyton* (diatoms usually found attached to surfaces of algal or flowering plant macrophytes), *epipelon* (diatoms usually found living on the surface of or in the interstices of mud), and *plankton* (diatoms usually found free-floating in the water column). The relative abundances of these are shown in Figure 17.5, and it is clear from this that plankton forms

the South Docks in February 1994 was terminated three months later when the plants were visibly disintegrating under a dense growth of epiphytic algae (Figure 17.3). This overgrowth could not be attributed to excessive nutrient concentrations in the dock, for phytoplankton production had reduced these to levels similar to those in coastal waters outside the estuaries throughout the period of the experiment (Dr J.W. Eaton, pers. comm.). The factor limiting the development of a macrophyte forest in the docks may well be the absence of turbulence. On intertidal rocky shores, wave action causes brown algal macrophytes to whip-lash, and this movement is likely to enhance the self-cleaning mechanisms of the plants (skin shedding). It would be interesting to see if perennial seaweeds, whose defences against epiphytes are chemical rather than physical (*Desmarestia* spp.?), could prove more successful in docks. Even without perennial species, docks have obvious potential as refuges for benthic algae, and the fact that dock gates

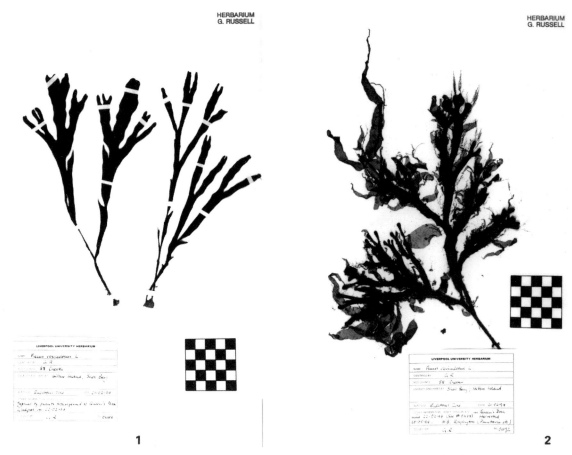

Figure 17.3. Herbarium preparations of *Fucus vesiculosus* from Hilbre Island.
1: plants from sample collected in February 1994 for introduction into Liverpool South Docks, and
2: heavily epiphytised plant after three months immersion. Scale grid = 5 x 5cm.

the most abundant group for much of the 80-year history of the marsh. It is evident also that there is considerable variation from year to year in all groups. Nevertheless, there is a significant change (p< 0.05) in all three after 1960, at which date there is a reduction in plankton and increases in both epiphyton and epipelon. While it might be tempting to ascribe this pattern to some change in water quality, the likeliest explanation is simply that the height of the marsh had reached a critical level at which tidal inundations became less frequent and of shorter duration. This would have reduced the incidence of plankton while the more stable conditions presumably proved more favourable for the development of epiphyton and epipelon.

The most abundant planktonic diatom in the monolith was *Skeletonema costatum* (Grev.) Clev. and its record of changing frequency is shown Figure 17.6. This species was also significantly less abundant after 1960 (p< 0.05), though varying from year to year throughout the period. This irregularity seems likely to be a natural phenomenon, rather than a sampling artefact, as numbers of *Skeletonema* in recent water samples are variable through time and from site to site in the same year (Figure 17.7). Figure 17.7 shows also that *Skeletonema* has a single annual peak of rather short duration from April to May. None of the important nutrients for diatom metabolism

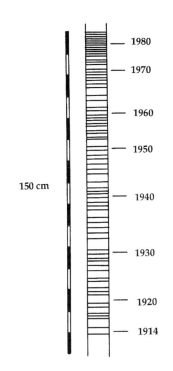

Figure 17.4. Diagram of exposed face of *Phragmites* marsh at Hale in November 1988. Transverse lines denote horizontal bands of leaf litter alternating with mud layers of different thickness. Redrawn from Jemmett (1991).

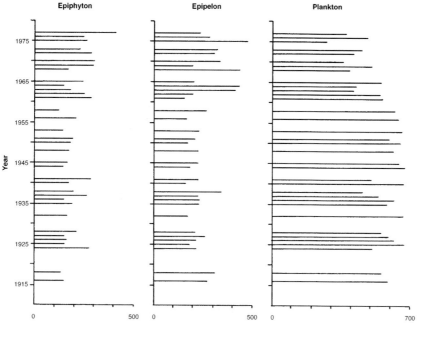

Figure 17. 5 (left) Mean abundances of subfossil diatoms in a Hale marsh monolith; epiphytic, epipelic and plank-tonic diatoms are treated separately. Drawn from raw data obtained by A.W.L. Jemmett. Mean values were calculated from three replicate samples from each year investigated. The numbers of frustules counted in each sample ranged from 956 to 1,114 but the final numbers were all adjusted to 1,000.

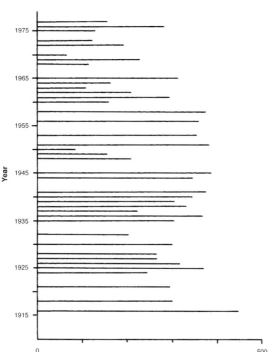

Figure 17.6 (above) Mean abundances of *Skeletonema costatum* in Hale marsh monolith. Drawn from raw data obtained by A.W.L. Jemmett.

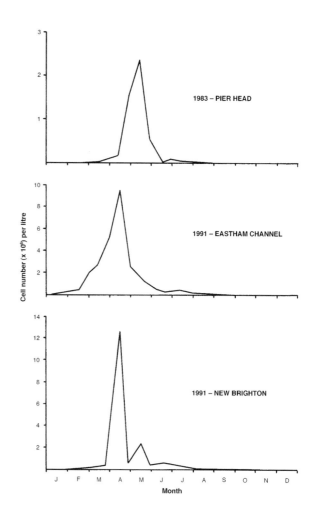

Figure 17.7 (right) Annual abundances of *Skeletonema costatum* in water samples taken from three sites in the Mersey Estuary: Pier Head (Narrows), Eastham Channel (Inner Estuary) and New Brighton (Outer Estuary). Note variation in sizes of spring peaks from year to year and at different sites in the same year. Note also consistent absence of autumn bloom. 1983 data obtained by G.R. 1991 graphs redrawn from Environmental Resources Ltd (1993).

follows this pattern (NRA 1995). The best fit is provided by silicate, levels of which fall away markedly in the inner estuary in late spring. However, silicate concentrations return to higher values in autumn when no corresponding *Skeletonema* bloom occurs. It is possible that zooplankton grazing prevents an autumn bloom, but a likelier explanation is that light is then insufficient to sustain growth. The turbidity of Mersey Estuary water has been quantified by Russell *et al.* (1983), who recorded Secchi disc extinction depths (i.e., depths at which a white disc lowered into the water disappears from view) that seldom exceeded 10cm.

Skeletonema is a common estuarine diatom and its abundance has been high throughout the period of marsh development. There is no evidence of any adverse effects of the increasing heavy-metal pollution at that time. A possible explanation of its resilience has been provided recently by Medlin *et al.* (1991) who compared four geographically isolated strains on the basis of morphology and by means of their ribosomal DNA nucleotide sequences. The strains proved to be diverse genetically and two were considered sufficiently distinct to belong to a new species. Thus, the success of *Skeletonema* in a polluted environment can be linked once again with genetic variability and with a broad-niche range.

Discussion

There is little doubt that the number of species of macroalgae at Hilbre has been in decline over the past 100 years, despite the fact that the Dee Estuary has not been and is not heavily polluted. Although there is always a danger in proposing single factor explanations of events of this kind, light reduction caused by an increase in suspended fine sediments seems likely to have been a major agent of change. Jemmett (1996) has emphasised that the cloudiness of Dee Estuary water is due to suspended silt and clay particles, and is a natural feature, which should not be misconstrued as evidence of poor water quality. However, a connection is possible in that the suspended material also includes a proportion of organic matter, some of which may well originate in domestic effluent.

In the River Mersey, where a definite but unquantifiable decrease in macroalgal diversity has also occurred, pollution has been on an altogether different scale. In 1982, the Secretary of State for the Environment referred to the river as '... an affront to the standards a civilised society should demand of its environment ...' (Head & Jones 1991). Yet it has proved possible for some naturally variable taxa to exploit even these demanding conditions and to achieve some success in vegetating the available intertidal substrate. As in the River Dee, turbidity has certainly been involved in the species losses, for clarification of impounded Mersey water in Liverpool docks has been followed quite rapidly by the appearance of several unexpected species of macroalgae. However, pollution and turbidity are again inextricably linked, for many of the chemical pollutants may be bound to the sediments as well as dissolved in the water.

Signs of amelioration of Mersey water have been detected beyond the confines of the restored docks. Concentrations of lead and mercury in the tissues of *Fucus vesiculosus* have declined since 1980 (Head & Jones 1991), reflecting a general reduction of most metals in the estuarine sediments (Mersey Basin Campaign 1995; Head & Jones 1991). Since 1981, North West Water has been engaged upon an ambitious programme to reduce domestic effluent in the River Mersey and its tributaries, with the aim of achieving a minimum quality of Fair throughout by the year 2000 (Head & Jones 1991). It has also been reported that Welsh Water (Hyder plc) have plans for improved sewage treatment, which ought to lead to better water quality in the Dee Estuary (Jemmett 1996).

There is considerable public concern about the quality of the two estuaries and good evidence that improvements are under way. Management plans have been published (Mersey Basin Campaign 1995; Jemmett 1996) detailing proposals for further environmental enhancement. It is difficult to guess what the impact of these on the macro- and microalgal vegetation might be, but the adaptability of algae should ensure that any new niches will be filled rather rapidly.

Acknowledgements

We are grateful to Gill Haynes for technical help over the years and to Peter Head for up-to-date advice on the state of the Mersey Estuary.

References

Allen, J.R., Wilkinson, S.B. & Hawkins, S.J. (1995). Redeveloped docks as artificial lagoons: The development of brackish-water communities and potential for conservation of lagoonal species. *Aquatic Conservation: Marine and Freshwater Ecosystems*, **5**, 299–309.

Bäck, S., Collins, J.C. & Russell, G. (1992a). Effects of salinity on growth of Baltic and Atlantic *Fucus vesiculosus*. *British Phycological Journal*, **27**, 39–47.

Bäck, S., Collins, J.C. & Russell, G. (1992b). Comparative ecophysiology of Baltic and Atlantic *Fucus vesiculosus*. *Marine Ecology Progress Series*, **84**, 71–82.

Batters, E.A.L. (1902). A catalogue of the British marine algae. *Journal of Botany, London*, **40** suppl., 1–107.

Burrows, E.M. & Lodge, S. (1951). Autecology and the species problem in *Fucus*. *Journal of the Marine Biological Association of the United Kingdom*, **30**, 161–76.

Davies, C.D. (1989). Wirral scheme. *Coastal Management* (Institute of Civil Engineering – Maritime Engineering Board), pp. 293–307. Thomas Telford, London.

Dring, M.J. (1987). Light climate in intertidal and subtidal zones in relation to photosynthesis and growth of benthic algae: a theoretical model. *Plant Life in Aquatic and Amphibious Habitats* (ed. R.M.M. Crawford), pp. 23–34. Blackwell Scientific Publications, Oxford.

Environmental Advisory Unit of Liverpool University Ltd. (1991). *The Mersey Oil Spill Project 1989–90*. Mersey Oil Spill Advisory Group, Liverpool.

Environmental Resources Ltd. (1993). *Stage IIIa Environmental Studies: E1 Plankton Studies in the Mersey Estuary*. Final

Report to Mersey Barrage Company.

Gibson, R.J.H. (1889). Report on the marine algae of the L.M.B.C. district. *Proceedings and Transactions of the Liverpool Biological Society*, **3**, 128–54.

Gibson, R.J.H. (1891). A revised list of the marine algae of the L.M.B.C. district. *Proceedings and Transactions of the Liverpool Biological Society*, **5**, 83–43.

Head, P.C. & Jones, P.D. (1991). The Mersey estuary: turning the tide of Pollution. *Environmental Pollution*, **1**, 517–28.

Holmes, E.M. & Batters, E.A.L. (1890). A revised list of the British marine algae. *Annals of Botany*, **5**, 63–107.

Jemmett, A.W.L. (1991). *An investigation into the heavy metals, sediment and vegetation of a Mersey estuary saltmarsh.* PhD thesis, University of Liverpool.

Jemmett, A.W.L. (1996). *The Dee Estuary Strategy Final Report: January 1996.* Dee Estuary Strategy, Metropolitan Borough of Wirral, Birkenhead.

Marrat, F.P. (1863–64). Several short hand-written lists in issues of the *Liverpool Naturalists' Scrap Book* for these years. Liverpool City Library, H580LIV.

Medlin, L.K., Elwood, H.J., Stickel, S. & Sogin, M.L. (1991). Morphological and genetic variation within the diatom *Skeletonema costatum* (Bacillariophyta): Evidence for a new species *Skeletonema pseudocostatum*. *Journal of Phycology*, **27**, 514–24.

Mersey Basin Campaign (1995). *Mersey Estuary Management Plan: A Strategic Policy Framework.* Liverpool University Press, Liverpool.

Nevo, E., Noy, R., Lavie, B., Beiles, A. & Muchtar, S. (1986). Genetic diversity and resistance to marine pollution. *Biological Journal of the Linnean Society*, **29**, 139–44.

NRA (1995). *The Mersey Estuary: A Report on Environmental Quality.* HMSO, London.

NRA (1996). *The River Dee Catchment Management Plan Action Plan:* 1996. National Rivers Authority, Bangor.

Reed, R.H. & Russell, G. (1978). Salinity fluctuations and their influence on 'bottle brush' morphogenesis in *Enteromorpha intestinalis* (L.) Link. *British Phycological Journal*, **13**, 149–53.

Reed, R.H. & Russell, G. (1979). Adaptation to salinity stress in populations of *Enteromorpha intestinalis* (L.) Link. *Estuarine and Coastal Marine Science*, **8**, 251–58.

Russell, G. (1972). Phytosociological studies on a two-zone shore. I Basic pattern. *Journal of Ecology*, **60**, 539–45.

Russell, G. (1973). Phytosociological studies on a two-zone shore. II Community structure. *Journal of Ecology*, **61**, 525–36.

Russell, G. (1982). The marine algae. *Hilbre the Cheshire Island: its history and natural history.* (ed. J.D. Craggs), pp. 65–74, Liverpool University Press, Liverpool.

Russell, G. (1995). Pyrolysis mass spectrometry: a fresh approach to old problems in brown algal systematics? *Marine Biology*, **123**, 153-157.

Russell, G., Hawkins, S.J., Evans, L.C., Jones, H.D. & Holmes, G.D. (1983). Restoration of a disused dock basin as a habitat for marine benthos and fish. *Journal of Applied Ecology*, **20**, 43–58.

South, G.R. & Tittley, I. (1986). *A Checklist and Distributional Index of the Benthic Marine Algae of the North Atlantic Ocean.* Huntsman Marine Laboratory and British Museum (Natural History), St Andrews and London.

Wilkinson, S.B. (1995). *The ecology of the benthos in Liverpool docks.* PhD thesis, University of Liverpool.

CHAPTER EIGHTEEN

The influence of atmospheric pollution on the lichen flora of Cheshire

B.W. FOX

Introduction

The earliest known records of lichens in Cheshire are probably contained in Dawson Turner and L.W. Dillwyn's account in *The Botanists Guide through England and Wales* printed in 1805, in which 26 species were recorded. Based on my own work in preparing *The Lichen Flora of Cheshire*, this total is much lower than is known today (over 260 species) in the county, but it includes large, very pollution sensitive species such as Lichen pulmonarius (*Lobaria pulmonaria*), Lichen scrobiculatus (*Lobaria scrobiculata*), found by John Bradbury (1768–1823) of Stalybridge, Cheshire on old oak trees in Stayley Rushes. (Lichen names in italics are those currently used in Purvis *et al.* 1992.)

Other species collected by Bradbury were Lobaria laetevirens (*Lobaria virens*), Parmeliella plumbea (*Degelia plumbea*) and Cornicularia triste (*Cornicularia normo-erica*), all of which are long since extinct from Cheshire and from this locality. Even the ultra sensitive *Usnea articulata* was recorded by him 'in plenty' in Lyme Hall parkland, in Cheshire.

Since some of the sites at which these were recorded were down-wind of the Manchester conurbation, one can conclude that any influence of aerial pollution originating from this source had not yet started to affect the lichen population at this time.

We also know that on the red sandstone rocks of the Bidston lighthouse area, on the west coast of the Wirral peninsula, Umbilicaria pustulata (*Lasallia pustulata*) was found in 'plenty' by Bradbury in 1805, together with *Umbilicaria deusta*. They were still present when Frederick Price Marrat (1820–1904) visited the site (Marrat 1860), and an excellent herbarium specimen, that was obviously growing well, is housed in the herbarium at Liverpool Museum.

In 1859, Leo Grindon in *The Manchester Flora* wrote of lichens.

Many pretty species [lichens] are to be found on the moors, and in the neighbouring woods and cloughs and in parks and old orchards in Cheshire; but the majority of those enumerated are not obtainable nearer [to Manchester] than on the high hills beyond Disley, Ramsbottom, Stalybridge and Rochdale, and even there a quantity has been lessened of late years, through the cutting down of old woods and the influx of factory smoke, which appears to be singularly prejudicial to these lovers of pure atmosphere.

It was thus reasonable to assume that the plume of pollution originating from the Manchester area had not yet penetrated much further than its outskirts represented by these towns.

A century ago, Wheldon recorded *Ochrolechia tartarea* in Eastham Wood, Wirral (Travis 1922), a species now apparently absent from Cheshire. It was not found on my visit to these woods in 1991, and judging from the present change in the surrounding environment, it is unlikely to be found there in the near future. The pollution pressure on this woodland is almost certainly due to surrounding encroaching urbanization and the consequent pollution from that source.

The overall impression gained therefore is that there has been a marked increase in the damaging effects of atmospheric pollution over the last century and that judging from sites like this one, no improvement has taken place in recent years.

Wheldon and Travis, in their *Lichen Flora of South Lancashire*, deal at length with the depredation of smoke pollution in these areas next to Cheshire. They suggested that maximum possible damage had occurred. Judging from Travis's collections in the Liverpool Museum, considerable decimation of *Umbilicarias* in the Bidston lighthouse area had also occurred. In *The Lichens of the Wirral* (1922) and later in his additions (1925), Travis noted this degradation, and that *Lasallia pustulata* only occurred in two places in a very stunted state. I visited the site in 1980, and found several plants of *Umbilicaria polyphylla*, apparently growing reasonably well, but no evidence of either *U. deusta* or *L. pustulata*. I visited the site again in 1991 and found only two specimens of *U. polyphylla*, growing reasonably well. However, the site is subject to a high level of public pressure, and this in addition to any influence of atmospheric pollution could cause the decrease in the species diversity of the lichen flora (Table 18.1).

Apart from sulphur dioxide, the effects of carbon

1805   *Lobarion* community recorded as well established.
1859   Leo Grindon noted influence of pollution on lichens in the Manchester area, *Lobarion* still present East of Manchester.
1860   Marrat collects fine specimens of *L. pustulata* at Bidston.
1897   *Ochrolechia tartarea* collected by Wheldon in Eastham woods.
1915   From Travis's collection of *Umbilicarias* in Liverpool herbarium, pollution had set in having considerable effect.
1922   *Lasallia pustulata* recorded by Travis at Bidston, but very depauperate, hardly recognisable. *Umbilicaria deusta* still present but damaged.
1977   The first record in Cheshire of *Usnea subfloridana* at Rostherne Mere, by G.M.A. Barker.
1980   Both *L. pustulata* and *U. deusta* gone from Bidston. Several specimens of *U. polyphylla* present. (B.W. Fox)
1983   *Usnea* spp. found at Foden's Flash by J. Guest.
1991   Only two small plants of *U. polyphylla* could be found by B.W. Fox at Bidston. *Usnea* spp. becoming widespread in Willow Carrs.

Table 18.1. The chronology of lichen observations in Cheshire.

pollution are less well known from a lichen point of view. This may be more important for monitoring other plants and animals however, and the excellent surveys of melanic forms of the Scalloped Hazel moth (*Gonodontis bidentata*) conducted during the early 1970s by J.B. Bishop and his co-workers may reflect this (Bishop & Cook 1975).

Sulphur dioxide pollution levels in Cheshire

Following the fateful smog of London in 1952, when over 4,000 people died as a direct result, the Clean Air Act of 1956 became law. It was not until the early sixties that this Act developed its teeth, and by the Clean Air Act 1968 and later Control of Pollution Act in 1974, the effect of pollution decrease was becoming obvious. No more dense smogs have been seen during climatic inversions

in autumn. One of the earliest systematic records of the sulphur dioxide levels of the county of Cheshire was prepared in 1961 by the Warren Springs Pollution Laboratories in Stevenage. Some 90 sites were established throughout Cheshire and the sulphur dioxide levels were monitored for varying lengths of time in each of the sites. Some were monitored from 1961, and at a few places there has been a continuous record of the mean winter levels until the mid eighties. Four of the sites for which long records are available are included in Figure 18.1, and it can be seen that the values have fallen from 300–400 micrograms per cubic metre to less than 20 in the mid 1980s when recordings ceased. Levels of over 2,000 micrograms per cubic metre occurred in the Widnes area in 1961 and Ellesmere Port in 1966. Even these very high levels were probably lower than those which may have occurred before systematic recording began. It is also interesting to note that for some reason a second peak occurred during the early seventies, especially in the Warrington area. The most recently recorded levels in 1985 averaged between 20 and 40 micrograms per cubic metre for most of these areas.

The effect of sulphur dioxide pollution on Lichens

Sulphur dioxide is a major component of industrial and urban emissions, and has been shown to be very toxic to lichens. *In vitro* work by Ferry & Baddeley in 1976 showed that the gas was very disruptive to the photosynthetic biochemistry of the plant. It is water soluble and forms sulphurous acid when combined with terrestrial water or rain. The main effect of this sulphurous acid is, at high concentrations, to disrupt photosynthesis itself, as well as nitrogen fixation. Recent work (Lange *et al.* 1989) shows that an even more sensitive effect of sulphur dioxide toxicity is on the transfer of carbohydrates from the algal to the fungal partner in the symbiont. Other unknown pollutants from traffic emissions seem to increase the level of chlorophyll present, thought to be caused by the stimulatory effects on the

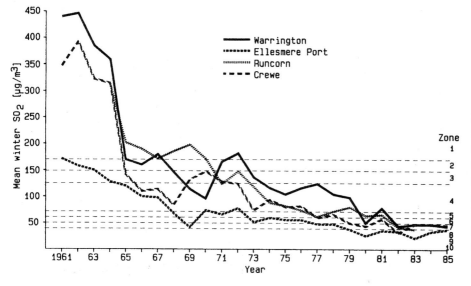

Figure 18.1. Mean winter sulphur dioxide levels (in micrograms per cubic metre) for four key sites in Cheshire. Adapted from data from the Warren Springs Pollution Laboratory, Stevenage. Superimposed are lines corresponding to Hawksworth & Rose's zone scales of pollution.

Zone	Sulphur dioxide levels µgm⁻³	Tree bark distribution	Zone	Sulphur dioxide levels µgm⁻³	Tree bark distribution
0	???	Nil	6	~50	*Parmelia caperata* (r)
1	>170	[*Pleurococcus*] base only			**Pertusaria albescens (r)**
2	~150	*Lecanora conizeoides* base only			**Pertusarias** (r)
3	~125	*Lepraria incana* base only			*Parmelia revoluta* (r)
		Buellia punctata			**Graphis spp.** (vr)
4	~70	*Parmelia sulcata*			*Pseuderv furfuracea.*
		Hypogymnea physodes			*Bryoria fuscescens*
		Lecanora expallens			*Parmelia revoluta* (r)
		Chaenotheca ferrug.(r)			**Parmelia easperatula (r)**
		Buellia canescens	7	~40	**Usnea subfloridana**
		Physcia adscendens			*Pertusaria hemisphaerica*
		Xanthoria parietina	8	~35	*Usnea ceratina*
5	~60	*Evernia prunastrii*			**Parmelia perlata** (vr)
		Ramalina farinacea			*Normandina pulchella*
		Physconia grisea (r)	9	<30	*Lobarias*
		Phaeophyscia orbicularis			*Dimerella lutea*
		Schismatomma decolorans			*Pachyphiale cornea*
		Xanthoria candelaria			*Usnea florida*
		Opegrapha spp.	10	0	All species.

Table 18.2. A brief listing of lichens employed in Hawksworth & Rose scale of pollution zones. Only bark species are used. Those in bold print are at present known to be rare in Cheshire.

nitrogen metabolism, which in turn stimulates chlorophyll synthesis (von Arb & Brunold 1989). It is not known how these possible pollutants modify lichen growth and survival.

One of the well known scales relating the frequency of lichens to the levels of sulphur dioxide in the environment was published by Hawksworth & Rose in 1970 (Table 18.2), and has been adapted in a number of different forms by authors throughout the world to provide a biological monitor for the study of this particular pollutant in the environment. In it, the authors defined a number of scales or zones from 0 (most polluted) to 10 (no pollution) which different species of lichens can tolerate, and by observing the mode of growth of these species on different substrates, some estimate of the prevailing sulphur dioxide concentrations can be determined. A 'mucky air map' was derived from information received from 15,000 schoolchildren as a result of the skilful application of a simplified version of this monitor by Gilbert in 1971 and reproduced by Richardson in his book *The Vanishing Lichens.*

Between 1958 and 1971 there was a 54% reduction in the ambient sulphur dioxide levels in the north-west region (North West Economic Planning Council 1974). These were still regarded, however, as the highest in the country outside the Greater London area. R. Bevan of the University of Liverpool was commissioned by Merseyside County Council to complete a biological survey of the Merseyside area, using lichens and a fungus, the well known Tar Spot (*Rhytisma acerinum*), which grows conspicuously on Sycamore (*Acer pseudocampestris*) and can be readily observed, and to some degree quantified.

In this study, the scale of indicators used was to cover the upper range of atmospheric pollution from 40 to 170 micrograms per cubic metre of sulphur dioxide. *Rhytisma* does not grow above an average pollution level

of 90 micrograms per cubic metre, and is thus a good median indicator for the range chosen for study. Three lichens occurring on specific habitats were chosen. It is important that the habitats were consistent throughout the survey, and these were sensibly defined at an early stage: *Lecanora conizeoides* on the lower two metres of tree trunks (representing an upper limit of 170 micrograms per cubic metre), *Xanthoria parietina* on asbestos roofs (representing an upper limit of 125 micrograms per cubic metre), and *Parmelia saxatilis* on acid sandstone (an upper limit of 60 micrograms per cubic metre). These were a good choice for this type of survey as they were easy to identify and record. However, one must bear in mind that the latter two lichens can locally be strongly influenced by the presence of eutrophication derived from bird activity and the proximity of sewerage farms generating ammoniacal gases. The result was an interesting and informative map based on this data (Figure 18.2a/b) showing the expected high concentrations centred around the principal towns in the area. The results were subsequently published by Vick & Bevan (1976).

A further similar survey, commissioned by the Merseyside County Council, was undertaken by Alexander and Henderson Sellers of the University of Liverpool Geography Department in 1980 (Alexander 1982). These results confirmed that the levels of pollution were decreasing over the intervening period and could be observed by the general movement of the inner limits of the zones into the town areas (Figure 18.2a/b).

In 1986, a further survey was undertaken, by Sewell and Ashton on behalf of Landlife,* using three 'indicator' species, viz., *Lecanora muralis*, based on a biological scale devised by Seaward (1976), *Xanthoria parietina* and *Parmelia saxatilis* on the same substrates. This system is

* Copy at Joint Countryside Advisory Service, Bryant House, Liverpool Road North, Maghull, Merseyside, L31 2PA.

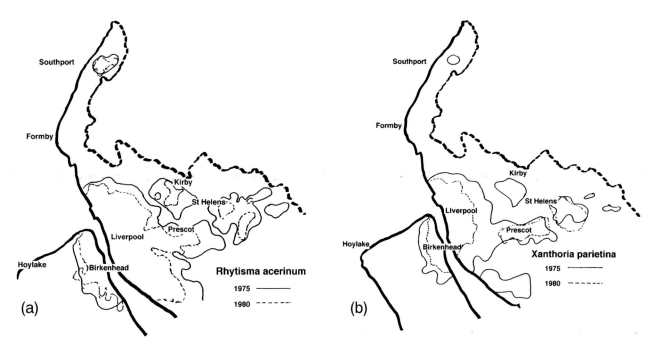

Figure 18.2. An approximate contour plot of the inner range of *Rhytisma acerinum (a)* and *Xanthoria parietina (b)*, derived from the work of Vick & Bevan (1975, solid line) and from Alexander (1980, dashed line).

based on absence or presence on substrates ranging from cement tile roofs to siliceous wall capstones, loosely based on relative pHs ranging from 9.6 to 5.25. This appears to work well for *L. muralis*, but has not been proven for *X. paretina* and *P. saxatilis*. Lack of substrate availability in the region, and the lack of attention to the age of the substrate exposure also reduced the value of this survey. The authors did indicate that some further improvement had occurred in the six years and correctly surmised that this type of study may have reached the limits of its usefulness. However, I strongly disagree with their conclusion that in general 'lichens as bioindicators of ameliorating conditions may be approaching the end of its viability as a survey method'. In my view, the subtlety of the changes occurring over a greater range of pollution sensitive species has only just begun to be appreciated. These taxa will, I am sure, provide a potentially powerful tool to recognise new pollutants in the future, when their subtle effects on lichens are more fully understood.

The lichens of Crack-willow (*Salix fragilis*) carrs in Cheshire

In 1977, G.M.A. Barker collected a small specimen of *Usnea subfloridana* in Shaw Green Willows in Rostherne Mere. This appears to be the first voucher-supported specimen of *Usnea* in Cheshire this century. In 1983 Mr Jonathen Guest drew my attention to further small specimens in Foden's Flash, a bird reserve. Since then there have been many reports of small plants of this species in many willow carrs, accompanied by a characteristic group of other lichens that appear to precede and succeed this species in this type of habitat.

This small community of species is characterised by a dominance of *Parmelia sulcata*, almost invariably associated with some *Parmelia subaurifera*, and/or *Parmelia glabratula*, *Hypogymnia physodes*, *Evernia prunastrii*, and more rarely by *Ramalina farinacea*, *Parmelia revoluta*, *Parmelia subrudecta*, *Parmelia caperata* and *Physcia aipolia*. It appears to be characteristic of Crack-willow (*Salix fragilis* – as distinct from other waterside *Salix* species) bark, and the community does not fit comfortably into any of the phytosociological groups described by James, Hawksworth & Rose in 1977. This community of lichens appears to be part of a pattern of recolonisation which has occurred over this period within many of the central industrial areas of England. The best site for observing this re-entry in most cases has been old Crack-willow overhanging standing water in partial shade. The preference for this habitat is almost certainly related, in part, to the favourable pH of this type of bark and the relatively constant high humidity content of the atmosphere surrounding it.

Their colonisation in Cheshire has also been observed on other substrates, such as boles of Sycamore trees, and was summarised by Guest in 1989. Several of the species have been observed as near as 15km from the centre of Manchester.

Dynamic patterns of lichen recolonisation

By observing a number of willow carrs in different parts of Cheshire, I was able to indicate (Fox 1988) that if their frequency within these habitats were compared, an order of entry could perhaps be established. The suggested sequence has been modified slightly following incorporation of further records from these areas. A more detailed study of the lichen population has

Figure 18.3. Typical Crack-willow carr, 1997. (*Photo: B.W. Fox*).

now been undertaken in readiness for the preparation of a full flora of all lichen species in Cheshire. Some indication of the present (*c.*1992) status of the *Parmelia sulcata* community can be deduced by studying the frequency of occurrence of different members of the *Parmelia sulcata* alliance in a selected number of 32 willow carr sites in Cheshire (Figure 18.5). This histogram gives some indication of the order of recolonisation of the different species of this community into these sites. It can be seen that *P. sulcata*, being present in almost all the sites, is a good indicator of the presence of this community at willow carr sites and appears to be the first coloniser of them. It would be valuable to survey these sites again now, some eight years later, to discover any further improvements in their status.

It is now well over a decade since levels of sulphur dioxide reached low winter mean levels of around 20–30 micrograms per cubic metre. Why therefore have we not seen the re-entry of *Graphis* species *Phlyctis argena*, more *Parmelia* species, *Nephromas*, and even some of the *Lobarion* species? There have been several accounts of small healthy looking tufts of *Usnea* species, which on later re-inspection have disappeared or degenerated, or at best halted in growth. In addition, the Hawksworth and Rose scale originally appeared to indicate that pollution levels corresponded with the more severe end, Zones 0 to 3, representing a severe level of pollution throughout the county, with sulphur dioxide levels in

Figure 18.4. Fallen branch of Crack-willow covered with *Parmelia saxatilis* and *Parmelia subaurifera*, in a Cheshire willow carr. (*Photo: B.W. Fox*).

excess of 125 micrograms per cubic metre (approximately .044 parts per million). The more recent appearance of such species as *Usnea subfloridana* and *Physcia aipolia* appear to be examples of the so-called zone skipping phenomenon, described by Hawksworth & McManus in 1989. This is considered to be due to the

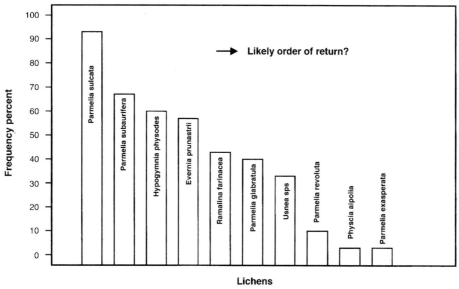

Figure 18.5. Relative abundance of different species derived from 32 Crack-willow carr sites, around 1992. The plot consists of the relative number of sites containing the species, and may represent the order of re-entry into these sites. The pattern refers to the situation around 1992, and a present-day survey would also include other species such as *Parmelia caperata* and *Parmelia perlata*.

preferential colonisation of some zone 6 and 7 species in the rapidly reducing sulphur dioxide environment, implying that the remaining species expected within these zones will appear in good time.

One can propose several other possible reasons which cause this impedance of the establishment of further new and more interesting species:

1. the levels of atmospheric sulphur dioxide may be rising again;

2. there may be some other lichen damaging contaminants from other sources, such as car emissions, oil and gas central heating systems, new industrial processes that may have caused new contaminants to enter the atmosphere which may be presently rising and are unmonitored;

3. there may be a greater influence of transient spikes of higher sulphur dioxide level caused by local climatic factors or occasional pollution accident.

This latter condition may be important as the figures for

Warrington (Figure 18.6), for example, show. The levels of sulphur dioxide rise episodically to high levels, and these could be highly damaging even for the short periods of exposure experienced. This kind of atmospheric pollution pressure could be responsible for the rise and fall observed in *Usnea* spp.

The monitoring of local spikes of pollution is not possible with the present distribution and numbers of monitoring equipment, and indeed to undertake this type of measurement may be too costly to apply generally. It would however be useful to be able to install monitors in those areas where intermittent growth of these species has been observed to see if any of these pollution spikes can still be detected.

I would like to propose that the factors favouring the successful return of the more sensitive species may be by establishing conditions which ensure that:

1. the development of the soredium on the substrate is largely unaffected by the atmospheric concentration of Establishment Hindering Chemicals (EHCs); and

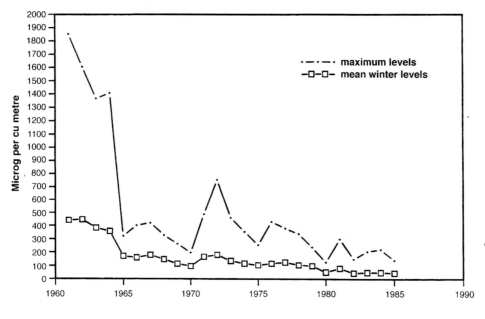

Figure 18.6. The maximum recorded sulphur dioxide levels for the Warrington area, compared with the annual means during the same years as in Figure 18.1. Note the enhanced ordinate scale.

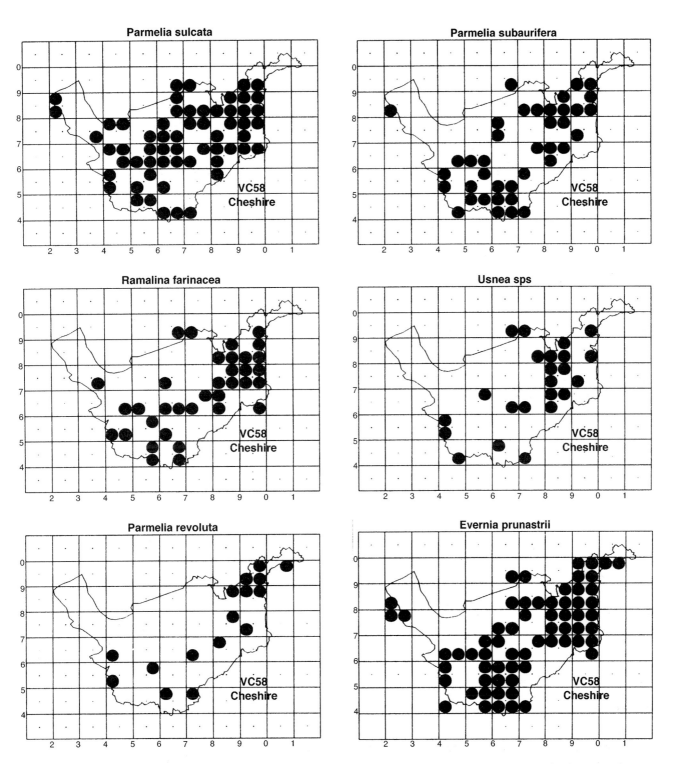

Figure 18.7. The distribution of six of the species used in the coincidence map, Figure 18.8. (Map prepared using software supplied by Dr Alan Morton, Imperial College, University of London, UK).

2. continuous growth of lichens once established on their substrate are not hindered by Continuous Growth Inhibiting Chemicals (CGICs). It may be the selective influence of these, as yet hypothetical pollutants on the different sensitivities of lichen species which is creating the zone skipping phenomenon. I would not be surprised to learn that the chemical nature of these two groups of pollutants will be found to be different, and will require detailed studies to determine their nature.

Using Alan Morton's DMAP software,* a coincidence map (Figure 18.8) of six members of this community can be plotted over Cheshire, and assuming that willow carrs can be found almost everywhere in the county, some indication of the geographical distribution of this

* Dr A. Morton, Department of Biology, Imperial College, University of London, Berks, UK

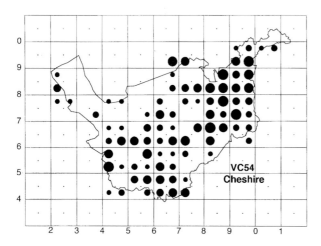

Figure 18.8. The coincidence distribution of the six species illustrated in Figure 18.7. Symbol size is directly related to the number of the species found together in the quadrant. It can be seen that clusters of enriched sites occur in the north-east and south-central areas of the county. A band of low frequency occurs from the industrial areas in the north-west to the south-eastern side of the county and could represent a plume of episodic pollution down-wind of the industrial areas in the north-west. The data applies to that observed in and around 1992. By 1996 this situation had almost certainly improved. (Dr Alan Morton's, DMAP software).

returning population can be deduced. It can be seen that the highest values occupy three broad areas, the north-eastern corner, the south and western corner, and a small area in the north-central region. This map was constructed from data up to 1991–92 and is likely to represent a transient status at that time. There appears to be a plume of slower recovery occurring across the county, down-wind of the industrial conurbations of Ellesmere Port and northern Wirral, which could lead to a quite different distribution following the recovery of lower pollution levels throughout the county. Only time will tell.

Conclusions

Since the pollution control measures of the 1960s, there has been a steady increase in the numbers and variety of lichens in all areas of Merseyside. More and more foliose and fruticose species are entering town parks in suitable sites and once established, an increasing proportion of young plants survive and continue to grow to larger, more visible lichens. Surveys undertaken in the 1970s and 1980s are reviewed and compared, and it is time to ask if the rapid changes occurring in the distribution of the taxa surveyed can provide new information about the changing quality of our atmosphere.

Of particular interest is the re-entry of a characteristic group of lichens dominated by *Parmelia sulcata* appearing on Crack-willow trees, usually overhanging wet areas in Cheshire. This data confirms that the consid-

erable decrease in the sulphur dioxide pollution of the atmosphere of Cheshire is reflected in a marked improvement in the quality of the lichen flora. There is evidence of 'zone skipping', considered to be the effect of a rapidly reducing sulphur dioxide level allowing selective growth of certain zone 6 and 7 species. What is inhibiting the more pollution sensitive species from appearing? Transient spikes of sulphur dioxide pollution occurring as a result of local climatic disturbance or occasional industrial problems is considered a possible cause. However, there is also the possibility that a further unknown series of pollutants may be affecting the establishment and continuing growth of certain species which would otherwise have started colonising selected sites without their influence. The need for more localised and selective sulphur dioxide monitoring sites is suggested, and further work to establish the possible nature of growth Inhibiting Chemicals (EICs and CGICs) is strongly recommended.

References

Alexander, R.W. (1982). The interpretation of lichen and fungal response to decreasing sulphur dioxide levels on Merseyside. *Environmental Education and Information*, **2 & 3**, 193–202.

Bishop, J.A. & Cook, L.M. (1975) Moths, melanism and clean air. *Scientific American* , **232(1)**, 90–99.

Ferry, B.W. & Baddeley, M.S. (1976). Sulphur Dioxide Uptake in Lichens. *Lichenology: Progress and Problems* (eds D.H. Brown, D.L. Hawksworth & R.H. Bailey), pp. 407–18. Academic Press, London.

Fox, B.W. (1988). Improvements in the lichen flora of a midland county. Abstract of AGM symposium. *Bulletin of the British Lichen Society*, **62**, 7.

Grindon, L.H. (1859). *The Manchester Flora*. William White, London.

Guest, J. (1989). Further colonization of Cheshire by epiphytic lichens. *Bulletin of the British Lichen Society*, **64**, 29–31.

Hawksworth, D.L. & McManus, P.M. (1989). Lichen colonization in London under conditions of rapidly falling sulphur dioxide levels, and the concept of zone skipping. *Botanical Journal of the Linnean Society*, **100**, 99–109.

Hawksworth, D.L. & Rose, F. (1970). Qualitative scale for estimating sulphur dioxide air pollution in England and Wales using Epiphytic Lichens. *Nature*, **227**, 145–48.

James, P.W., Hawksworth, D.L. & Rose, F. (1977). Lichen Communities in the British Isles: A preliminary conspectus. *Lichen Ecology* (ed. M.R.D. Seaward), pp. 296–413. Academic Press, London.

Lange, O.L., Herber, U., Schulze, E.D. & Ziegler, H. (1989). Atmospheric pollutants and plant metabolism. *Ecological Studies*, **7**, 238–73.

Marrat, F.P. (1860). Hepatics and Lichens of Liverpool and its vicinity. *Proceedings of the Literary and Philosophical Society of Liverpool*, **14**, (Appendix), 3–14.

North West Economic Planning Council (1974). *Strategic Plan for the North West*. HMSO, London.

Purvis, O.W., Coppins, B.J., Hawksworth, D.L., James, P.W. & Moore, D.M. (eds) (1992). *The Lichen Flora of Great Britain and Ireland*. The British Lichen Society, The Natural History Museum, London.

Richardson, D. (1975). *The Vanishing Lichens*. Douglas David and Charles Ltd, Canada.

Seaward, M.R.D. (1976). Performance of *Lecanora muralis* in an urban environment. *Lichenology: Progress and Problems*, (eds

D.H. Brown, D.L. Hawksworth & R.H. Bailey), pp. 323–57. Academic Press, London.

Travis, W.G. (1922). The Lichens of the Wirral. *Lancashire and Cheshire Naturalist*, **14**, 177–90.

Travis, W.G. (1925). Additions to the Lichen Flora of the Wirral. *Lancashire and Cheshire Naturalist*, **17**, 152–54.

Turner, D. & Dillwyn, L.W. (1805). *The Botanists' Guide through England and Wales*. Philips and Farndon, London.

Vick, C.M. & Bevan, R. (1976). Lichens and Tar Spot Fungus (*Rhytisma acerinum*) as indicators of sulphur dioxide pollution on Merseyside. *Environmental Pollution*, **11**, 203–16.

von Arb, C. & Brunold, C. (1989). Lichen physiology and air pollution. 1 Physiological responses in situ *Parmelia sulcata* among air pollution zones within Biel, Switzerland. *Canadian Journal of Botany*, **68**, 35–42.

Wheldon, J.A. & Travis, W.G. (1915). The lichens of South Lancashire. *Linnean Society Journal of Botany*, **43**, 87–136.

CHAPTER NINETEEN

Vascular plants: a game of chance?

E.F. GREENWOOD

Introduction

The Mersey Basin region is a varied one, made up of extensive coastal areas, mainly of saltmarsh and sand dune, but with some clay cliffs, intensive arable and pastoral agricultural areas, and on the eastern margins large upland areas over 500m altitude. Formerly important habitats, such as the many lowland raised bogs, shallow meres and sandstone headlands on the tidal River Mersey, have largely gone through developments of various kinds. Perhaps the most significant feature today is the presence of the large urban and industrial conurbations of Greater Manchester and Merseyside – a powerhouse of the industrial revolution for over 200 years. Human intervention in the area has been massive and nothing of the truly natural habitats remains. Nevertheless, large areas of wildscape and semi-natural areas survive or have been created in the last 200 years as a consequence of human intervention.

What has been happening to the flowering plants and ferns which, with other plant and animal species including humans, inhabit this area? The changes that have occurred may look as if it is all a matter of chance, but is it? Unfortunately the poor quality of the available data limits precise conclusions, but interesting generalisations are certainly possible.

The Mersey Basin area is defined as the catchment area of the River Mersey. (The Mersey Basin Campaign area also includes the Leeds & Liverpool Canal corridor.) It includes parts of the modern administrative areas of Cheshire, Derbyshire, Greater Manchester, Lancashire and Merseyside. However, botanical recording is based on Watsonian vice-counties, which are sub-divisions of the administrative counties of the late 19th century. Most of VC58 and 59 (Cheshire and South Lancaster) contain substantial parts of the Mersey Basin of which only a small upland area falls outside (in Derbyshire VC57) these vice-counties. This chapter relates primarily to vice-counties 58 and 59 (the Mersey Basin region) together with vice-county 60 (West Lancaster or most of Lancashire north of the River Ribble) where detailed recording has taken place over the last 30 years.

The first local floras (Hall 1839; Buxton 1849) are really no more than rough indications of what grew in the region 150 or so years ago. Since then a succession of local floras has been published, but it was not until the advent of the 'Atlas' (Perring & Walters 1962) that a quantitative approach to recording was developed at a 10 x 10km square level. Unfortunately, in the Mersey Basin only Cheshire, at a 5 x 5km square level, was recorded systematically (Newton 1971). Further north, in VC60, systematic recording on a 2 x 2km square basis was undertaken over the last 30 years, giving a much more detailed analysis than is available further south. However thorough, recording the presence or absence of species in an area as large as a 2 x 2km square is a somewhat chancy affair. The Botanical Society of the British Isles (BSBI) monitoring exercise in 1987–89 (Rich & Woodruff 1990) demonstrated the difficulties of trying to make assessments of change since 1962 on a quantitative basis. Despite these difficulties it is believed that over 100 years, at least, it is possible to make some valid generalisations. In compiling this account the main works used were those by de Tabley (1899), Newton (1971, 1990) (for Cheshire), Savidge, Gordon & Heywood (1963) (for South Lancaster) and Wheldon & Wilson (1907), and results of survey work compiled in the last 30 years (for West Lancaster). Where possible the results of recent recordings have been included.

Changes to the vascular flora over 200 years

Native species

These are defined as those which are present without intervention by humans, whether intentional or unintentional, having come from an area in which they are native. In practice it is impossible to determine the status of some species, and in this account all species which are known to have been established in the area for 500 or more years are regarded as native. This still leaves a few species whose status is uncertain and inevitably, in the absence of definitive information, decisions regarding the status of individual species are subjective.

Table 19.1 analyses the composition of the flora of the Mersey Basin region (VC58 and 59) along with floras of West Lancaster (60), Durham (66) and Northumberland (67 and 68), all of which contain large conurbations,

	*Total no. of species and hybrids (incl. aliens)	Total no. of native species and hybrids	Total no. of all aliens		Total no. of casual aliens		Total no. of extinct native species	
			No.	% of total no. of species and hybrids	No.	% of aliens	No.	% of total of native species
VC58 Cheshire	1,231	868	363	30	138	38	82	9.5
VC59 S. Lancaster	1,516	831	685	45	419	61	100	12.0
VC60 W. Lancaster	1,579	922	657	42	428	65	38	4.0
VC66 Durham[1]	1,656	1,000	656	40	226	34.5	96	9.6
VC67 & 68 Northumberland[2]	1,575	949	626	40	347	55	23	2.4

*Excluding critical micro species of *Rubus, Hieracium* and *Taraxacum*   [1] Graham (1988)   [2] Swann (1993)

Table 19.1. Analysis of the flora of five northern vice-counties in England.

upland and coastal regions, and compares them in terms of native, established and casual alien and extinct species. Extinct species are those which have not been recorded for a substantial time (say 30 years) or where the last established locality for a species is known to have been lost. Caution should be exercised when considering if a species is extinct, as some have only occurred at a few sites and then irregularly over long periods of time, e.g., Lesser Twayblade (*Listera cordata*). Those consid-

ered extinct in VCs 58–60 are listed in Table 19.2.

Table 19.1 shows that in the Mersey Basin region approximately 844 native species have been recorded (average for the five areas listed in the table: 913). This may be too low, and with more intensive fieldwork it should be possible to increase this number substantially – say by 30 species.

Of those native species recorded approximately 10.8% (or 91 species) have become extinct (listed in Table 19.2).

Table 19.2. Extinct native species in Cheshire and Lancashire (VC58-60) with dates of last known occurrence when known.

Species	VC58	VC59	VC60	BSBI *
Agrostemma githago	×	1964	✓	–49 S
Alopecurus bulbosus	×	1864	–	+91 I
Anagallis minima	×	1859	✓	–73 S
Antennaria dioica	–	1954	1909	–5 I
Anthriscus caucalis	✓	✓	1905	–21 I
Asplenium marinum	✓	1908	✓	–10 I
Asplenium viride	–	1915	–	–22 I
Bupleurum tenuissimum	×	–	–	–42 S
Calamagrostis canescens	✓	1892	–	4 I
Calystegia soldanella	×	✓	✓	–37 S
Carduus tenuiflorus	✓	1907	✓	–32 S
Carex diandra	×	1850	✓	–33 I
Carex dioica	×	✓	✓	–11 I
Carex divulsa	×	–	–	+28
Carex elata	✓	1880	✓	–20 I
Carex elongata	✓	1900	–	+35 I
Carex filiformis	1867	1859	–	–
Carex lasiocarpa	×	×	✓	–28 I
Carex limosa	✓	–	1912	–26 I
Carex strigosa	✓	1927	✓	+10 I
Carex viridula ssp. brachyrrhyncha	×	1964	✓	–2 I
Carex viridula ssp. viridula	×	✓	✓	–32 ?
Catapodium rigidum	×	1939	✓	–1 I
Centaurium latifolium	–	1872	–	–
Centaurium littorale	×	✓	✓	–100 S
Cephalanthera longifolia	–	–	1898	–
Ceratophyllum submersum	✓	1870	✓	+17 I
Cicuta virosa	✓	1861	–	–5 I
Clinopodium acinos	×	1859	✓	–21 I
Cochlearia officinalis	✓	1868	✓	–17 ?
Coeloglossum viride	×	1914	✓	–40 S
Cryptogramma crispa	×	×	✓	–12 I
Cuscuta spp. and C.epithymum	×	✓	–	–32 S
Daphne laureola	✓	1866	✓	–13 I
Descurainea sophia	×	✓	✓	+4 I
Dianthus deltoides	×	1887	–	–41 I
Diphasiastrum alpinum	–	1850	1907	–57 I
Draba muralis	–	1851	✓	+17 I
Drosera intermedia	1980s	1965	1858	–4 I
Drosera longifolia	×	1868	1902	–33 I
Dryopteris oreades	–	1957	–	+100 S
Dryopteris cristata	×	1851	–	–3 ?

* BSBI Monitoring Scheme changes (%) 1962–1987/88 based on sample 10 × 10km squares of the UK National Grid

Eleocharis multicaulis	✓	1885	✓	−11 I
Eleogiton fluitans	✓	1917	1925	−4 I
Elytrigia atherica	✓	1891	✓	0 I
Epipactis phyllanthes	×	✓	✓	+35 I
Erodium moschatum	1962	1929	1931	−32 I
Euphorbia exigua	×	✓	✓	−17 S
Euphrasia micrantha	−	−	1907	+5 ?
Filago minima	×	1914	✓	−29 S
Filago vulgaris	×	1930	✓	−33 S
Galeopsis angustifolia	×	1907	✓	−51 S
Genista anglica	×	✓	✓	−40 S
Gentiana pneumonanthe	✓	✓	1941	−17 I
Gentianella campestris	×	✓	✓	−28 I
Geranium columbinum	×	1913	✓	−13 I
Gnaphalium sylvaticum	×	1928	✓	−61 S
Groenlandia densa	×	1918	✓	−24 S
Gymnocarpium robertianum	−	1860	✓	+10 I
Gymnocarpium dryopteris	×	1964	✓	−43 S
Hammarbya paludosa	×	1878	−	+65 I
Helleborus viridis	−	1963	✓	−18 I
Huperzia selago	×	1880	✓	−32 I
Hymenophyllum tunbrigense	−	1840	1907	+100 I
Hymenophyllum wilsonii	−	1860	✓	−5 I
Hypericum elodes	✓	1873	1931	−4 I
Hypochaeris glabra	×	1928	−	−47 S
Impatiens noli-tangere	x	1903	✓	−22 I
Isoetes sp.	−	−	Pre-historic	−
Juncus balticus	−	✓	Early 1960s	−
Lamium confertum	−	1899	1907	−26 I
Lathyrus sylvestris	×	✓	−	−25 I
Lepidium ruderale	×	1959	−	−4 I
Limonium vulgare	×	1860	✓	−22 ?
Limosella aquatica	✓	✓	1964	−5 I
Listera cordata	×	✓	✓	−45 I
Lithospermum arvense	×	1936	✓	−26 S
Lycopodiella inundata	×	1880	−	−17 I
Lycopodium clavatum	×	✓	✓	−45 S
Maianthemum bifolium	−	1597	−	−
Marrubium vulgare	×	1926	−	−54 S
Mentha pulegium	×	✓	−	−33 I
Mertensia maritima	−	−	1941	−100 S
Meum athamanticum	−	1914	−	−33 I
Moenchia erecta	×	1915	−	−18 I
Monotropa hypopitys	×	✓	✓	−13 I
Myosurus minimus	✓	−	1857	−12 I
Ophrys insectifera	−	1850	✓	−29 I
Orchis ustulata	−	−	1930s	−66 S
Orobanche rapum-genistae	×	1850	−	−100 S
Papaver argemone	×	×	✓	−30 S
Peucedanum palustre	−	1870	−	−
Peucedanum ostruthium	×	1908	−	−22 I
Phegopteris connectilis	×	✓	✓	−18 I
Pilularia globulifera	×	1874	−	−5 I

Poa palustris	×	✓	−	−100 I
Polystichum setiferum	✓	1850	✓	+30 S
Potamogeton coloratus	×	1827	✓	−15 I
Potamogeton gramineus	✓	✓	1881	−28 I
Potamogeton praelongus	×	×	−	−100 S
Pseudorchis albida	×	1900	Early 20thC	−100 I
Puccinellia rupestris	×	1888	−	−4 I
Pyrola media	×	1860	−	+1 I
Pyrola minor	×	✓	−	−45 I
Radiola linoides	×	1915	1949	−35 S
Ranunculus arvensis	×	1934	1963	−52 S
Ranunculus parviflorus	×	×	−	−32 I
Ranunculus sardous	×	✓	✓	+13 I
Ranunculus x *bachii*	−	1914	1914	−
Rhynchospora alba	✓	1900	✓	−22 I
Ribes spicatum	−	1903	−	−45 I
Ruppia cirrhosa	−	1863	−	−45 I
Sagina subulata	×	−	−	−4 I
Salix phylicifolia	−	1888	✓	−19 I
Salvia verbenaca	✓	×	1901	+35 I
Saxifraga hirculus	×	−	−	−
Scandix pecten-veneris	×	×	✓	−72 S
Scheuchzeria palustris	×	Pre-historic	Pre-historic	−
Schoenus nigricans	×	✓	✓	−18 I
Sedum anglicum	✓	−	1885	−6 I
Selaginella selaginoides	×	1942	✓	−26 I
Seriphidium maritimum	1974	1859	✓	−23 I
Serratula tinctoria	✓	×	✓	−14 I
Silaum silaus	−	1888	1901	−8 I
Sparganium natans	×	−	✓	−45 I
Spergularia rupicola	✓	1927	1925	+6 I
Spiranthes spiralis	×	1905	✓	−50 S
Stellaria palustris	−	1926	1974	−9 I
Tephroseris palustris	−	−	1858	−
Torilis nodosa	×	1963	✓	−24 S
Trifolium micranthum	✓	1956	✓	−9 I
Trifolium squamosum	−	1811	−	−24 I
Trifolium subterraneum	×	1948	−	+17 I
Trifolium suffocatum	−	1964	1860	+91 I
Trollius europaeus	×	✓	✓	−16 I
Utricularia intermedia	−	1838	−	−100 I
Utricularia minor	✓	1956	1944	−37 I
Valerianella dentata	×	1903	1957	−39 S
Veronica spicata ssp. *hybrida*	−	−	1863	−
Vicia lutea	×	✓	−	+91 I
Viola tricolor ssp. *curtisii*	×	✓	✓	−
Wahlenbergia hederacea	✓	1908	1880	−38 S
Zostera marina	1851	1900	−	−57 I
Total	**82**	**100**	**38**	

S = Significant
I = Insignificant
− = No record
✓ = Not extinct
× = Believed extinct (no date)

The average for the five areas analysed in Table 19.1 suggests the loss should only be 7.5% (or 68 species). More intensive field survey might reveal that some supposedly extinct native species are still growing in the area. Nevertheless, it is likely that human intervention has adversely affected the survival of native species in the more urbanised areas. It is also interesting that the lists of extinct species for the three Cheshire and Lancashire vice-counties vary considerably. Often this is because the species which have become extinct were always rare and therefore vulnerable, being known to occur in only a few localities. Whilst reasons for extinction can be readily related to the destruction of habitats through urbanisation and agricultural improvements, habitat changes may be more indirectly caused by eutrophication of soils. However, climatic changes and other factors may be involved.

In an effort to obtain a better understanding of what might be happening, a more detailed analysis of changes to the native flora was undertaken in West Lancaster, which is better known to the author. In VC60 extinction of the native flora accounts for only 4.0% (38) of the total, but new native species discovered in the last 30 years accounted for 4.7% (43) of the flora, Table 19.3. Of these, 11 species or 1.2% of the native species were added as a consequence of a better understanding of more difficult

or critical groups. At least seven of the species have spread into the area (colonist) during the last 100 years whilst the remainder were probably overlooked by earlier recorders.

Wheldon & Wilson's *Flora* (1907) is a model for its time and reflects careful study and extensive fieldwork over 30 years at the end of the 19th century. With this in mind, an attempt was made to identify species that might be becoming more common (114 species) or less common (48 species), Tables 19.4 and 19.5. The figure for decreasing species is likely to be fairly accurate, but the much larger number of increasing species is due at least in part to the more thorough recording in the last 30 years.

In Table 19.6 the changes to the native flora of VC60 are summarised. This suggests that the numbers of new and extinct taxa are roughly equal, but that more native species are extending their range than are becoming more restricted.

This is perhaps a surprising conclusion. However, it is likely that the more thorough surveys of the last 30 years account for the present apparently optimistic picture. Unfortunately, not many attempts have been made either nationally or locally to assess change in this way. Nevertheless Braithwaite (1992), examining the largely rural county of Berwickshire on the east coast and

Species	Distribution preferences		BSBI†
	Geographical	Habitat	
Alisma lanceolatum	General	Aquatic	+1 I
Anthemis arvensis	General	Ruderal	–28 S
Atriplex longipes	Northern	Coastal	–
Bromopsis benekenii	General	Woodland	–
Bromopsis erecta	Southern	Grassland	–1 I
Bromus lepidus	General	Ruderal	–49 S
Callitriche brutia	Southern	Aquatic	+100 ?
Carex ericetorum	Continental	Grassland	–
Carex lasiocarpa	Northern	Heath	–28 I
Carex strigosa	Southern	Woodland	+10 I
Ceratophyllum demersum	General	Aquatic	+16 I
Ceratophyllum submersum	Southern	Aquatic	+17 I
Crepis biennis	Northern	Woodland	–20 ?
Dactylorhiza praetermissa	Southern	Grassland	+22 S
Daphne mezereum	General	Woodland	–
Deschampsia cespitosa ssp. *parviflora*	General	Woodland	–
Dryopteris aemula	Western	Woodland	–12 I
Epilobium alsinifolium	Northern	Aquatic	–22 I
Epipactis leptochila var. *dunensis*	Endemic	Coastal	–
Epipactis phyllanthes	Western	Coastal	+35 I
Eriophorum latifolium	General	Aquatic	+13 I
Festuca filiformis	Western	Heath	–11 ?
Filago minima	General	Ruderal	–29 S
Galeopsis bifida	General	Ruderal	–
Galium palustre ssp. *elongatum*	General	Aquatic	+99 ?
Glyceria declinata	Western	Aquatic	+20 S
Hordelymus europaeus	Southern	Woodland	–49 I
Hordeum secalinum	Southern	Grassland	–6 I
Ledum palustre ssp. *groenlandicum*	American	Heath	–
Lotus glaber	Southern	Grassland	–4 I
Myosotis stolonifera	Western	Aquatic	+65 S
Persicaria laxiflora	Southern	Aquatic	–16 I
Persicaria minor	General	Aquatic	–13 I
Poa angustifolia	General	Grassland	–
Poa humilis	Northern	Grassland	–
Potamogeton coloratus	General	Aquatic	–15 I
Puccinellia distans	General	Coastal	+40 S
Sorbus torminalis	Southern	Woodland	+35 S
Spartina anglica	General	Coastal	+3 I
Stellaria pallida	Continental	Coastal	–34 S
Trifolium micranthum	Southern	Grassland	–9 I
Veronica catenata	General	Aquatic	+16 ?
Vulpia myuros	Southern	Ruderal	+28 S

* Critical species
S = Significant
I = Insignificant
† BSBI Monitoring Scheme changes (%) 1962–1987/88 based on sample 10 × 10km squares of the UK National Grid

Table 19.3. Native species discovered in VC60, in the period 1907–1995.

Species	Geographical	Habitat	BSBI†
Acer campestre	Continental	Marginal	+4 I
Agrostis gigantea	General	Ruderal	+33 S
Aira caryophyllea	Southern	Ruderal	–15 S
Anchusa arvensis	General	Ruderal	–8 I
Atriplex littoralis	General	Coastal	–6 I
Berula erecta	General	Aquatic	0 I
Beta vulgaris ssp. *maritima*	Western	Coastal	+282 ?
Bidens cernua	General	Aquatic	–18 I
Bidens tripartita	General	Aquatic	–5 I
Blysmus rufus	Northern	Coastal	–5 I
Brassica nigra	Southern	Ruderal	–1 I
Cakile maritima	Western	Coastal	–18 I
Calamagrostis epigejos	General	Marginal	+4 I
Callitriche hamulata	Northern	Aquatic	–11 ?
Carex curta	General	Aquatic	–4 I
Carex distans	General	Coastal	–23 I
Carex elata	General	Aquatic	–20 I
Carex pilulifera	General	Heath	–16 S
Catabrosia aquatica	General	Aquatic	–13 I
Catapodium rigidum	Southern	Ruderal	–1 I
Centaurium pulchellum	General	Coastal	–7 I
Cerastium arvense	General	Marginal	–18 S
Chaenorhinum minus	General	Ruderal	+2 I
Chamerion angustifolium	General	Ruderal	+1 I
Coincya monensis ssp. *monensis*	Endemic	Coastal	–22 I
Conium maculatum	General	Ruderal	–1 I
Coronopus squamatus	General	Ruderal	+6 I
Crambe maritima	Western	Coastal	–23 I
Dipsacus fullonum	Southern	Marginal	+8 S
Echium vulgare	General	Ruderal	–19 S
Eleocharis quinqueflora	General	Aquatic	–12 I
Eleocharis uniglumis	General	Aquatic	–4 I
Elytrigia atherica	Western	Coastal	0 I
Epilobium roseum	General	Aquatic	+4 ?
Equisetum variegatum	Northern	Coastal	–22 I
Erodium cicutarium	General	Ruderal	–12 S
Festuca altissima	Northern	Woodland	+51 I
Festuca arundinacea	General	Marginal	+23 ?
Fumaria purpurea	Endemic	Ruderal	–33 I
Gagea lutea	General	Woodland	–17 I
Galium uliginosum	General	Aquatic	–18 S
Glyceria maxima	General	Aquatic	+4 I
Glyceria notata	General	Aquatic	+16 S
Helictotrichon pubescens	Northern	Grassland	+8 I
Hordeum murinum	Southern	Ruderal	+1 I
Juncus subnodulosus	General	Aquatic	+17 ?
Lamium album	General	Marginal	–3 I
Lamium amplexicaule	General	Ruderal	+2 I
Lamium hybridum	General	Ruderal	+25 S
Lathraea squamaria	General	Woodland	–23 I
Lemna gibba	General	Aquatic	–13 I
Lepidium campestre	General	Ruderal	–28 S
Lepidium heterophyllum	Western	Ruderal	–31 S
Leymus arenarius	Western	Coastal	+8 I
Limonium humile	Western	Coastal	–22 ?
Limonium vulgare	Western	Coastal	–22 ?
Milium effusum	General	Woodland	–1 I
Myosotis ramosissima	General	Coastal	–20 S
Myosotis secunda	Western	Aquatic	+16 I
Myosotis sylvatica	General	Marginal	+62 S
Myriophyllum alterniflorum	General	Aquatic	–15 I
Ophrys apifera	Southern	Marginal	0 I
Ornithogalum umbellatum	Southern	Marginal	+14 I
Ornithopus perpusillus	Western	Grassland	–25 I
Orobanche minor	Southern	Ruderal	–20 S
Parapholis strigosa	Western	Coastal	–1 I
Phleum arenarium	Southern	Coastal	–16 I
Picris echioides	Southern	Marginal	+4 I
Plantago coronopus	Southern	Coastal	–9 I
Poa compressa	General	Ruderal	+27 S
Polygonum oxyspermum ssp. *raii*	Western	Coastal	–24 ?
Polystichum setiferum	Southern	Woodland	+30 S
Potamogeton alpinus	General	Aquatic	–49 I
Potamogeton pectinatus	General	Aquatic	+42 S
Potentilla palustris	General	Aquatic	–16 I
Ranunculus lingua	General	Aquatic	+40 S
Ranunculus sardous	General	Ruderal	+13 I
Ranunculus sceleratus	General	Aquatic	+2 I
Raphanus raphanistrum ssp. *maritimus*	Western	Coastal	–11 I
Raphanus raphanistrum ssp. *raphanistrum*	General	Ruderal	–7 I
Ribes nigrum	General	Marginal	+22 S
Ribes rubrum	General	Marginal	+21 ?
Rorippa palustris	General	Aquatic	+12 I
Rorippa sylvestris	General	Aquatic	+3 I
Rumex hydrolapathum	General	Aquatic	–7 I
Sagina maritima	Western	Coastal	–28 I
Salix viminalis	General	Aquatic	+10 S
Samolus valerandi	General	Coastal	–15 I
Saxifraga granulata	General	Marginal	–15 I
Schoenoplectus tabernaemontani	General	Aquatic	+11 I
Scrophularia auriculata	Southern	Aquatic	–2 I
Senecio erucifolius	Continental	Marginal	+3 I
Senecio viscosus	General	Ruderal	+8 I
Seriphidium maritimum	Western	Coastal	–23 I
Solanum nigrum	General	Ruderal	+4 I
Spirodela polyrhiza	General	Aquatic	–11 I
Stachys arvensis	Southern	Ruderal	–15 S
Stellaria neglecta	General	Woodland	+23 I
Tanacetum vulgare	General	Marginal	+1 I
Thalictrum flavum	General	Aquatic	–14 I
Thlaspi arvense	General	Ruderal	–1 I
Tilia cordata	Continental	Woodland	+19 I
Trifolium fragiferum	Western	Coastal	–15 I
Typha angustifolia	General	Aquatic	+14 I
Valerianella locusta	General	Ruderal	–16 S
Veronica anagallis-aquatica	General	Aquatic	–8 ?
Veronica hederifolia	General	Ruderal	+6 I
Veronica scutellata	General	Aquatic	–18 S
Vicia hirsuta	General	Ruderal	+1 I
Vicia lathyroides	General	Coastal	+26 I
Vicia tetrasperma	General	Ruderal	+10 I
Viola palustris	General	Aquatic	–5 I
Viola reichenbachiana	Southern	Woodland	–2 I
Vulpia bromoides	Southern	Ruderal	–5 I

S = Significant
I = Insignificant
† BSBI Monitoring Scheme changes (%) 1962–1987/88 based on sample 10 × 10km squares of the UK National Grid

Table 19.4. Species showing an apparently increasing frequency of occurrence in VC60, 1907–1992.

Species	Distribution preferences		BSBI†
	Geographical	Habitat	
Actaea spicata	General	Marginal	–
*Agrostemma githago	Southern	Ruderal	–49 S
Asplenium marinum	Western	Coastal	–10 I
Asplenium viride	Northern	Rocks	–22 I
Baldellia ranunculoides	General	Aquatic	–37 S
Botrychium lunaria	General	Grassland	–19 I
Cirsium heterophyllum	Northern	Marginal	–22 I
Coeloglossum viride	General	Grassland	–40 S
Cryptogramma crispa	Northern	Rocks	–12 I
Cynoglossum officinale	General	Marginal	–34 S
Drosera rotundifolia	General	Heath	–13 I
Filago vulgaris	Continental	Ruderal	–33 S
Gentianella campestris	General	Grassland	–28 I
Gnaphalium sylvaticum	General	Grassland	–61 S
Groenlandea densa	General	Aquatic	–24 S
Helleborus viridis	General	Woodland	–18 I
Huperzia selago	General	Rocks	–32 I
Hymenophyllum wilsonii	Western	Rocks	–5 I
Jasione montana	General	Heath	–22 S
Juniperus communis	General	Marginal	–11 ?
Limonium britannicum ssp. celticum	Endemic	Coastal	–11 ? (for agg).
Listera cordata	Northern	Heath	–45 I
Lithospermum arvense	General	Ruderal	–26 S
Lycopodium clavatum	General	Heath	–45 S
Malva neglecta	General	Ruderal	–3 I
Ophrys insectifera	General	Marginal	–29 I
Orchis morio	General	Grassland	–32 S
Osmunda regalis	Western	Aquatic	+23 I
Paris quadrifolia	General	Woodland	–24 S
Pedicularis palustris	General	Aquatic	–25 S
Pinguicula vulgaris	General	Aquatic	–14 I
Platanthera bifolia	General	Grassland	–34 S
Platanthera chlorantha	General	Grassland	–13 I
Primula farinosa	General	Aquatic	–5 I
Pyrola rotundifolia	Western	Coastal	+100 ?
Ranunculus baudottii	Western	Coastal	–4 I
Ranunculus circinatus	General	Aquatic	–27 S
Ranunculus peltatus	Western	Aquatic	+41 ?
Salix myrsinifolia	Northern	Marginal	–17 ?
Salix phylicifolia	Northern	Marginal	–19 I
Selaginella selaginoides	Northern	Aquatic	–26 I
Sherardia arvensis	General	Ruderal	–13 S
Sparganium natans	General	Aquatic	–45 I
Trollius europaeus	Northern	Marginal	–16 I
Utricularia vulgaris	General	Aquatic	–16 I
Verbena officinalis	General	Marginal	–15 I
Veronica agrestis	General	Ruderal	+7 I
Veronica polita	General	Ruderal	+4 I

S = Significant
I = Insignificant
† BSBI Monitoring Scheme changes (%) 1962–1987/88 based on sample 10 × 10km squares of the UK National Grid
* Probably of ancient introduction

Table 19.5. Species showing an apparently decreasing frequency of occurrence in VC60, 1907–1997.

	Numbers	% of native VC60 flora	National changes to VC60 species			
			Decreasing		Increasing	
			No	% of VC60 species	No	% of VC60 species
New taxa	43	4.7	17	40	16	37
Increasing taxa	114	12.4	65	57	46	40
Total	157	17.1	82	52	62	40
Extinct taxa	37	4.0	25	68	5	14
Decreasing taxa	48	5.2	42	88	5	10
Total	85	9.2	67	79	10	12

Table 19.6. Summary of changes in the native flora of VC60, 1907–1995 compared with possible changes observed nationally 1962–1988.

over a much shorter time period, concluded that 11.5% of the flora was declining and that only 5.5% was showing any increase – almost the exact reverse of the results reported here.

At a national level the BSBI instigated a sample survey, or monitoring scheme, of which one of the objectives was to assess the change in frequency of species in the British Isles between the publication of the 'Atlas' (Perring & Walters 1962) and their survey period of 1987/88. A report was produced (Rich & Woodruff 1990)

which described the project and presented the results. In doing so the authors described the difficulties of making any accurate conclusions. In particular the problems of comparing like with like proved largely insuperable. As a consequence in only a few cases was it possible to determine change in frequency with any confidence. Nevertheless the report (although not reproduced in Palmer & Bratton 1995) does indicate the percentage changes noted for each species and, using the data for England, included in Tables 19.4 and 19.5. Table 19.6

summarises the changes nationally for VC60 species. This shows that, for both extinct/decreasing species and new/increasing species, most appear to be declining in frequency nationally, but there is a difference between the two groups. In the new/increasing group 40% are increasing nationally whilst 52% are decreasing. However, in the extinct/decreasing group only 12% are increasing whilst 79% are decreasing nationally. However, the BSBI survey was assessing change over only the last 30 years or so, rather than the last 90 years for VC60. Nevertheless, despite the inadequacies of the data and difficulties of comparing like with like, it may be that the trends reported here for VC60 are reflected, on a shorter time-scale, nationally.

In Tables 19.7 and 19.8 an attempt is made to reflect those species showing change in terms of broad European phytogeographical affinities and habitat preferences. These tables show that generally distributed species or species with a southern distribution tend to be increasing, whilst northern species are declining. In terms of habitat preferences aquatic, ruderal and coastal species are all increasing, but heath and grassland species appear particularly vulnerable to change.

	Decreasing and extinct	Increasing and new	Net loss (-) or gain (+)
Northern	18	10	-8
Southern	5	28	+23
Western	14	21	+7
Continental	2	5	+3
General	46	89	+43
American	0	1	+1
Endemic	0	3	+3
Total	85	157	

Table 19.7. Changing numbers of native species in VC60, 1907–1992 according to European phytogeographical preferences (number of species).

	Decreasing and extinct	Increasing and new	Net loss (-) or gain (+)
Heath	13	4	-9
Aquatic	19	46	+27
Rock	6	0	-6
Ruderal	11	37	+26
Coastal	7	30	+23
Grass	16	9	-7
Wood	4	16	+12
Marginal	9	15	+6
Total	85	157	

Table 19.8. Changing numbers of native species in VC60, 1907–1992 according to broad habitat preferences (number of species).

These changes accord well with observed changes in environmental data where available. In 1991 Lancashire County Council published 'A Green Audit' (Anon 1991) which summarised these changes. In terms of climatic change over a period of 210 years, 1750–1960, summer temperatures have remained more or less unchanged but in winter January temperatures have risen 1.7°C, in February 0.5°C, in November 0.8°C and in December 1.1°C. Thus, winters have become warmer, but it should be noted that this warming started before the full onset of the industrialisation of the Mersey Basin region in the 19th century. Figures were not calculated for the years since 1960, but they include some of the coldest and mildest winters on record as well as some of the warmest summers. There is no doubt that there has been an overall warming of the climate, (Figure 19.1), which favours species of a more southerly distribution, whilst northern species are disadvantaged. Whether the warming is caused or influenced by human intervention is still uncertain, but recent evidence suggests that it is (see chapter one, this volume).

Rainfall data is available for Preston covering the 140 years from 1850 (Figure 19.2). Rainfall is highly variable,

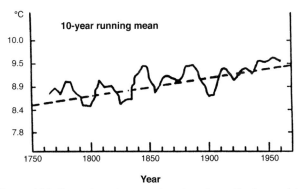

Figure 19.1. Long-term temperature trends on the Lancashire plain, 1750–1960.
(Source: Savidge, Heywood & Gordon 1963, and reproduced by permission of Liverpool Botanical Society.)

but overall the trend is for the climate to get wetter – by perhaps 100mm over the period of 140 years. This would favour wetland species where drainage has not taken place. However, the dry period starting in 1995 is exceptional, but it is still too early to determine its significance.

Data for habitat changes are poor. Woodland has probably changed little in extent, but over the last 200 years mainly broad-leaved plantations were planted in the Fylde (and in South Lancaster), and in the last 50 years conifer woods were planted more widely. Overall there is probably more woodland today than at any time in recent history, but this masks the destruction of old, more natural and therefore species rich woodland.

Hedgerows can be described as woodland edge habitats and can form important linear habitats. In the lowland areas in the west of Lancashire (and in

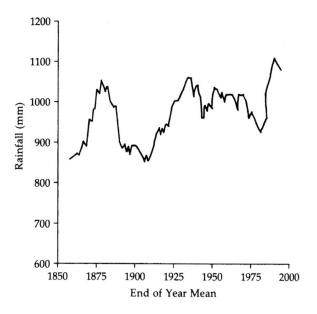

Figure 19.2. Rainfall at Preston, 1849–1989. (Source: Lancashire Polytechnic 1989, and reproduced by persmission of Liverpool Botanical Society).

Merseyside) most of the hedges were planted in the last 200 years. Nevertheless, it is estimated that 10% of the county's hedges are species rich. In a comparative survey between 1963 and 1988 about 25% of the hedges seem to have been lost, but often in localised areas in the west where few and comparatively recently planted hedges existed. In Cheshire 60% of hedges seem to have been lost in the period 1850–1987 (Anon 1993).

Upland heaths have probably not changed greatly in extent but moor burning is a characteristic feature of land management; occasionally more devastating fires have occurred, e.g., in 1947 in Bowland. However, moor burning, increased grazing and atmospheric pollution, especially of nitrogen, have probably changed the nature of the upland areas, reducing the extent of dwarf shrub communities and favouring the spread of grasslands (chapter eleven, this volume; de Smidt 1995). In lowland areas 98% of the Lancashire peat mosslands have been lost, a situation certainly reflected further south in the Mersey Basin. There were probably only a few lowland dry heaths in Lancashire and they have largely been lost, but a few persist south of the River Mersey, especially in Wirral. However, these have seen a considerable loss of species diversity during the last 100 years or so.

There have been major changes to aquatic habitats, but it is possible that overall there has been an increase in area. In the north of West Lancaster (VC60) the extensive swamps and fens at Leighton Moss containing 80ha of reedbed are entirely new, having been formed in the last 80 years or so. Similarly, numerous gravel pits, salt subsidence pools and reservoirs have greatly added to the extent of aquatic habitats over the last 100 years. On the other hand, marl pits have been reduced by some 63% (10,758 ponds) between 1845 and 1988 in Lancashire as a whole. However, half of this loss was between 1965

and 1988. In Cheshire (Anon 1993) the loss in a similar period was also 63% (29,500 out of 47,000).

In Lancashire there have been major changes in coastal habitats. Destruction of cliffs, saltmarshes and sand dunes occurred through urbanisation, industrialisation and land reclamation. Natural accretion (and erosion) has also taken place, so that overall there is perhaps an increase in the extent of coastal habitats with perhaps larger areas of saltmarshes (Berry 1967; Pringle 1987; Gray & Scott 1987), but smaller areas of sand dunes. However, change in the extent of coastal habitats may be cyclical as the relatively recently formed saltmarshes at Silverdale are now (1996) eroding (Peter 1994).

No figures exist for changes in Lancashire grasslands, but nationally it is estimated that 95% of species rich lowland, neutral grassland has been lost. It is believed that the local situation in the Mersey Basin region and in West Lancaster reflects the national position.

Comparable water quality data for Lancashire is not available, but it is clear that in recent years there has been considerable nutrient enrichment affecting in particular aquatic, heath and grassland habitats. Indeed it is known that in Lancashire EC standards are frequently exceeded, especially with high levels of nitrogen. During the last 20 years or so there have been heavy applications of nitrogenous fertilisers to agricultural land, and enrichment has also been caused by higher levels of atmospheric nitrogen (de Smidt 1995).

Rocky habitats in Lancashire have always been limited. However, the development of Heysham Harbour and public pressure at Silverdale removed or severely degraded this habitat. Similar habitats on the Mersey Estuary were also lost through urbanisation in the 19th century. Inland, peat erosion has revealed new areas of rock and scree in upland areas but they are nutrient poor and inhospitable to plant growth. Any plants that do colonise these areas are also liable to be eaten, as they are within heavily grazed areas the intensity of which has increased considerably in recent years.

Data for urbanisation of the landscape has not been compiled. However, by the beginning of the 19th century the population of Lancashire had already increased from a mainly rural (85%) population of 196,100 in 1690 to 672,700 in 1801. By 1994 the largely urban (75%) population of 1,424,000 was found in greatly enlarged conurbations. Much larger increases in population occurred in Greater Manchester and Merseyside with a population in 1994 of 2,578,000 and 1,434,400 respectively (Church 1996).

To accommodate this enlarged population and its associated industries considerable areas of agricultural land and semi-natural habitats were used by extending existing towns and cities, and in some areas by creating new ones (Freeman 1962; Lawton 1982).

Despite the quality of the available data, it would appear that the floristic changes observed in West Lancaster (VC60) are consistent with the habitat changes described above for a wider area.

The marked increase in aquatic species is consistent

with a wetter climate and the general increase in fens and man-made habitats of gravel pits and reservoirs, etc., despite the loss of marl pits and some drainage. However, closer examination shows that losses are of species requiring nutrient poor water, e.g., Lesser Bladderwort *(Utricularia minor),* whilst the increasing ones favour nutrient rich waters, e.g., Celery-leaved Buttercup *(Ranunculus sceleratus),* Bur-marigolds *(Bidens* spp.), etc., which is consistent with increased eutrophication.

Similarly, the increase in ruderal species reflects the general increase in the size of the urban habitat and the creation of open habitats through man-made activities. The significance of urban habitats will be discussed further in relation to alien species. Some reduction in arable farmland weeds might be expected, but this was scarcely if at all detected.

The increasing spread of coastal species is consistent with the developing saltmarshes and although the extent of the sand dunes is reduced, developing dunes are still found on the Fylde coast.

There seems to be a slight, perhaps insignificant, increase in species favouring woodland and marginal habitats. This is difficult to explain as there has been a reduction of older more natural and species rich woodlands.

The decrease in species favouring heaths, grasslands and rocky places correlates well with the known decrease in the extent of these habitats and with increasing soil nitrogen levels.

Tables 19.4–8 inclusive, which form the basis of this analysis, were compiled by comparing data for 1907 with data for 1992. Those species which have had short-term changes in abundance or where localities were not listed

Figure 19.3. Danish Scurvygrass on a Merseyside roadside *(Photo: H. Ash).*

by Wheldon & Wilson (1907) do not appear in the tables and yet can be very interesting. There has been an explosive spread of Danish Scurvygrass (*Cochlearia danica*) along motorways and A class roads in the Mersey Basin since 1992. In West Lancaster this spread was not noted until 1995 when the central reservation of the M6 suddenly became carpeted with this species from its junction with the M55 north of Preston to Carnforth and along the M55 itself towards Blackpool. The spread of this plant has been noted throughout England, and several notes have appeared in *BSBI News* (e.g., Allen 1996; Pinkiss 1996) reporting its spread and postulating reasons why this might have occurred. The most favoured reason is that the use of rock salt as a winter de-icing agent produces open habitats with saline conditions favoured by this species. However, it has also been noted to colonise non-saline but open habitats, and in Wirral it has colonised bare ground following roadside herbicide treatments.

Less easy to document has been the apparent spread of Sticky Mouse-ear (*Cerastium glomeratum*). In 1907 it was too common for localities to be listed by Wheldon & Wilson (1907), but qualitatively it seems to have increased in abundance not only in West Lancaster but generally; this was one of the few species where Rich & Woodruff (1990) noted a significant increase. It is a ruderal and grassland species favouring nitrogen enriched soils.

Western Gorse (*Ulex gallii*) is also not included in the tables, although for a period of several years in the 1970s it became particularly abundant for several miles on the roadside banks of the M6 near Lancaster. Heathland habitats, which it favours, are rare in VC60, but a suitable nearby seed source occurred at Galgate and the newly built motorway provided a suitable habitat. However, by 1996 it had disappeared.

These changes suggest that complex forces are at work in determining the changing abundance and distribution of species. Some of the changes are very short-term, but others appear long-lasting. Whilst the causes of change for some species may be clearly indicated, e.g., loss of habitat, for many species the reasons remain unknown.

The climatic and habit changes described are appropriate to most of the Mersey Basin region, with perhaps a greater emphasis there on habitat destruction through urbanisation. There is no reason to believe that, whilst this analysis is based largely on West Lancaster data, similar floristic changes and the processes involving change are not also taking place in the region as a whole.

Alien species

If the analysis of change to the native flora is somewhat inconclusive, an analysis of alien species presents further problems. Alien species are defined as those which are brought into the area by humans either intentionally or unintentionally, but which may be native in some other parts of the British Isles; or have come to the area without human intervention, but from an area in which it is alien.

As was discussed earlier, applying this definition in practice often causes difficulties which are compounded by trying to determine if a species is naturalised or not. A naturalised (or established) species is one which has been present at a site in the wild (i.e., without human intervention) for at least five years, and is spreading vegetatively or is effectively reproducing by seed. This definition also includes persistent species which are present in the wild for at least five years, but are neither spreading significantly vegetatively nor by seed. An attempt was made to apply these definitions to the flora of the Mersey Basin region and VC60. Using Newton (1971, 1990); Savidge, Heywood & Gordon (1963) and recent data, especially for VC59 and VC60, an attempt was made to determine the status of alien species on an individual basis. In making this analysis personal judgement was used, and as such the results might well vary if other individuals or a committee approach were taken to assess status.

Nevertheless, Table 19.1 shows that alien species form an important component of the total flora ranging from 30% in Cheshire to 45% in South Lancaster. In Cheshire the low figure is probably due to recorders not noting many clearly introduced species of casual occurrence or not considering species were 'wild'. However, if the number of naturalised species is considered, the numbers show much less variation ranging from 15% in West Lancaster to 18% in Cheshire (Table 19.1 and 19.9).

There is no doubt that, whilst there may be a small net loss of native species to the region, the total number of established taxa growing in the area is now greater than at any time in recent history. At any one time this total is further increased by a large number of casual species. This increase is due almost entirely to human intervention of some kind.

In order to understand further when and how the naturalised aliens came to be established, an analysis of the local floras of the region was undertaken. Table 19.10 lists the established alien species for Cheshire and Lancashire (VC58, 59 and 60) with an approximate date of when a species was first recorded or noted. No data is given where a species is not considered naturalised, even though it may occur in the vice-county.

Table 19.11 groups the date of first record into 50 year categories (chosen on the basis of the dates of local flora publications). This demonstrates that the numbers of species becoming established on an annual basis remains remarkably constant over nearly 170 years, averaging one to two species per annum, but the figures also suggest that during the present century the rate of naturalisation has increased.

Using local floras and Clement & Foster (1994), the main means of introduction for each species can be assessed (Table 19.12). Despite the influence of the major ports on the River Mersey, most species naturalised in the region are garden escapes, which, together with medicinal and culinary herbs and plants grown in aquaria, account for some 68% of naturalised alien species. Much less important are species introduced with crops or escaped crop and amenity plantings and accidental introductions. Also amongst the new taxa are a small number that are new to science. Examples include Common Cord-grass (*Spartina anglica*) and Montbretia (*Crocosmia* x *crocosmiflora*).

It may seem surprising that the horticultural industry should be so important, but the region has a long history of horticultural endeavour extending well over 200 years. This involved all classes of society from the 18th century (Secord 1994a & b). It also saw the establishment of the Liverpool Botanic Garden in 1802, the University

	Total no. Aliens	Casual		Naturalised	
		No.	% of total aliens	No.	% of total aliens
VC58	363	138	38	225	62
VC59	685	419	61	266	39
VC60	657	428	65	229	35

Table 19.9. Number of casual and alien species in Lancashire and Cheshire VC58–60.

Date Class	Before 1830 (50 years)	1830–1879 (50 years)	1880–1929 (50 years)	1930–1979 (50 years)	After 1980
Vice-county					
Cheshire 58	8%	32%	12.5%	35%	13%
S. Lancs 59	8%	25%	25%	35%	7%
W. Lancs 60	5%	14%	28%	44%	9%
Average	7%	23%	22%	38%	10%
Equivalent no. of species	17	57	53	91	24 (80 extrapolated over 50 years)
No. per year (approx.)	–	1	1	2	2

Table 19.11. Date of first record of naturalised aliens in Cheshire and Lancashire (VC58–60) (% of total naturalised aliens).

Table 19.10. Naturalised alien species in Cheshire and Lancashire (VC58–60)

Name of Plant	VC58	VC59	VC60	Means of Introduction	Region of Origin
Acer platanoides	1971	Post 1963	1964	Amenity	Eurasia
Acer pseudoplatanus	Before 1830?	15th Century	Before 1830?	Amenity	Eurasia
Aconitum napellus	–	1963	–	Garden	Europe
Acorus calamus	1670	1872	1837	Garden	E.Asia
Adiantum capillus-veneris	1971	–	–	Unknown	BI. World-wide
Aesculus hippocastanum	Before 1830?	Before 1830?	Before 1830?	Amenity	Europe
Agrostemma githago	1868	1839	1775	Grain	Mediterranean
Agrostis scabra	–	1966	–	Grain	Europe
Allium carinatum	1976	1995	1948	Garden	Europe, Asia
Allium oleraceum	1899	1865	1901	Accidental	Eurasia
Allium paradoxum	–	1983	1989	Garden	C.Asia
Allium scorodoprasum	1899	1899	1908	Accidental	Eurasia
Alnus incana	–	1960	1987	Amenity	Eurasia
Alopecurus myosuroides	1835	1838	1883	Accidental/Natural Spread?	Eurasia, Africa
Althaea officinalis	–	1826	–	Accidental?	Eurasia, Africa
Ambrosia psilostachya	–	1903	1902	Grain	N.America
Anaphalis margaritacea	1971	1907	1907	Garden	Eurasia, N.America
Angelica archangelica	1899-1933	1887	1982	Medicinal	Eurasia
Antirrhinum majus	1933	1807	1901	Garden	Mediterranean
Aponogeton distachyos	–	1976	–	Aquarium	S.Africa
Aquilegia vulgaris	1859	Native?	Native	Garden or natural spread	Eurasia, Africa
Arenaria balearica	1931	1933	1942	Garden	Mediterranean
Armoracia rusticana	1597	1933	1891	Medicinal	Eurasia, Africa
Arum italicum	–	1994	–	Garden	Europe
Asparagus officinalis ssp. *officinalis*	1838	1933	1829	Agriculture	Eurasia
Aster novi-belgii s.l.	1880	1837	1964	Garden	N.America
Astrantia major	–	1960	1987	Garden	Europe
Avena fatua	1805	1840	1900	Grain	Europe
Azolla filiculoides	1939	1976	1985	Aquarium	Americas
Barbarea intermedia	1859	1862	1967	Accidental	Europe, Africa
Berberis vulgaris	1808	1851	1858	Garden	Europe
Bidens frondosa	1989	1913	–	Garden	N. & S.America
Borago officinalis	–	1910	–	Garden	Europe
Brassica napus	1860s	1962	1962	Agriculture	?
Brassica rapa ssp. *campestris*	1850s	1963	1874	Agriculture	Europe
Bromopsis inermis	1990	1940–49	1948	Agriculture	Europe
Buddleja davidii	1971	Post 1963	1966	Garden	China
Calla palustris	1970	1994	–	Garden	Eurasia & N.America
Calystegia pulchra	1968	1951. 1st British	1964	Garden	N.E.Asia
Calystegia silvatica	1962	1863. 1st British	1952	Garden	S.W.Asia
Campanula persicifolia	–	1939	–	Garden	Eurasia
Campanula rapunculoides	1859	1871-1900	1881	Garden	Eurasia
Carpinus betulus	1859	1851	1907	Amenity	Eurasia
Castanea sativa	1851	1851	1907	Amenity	Eurasia, Africa
Centaurea montana	1990	1918	1966	Garden	Europe
Centranthus ruber	1851	1851	1962	Garden	Mediterranean, Asia
Cerastium tomentosum	1933	1933	1931	Garden	Europe
Ceratochloa carinata	1970	1958	1989	Accidental	N.America
Chamaecyparis lawsoniana	c.1990	–	–	Garden	N.America
Chelidonium majus	1839	1839	1858	Medicinal	Eurasia
Chenopodium bonus-henricus	1839	1839	1874	Medicinal	Eurasia
Chenopodium murale	1839	1839	1900	Accidental	Eurasia, Africa
Cicerbita macrophylla	1971	1960	1942	Garden	Eurasia
Cichorium intybus	1839	1914	1860	Garden	Eurasia
Claytonia perfoliata	1860s	1916	1946	Garden	N.America
Claytonia sibirica	1962	1876	1902	Garden	N.America
Clematis vitalba	1801	1850	1899	Garden or natural spread	Eurasia, Africa
Conyza canadensis	1925	1913	1962	Accidental	N.America

Name of Plant	VC58	VC59	VC60	Means of Introduction	Region of Origin
Cornus sericea	1990	1940	1986	Amenity	N.America
Coronopus didymus	1847–59	1880	1915	Accidental	S.America
Corrigiola litoralis	–	1928	1965	Accidental	Eurasia, Africa
Cotoneaster horizontalis	1990	1955	1967	Garden	W.China
Cotoneaster microphyllus/ integrifolius	–	–	1902	Garden	Himalayas, W.China
Cotoneaster salicifolius	–	1991	–	Garden	China
Cotoneaster simonsii	–	1961	1967	Garden	Himalayas
Cotula coronopifolia	1880	1970s	–	Garden	S.Africa, Australasia
Cotula squalida	–	1960	–	Garden	New Zealand
Crambe cordifolia	–	1988	–	Garden	S.E.Asia
Crassula helmsii	1990	1976	1984	Aquarium	Australasia
Crataegus laciniata	–	1964	–	Garden	Europe, Asia
Crepis vesicaria	1941	1930	1979	Accidental	Mediterranean, Asia
Crocosmia x crocosmiiflora	1970	1963	1964	Garden	Cultivated
Crocus nudiflorus	1830	1830	1907	Garden	Europe
Crocus tommasinianus	c.1990	–	–	Garden	Europe
Crocus vernus	1859	1808	1885	Garden	Europe
Cymbalaria muralis	1851	1851	1874	Garden	Europe
Cymbalaria pallida	1990	1962	1948	Garden	Europe
Cyperus longus	–	1952	–	Garden	Europe
Descurainia sophia	1838	1838	1837	Accidental	Widespread
Diplotaxis muralis	1860	1872	1899	Grain, ballast	Mediterranean
Diplotaxis tenuifolia	1670	1927	1920s	Grain	Europe
Doronicum pardalianches	1855	1930	1941	Medicinal	Europe
Doronicum plantagineum	1989	–	–	Garden	Europe
Egeria densa	–	1955	–	Aquarium	S.America
Elodea canadensis	1859	1859	1864	Aquarium	N.America
Elodea nuttallii	1987	1979	1976	Aquarium	N.America
Epilobium brunnescens	1933	1937	1931	Garden	New Zealand
Epilobium ciliatum	1965	1970s?	1964	Timber	N.America
Epilobium pedunculare	1938	–	–	Garden	New Zealand
Erinus alpinus	–	1872	1942	Garden	Europe
Erysimum cheiranthoides	1851	1851	1875	Accidental	Widespread
Erysimum cheiri	1850s	1851	1864	Garden	Europe
Euphorbia cyparissias	1895	1941	1897	Garden	Europe
Euphorbia esula and Euphorbia x pseudovirgata	–	1946	1966	Grain	Eurasia
Fagus sylvatica	Before 1830	15th Century	18th Century	Amenity	Eurasia
Fallopia baldschuanica	–	1939	1962	Garden	Japan
Fallopia japonica	1962	1933	1933	Garden	Japan
Fallopia japonica x sachalinensis = F. x bohemica	–	1992	1985	Accidental	Unknown
Fallopia sachalinensis	1962	1953	1963	Garden	Japan
Festuca heterophylla	1968	–	1987	Garden	Europe
Ficus carica	1979	1913	1968	Accidental, garden	S.W.Asia
Foeniculum vulgare	1851	1838	1907	Medicinal	Widespread
Fragaria x ananassa	1971	19th Century	1965	Garden	Cultivation
Galanthus nivalis	1851	1887	1858	Garden	Eurasia
Galega officinalis	1971	1912	–	Garden	Eurasia
Galinsoga parviflora	1928	1874	1964	Garden	S.America
Galinsoga quadriradiata	1940	1948	1962	Horticultural seed	Tropical America
Gaultheria mucronata	–	1956	1966	Garden	S.America
Gaultheria shallon	1971	Post 1963	–	Amenity	N.America
Geranium endressii	1971	1963	1964	Garden	Europe
Geranium phaeum	1839	1851	1858	Garden	Europe
Geranium pyrenaicum	1873	1885	1882	Garden	Eurasia, Africa
Geranium x magnificum	–	–	1966	Garden	Cultivation
Gunnera tinctoria	1970	–	–	Garden	S.America
Hebe sp.	c.1990	–	–	Garden	Australasia
Heracleum mantegazzianum	1962	1952	1962	Garden	S.W.Asia
Hesperis matronalis	1801	1851	1837	Garden	Eurasia

Name of Plant	VC58	VC59	VC60	Means of Introduction	Region of Origin
Hieracium grandidens	1971	1948	1966	Accidental	Europe
Hippophae rhamnoides	1872	1899	1905	Amenity	Eurasia
Hirschfeldia incana	1920	1939	1966	Grain	Mediterranean
Hyacinthoides hispanica and hybrids	1971	1928	1965	Garden	Europe
Hypericum calycinum	1971	–	1965	Garden	Eurasia
Illecebrum verticillatum	–	1976	–	Accidental	Europe
Impatiens capensis	1938	1905	–	Garden	N.America
Impatiens glandulifera	1932	1913	1938	Garden	Asia
Impatiens parviflora	1917	1855-63	1966	Timber	Asia
Inula helenium	1805	1963	1805	Garden	Eurasia
Iris foetidissima	1990	–	–	Garden	Europe
Juncus tenuis	1927	1903	1925	Grain	N.America
Kniphofia uvaria	–	1990s	1987	Garden	S.Africa
Lagarosiphon major	1971	1953	–	Aquarium	S.Africa
Lamiastrum galeobdolon ssp. *argentatum*	1990	Post 1963	1987	Garden	?
Lamium maculatum	1859	1838	1858	Garden	Eurasia
Larix decidua	Before 1830?	Before 1830?	Before 1830?	Forestry	Europe
Lathyrus aphaca	–	1854	–	Impurity	Eurasia
Lathyrus latifolius	1971	1902	1965	Garden	Europe, Africa
Lathyrus tuberosus	–	1914	–	Grass, bird seed	Eurasia
Ledum palustre ssp. *groenlandicum*	1971	1917	1972	Accidental or natural	N.America
Lemna minuta	1989	1995	–	Aquarium	N. & S.America
Lepidium draba ssp. *draba*	1899	1929	1899	Grain	Eurasia
Lepidium latifolium	1840	1633	1994	Medicinal	Eurasia, Africa
Leucanthemum maximum and *L.* x *superbum*	1971	1963	1964	Garden	Cultivated
Ligustrum ovalifolium	1971	1963	1964	Garden	Eurasia
Lilium martagon	1965	1964	1949	Garden	Eurasia
Lilium pyrenaicum	–	1986	1965	Garden	Europe
Linaria purpurea	1930	1976	1964	Garden	Mediterranean
Linaria repens	–	1934	1907	Accidental	Europe
Linaria vulgaris x *L. repens* = *L.* x *sepium*	–	1979	–	Accidental	Europe
Lobularia maritima	1962	1872	1903	Garden	Mediterranean, Asia
Lolium multiflorum	1861	1933	1893	Agriculture	Europe
Lolium temulentum	1858	1849	1899	Grain	Eurasia
Lonicera involucrata	–	–	1965	Garden	N.America
Lonicera xylosteum	1870	1838	1900	Garden	Eurasia
Ludwigia palustris	–	1927		Accidental	Eurasia, Africa
Lunaria annua	–	1991	1931	Garden	Europe
Lupinus arboreus	1971	1926	1966	Garden	N.America
Lupinus nootkatensis, L. x *regalis* and *L. polyphyllus*	1933	19th Century	1899	Garden	N.America
Luzula luzuloides	–	1920	–	Garden	Europe
Lychnis coronaria	c.1990	–	–	Garden	Eurasia
Lycium barbarum and *L. chinense.*	1899	1929	1907	Garden	China
Lysimachia punctata	1971	1962	1966	Garden	Eurasia
Mahonia aquifolium	1970	1948	1967	Garden	N.America
Matricaria discoidea	1904	1902	1901	Grain	N.E.Asia
Meconopsis cambrica	–	1963	1863	Garden	Europe
Medicago arabica	1851	1854	–	Grass	Europe, Africa
Medicago polymorpha	1857	1850s	1887	Grass	Eurasia, Africa
Medicago sativa	1845	1851	1901	Grass	Eurasia, Africa
Melilotus altissimus	1851	1828	1858	Accidental	Europe
Melilotus officinalis	1861	1903	1904	Bird Seed	Eurasia
Melitotus albus	1863	1828	1903	Accidental	Eurasia
Mentha spicata	1971	1926	1900	Herb	?
Mentha spicata x *M. longifolia* = x *villosonervata*	1971	–	1982	Herb	?
Mentha spicata x *M. suaveolens* = *Mentha* x *villosa*	1971	–	1960	Herb	?
Mimulus guttatus s.l.	1808	1871	1887	Garden	N.America

Name of Plant	VC58	VC59	VC60	Means of Introduction	Region of Origin
Mimulus moschatus	1970	1933	1959	Garden	N.America
Myrrhis odorata	1825	1828	1858	Medicinal	Europe
Narcissus pseudonarcissus (and horticultural forms)	c.1990	–	1960s	Garden	Cultural
Nectaroscordum siculum	1990	–	–	Garden	Europe, Asia
Nepeta x faassenii	1990	–	–	Garden	Cultivation
Nymphoides peltata	1810	1844	1985	Garden	Eurasia
Oenothera biennis	1851	1801	1898	Garden	N.America?
Oenothera x fallax	–	1892	1907	Garden	Local
Oenothera glazioviana	1893	1881	1902	Garden	N.America
Onobrychis viciifolia	1990	1845	1966	Agriculture	Europe
Onopordum acanthium	–	1903	–	Garden	Eurasia
Ornithogalum angustifolium	1859	1864	1775	Garden	Europe
Oxalis articulata	–	1963	1966	Garden	S.America
Oxalis corniculata	1840	–	1922	Garden	Widespread
Oxalis stricta	–	1848	1964	Garden	N.America
Papaver somniferum ssp. somniferum	1899	1962	1941	Garden	W.Asia
Parthenocissus tricuspidata	–	Post 1963	1965	Garden	E.Asia
Pentaglottis sempervirens	1849	1838	1858	Garden	Europe
Persicaria campanulata	1971	1980	–	Garden	Himalayas
Persicaria wallichii	1990	1953	1965	Garden	Himalayas
Petasites albus	1962	1910	1945	Garden	Eurasia
Petasites fragans	1872	1887	1900	Garden	Mediterranean
Petasites japonicus	1940	1988	–	Garden	E.Asia
Peucedanum ostruthium	1918	1806	1888	Medicinal	Europe
Pilosella aurantiaca ssp. carpathicola	1899	c.1900	1903	Garden	Europe
Pinus nigra	–	1890s	–	Forestry	Europe
Pinus sylvestris	Before 1830?	Before 1830?	Before 1830?	Forestry	Eurasia
Poa chaixii	1970	–	–	Garden	Europe
Populus alba	1838	1838	1907	Garden	Europe
Populus candicans group	1971	1933	1964	Amenity	?
Populus nigra hybrids	1838	1838	1893	Garden	Cultivation
Populus x canescens	1899	1849	1963	Garden	Europe
Potamogeton epihydrus	–	1959	–	Aquarium	Europe/N.America
Potentilla argentea	1861 (Native?)	1902 (Native?)	1987	Accidental/Natural	Widespread
Potentilla norvegica	1971	1898	1941	Garden, bird seed	Widespread
Potentilla recta	1971	1972	1962	Garden, bird seed	Eurasia, Africa
Prunus cerasifera	1990	–	1965	Garden	Eurasia
Prunus cerasus	1838	1865	1965	Garden	Eurasia
Prunus domestica ssp. domestica & ssp. institia	1838	1838	1962	Garden	Eurasia?
Prunus laurocerasus	1971	1963	1975	Garden	Eurasia
Pseudofumaria lutea	1805	1858–65	1860	Garden	Europe
Pyrus communis and P. pyraster	1851	1851	1860	Garden	Eurasia
Quercus cerris	1971	1962	1964	Amenity	Eurasia
Quercus ilex	1971	1962	1964	Garden	Mediterranean
Reseda alba	1980	1861	–	Garden	Mediterranean, Asia
Rheum officinale and R. x hybridum	–	1954	1987	Garden	Asia & Cultivation
Rhododendron ponticum	1874	1962	1962	Amenity	Eurasia
Ribes rubrum	1846	1842	1858	Garden	Europe
Ribes sanguineum	1990 Self sown	?19th Century	1964	Garden	N.America
Rosa rubiginosa	1899	1851	1906	Amenity	Eurasia
Rosa rugosa	1995	1927	1946	Amenity	E.Asia
Rubus armeniacus	–	1963?	–	Garden	Eurasia
Rubus tricolor	–	Post 1970	1989	Garden	China
Rumex pseudoalpinus	–	1859	–	Garden	Eurasia
Rumex scutatus	–	1907	1972	Garden	Eurasia, Africa
Ruscus aculeatus	1971	1963	1967	Garden	Mediterranean
Sagittaria sagittifolia	1859	1923	1875	Accidental/Natural Spread	Eurasia
Salix daphnoides	–	1951	1967	Amenity	Europe
Salix x rubens and Salix x rubra	1971	?	1986	Amenity	Europe

Name of Plant	VC58	VC59	VC60	Means of Introduction	Region of Origin
Sambucus ebulus	1838	1838	1946	Garden	Eurasia
Sambucus racemosa	1970	1949	–	Garden	Eurasia
Sanguisorba minor ssp. *muricata*	1971	1907	1989	Agriculture	Europe
Saponaria officinalis	1839	1801	1718	Garden	Eurasia
Saxifraga x *polita*	–	1985	1857	Garden	?
Saxifraga x *urbium*	1836	1801	1899	Garden	Cultivation
Securigera varia	1855	1876	1910	Garden	Eurasia
Sedum album	1928	1881	1891	Garden	Eurasia, Africa
Sedum spurium	–	1959	1968	Garden	C.Asia
Sempervivum tectorum	1839	1839	1837	Garden	Europe
Senecio fluviatilis	1810	1849	1858	Garden	Europe
Senecio ovatus	–	1963	1966	Garden	Eurasia
Senecio squalidus	1926	1931	1948	Garden	Europe
Senecio viscosus	1867	1858	1688	Accidental	Eurasia
Senecio x *albescens*	1995	1993	1985	Garden	?
Sigesbeckia serrata	–	1928	–	Grain	C. & S.America
Sinapis alba	1839	1963	1874	Agriculture	Eurasia
Sisymbrium altissimum	1903	1923	1901	Grain	Eurasia
Sisymbrium orientale	1962	1909	1910	Grain	Mediterranean
Sisymbrium strictissimum	–	1890	Unknown	?	Europe
Sisyrinchium bermudiana/ S. montanum?	1985	1911	1900	Garden	N.America
Smyrnium olusatrum	1837	1903	1858	Medicinal	Widespread
Soleirolia soleirolii	1971	–	1965	Garden	Mediterranean
Solidago canadensis	1925	1963	1964	Garden	N.America
Solidago gigantea	1988	1927	1971	Garden	N.America
Sorbus aria s.l.	1851	1851	Native	Garden	Europe
Sorbus croceocarpa	–	–	1990	Garden	?
Sorbus intermedia	1968	1926	1964	Garden	Europe
Spartina anglica	1925	1930	1932	Amenity/natural spread	Accidental
Spiraea salicifolia and spp.	1930	1908	1963	Garden	N.America & cultivated
Stratiotes aloides	1858	1805	1858	Garden/accidental	Eurasia
Symphoricarpos albus	1933	1933	1933	Amenity	N. America
Symphytum x *uplandicum*	1968	1901	1968	Agriculture	Cultivated
Syringa vulgaris	–	Post 1963	1964	Garden	Europe
Tanacetum balsamita	–	1801	–	Garden	S.W.Asia
Tanacetum parthenium	1851	1938	1907?	Medicinal	Eurasia
Taxus baccata	1844	Native?	Native	Amenity	Eurasia, Africa
Tellima grandiflora	1941	–	1975	Garden	N.America
Thlaspi arvense	1867	1872	1883	Grain	Eurasia, Africa
		Possibly native			
Tilia x *vulgaris*	1851	1851	1907	Amenity	Europe
Tolmiea menziesii	1971	1928	1964	Garden	N.America
Tragopogon porrifolius	–	1597	–	Garden	Europe
Trifolum hybridum	1860	1859	1887	Agriculture	Eurasia
Trifolium pannonicum	–	1916	–	Impurity	Europe
Tulipa sylvestris	1836	–	1870	Garden	Mediterranean
Valeriana pyrenaica	1970	1963	–	Garden	Europe
Vallisneria spiralis	–	1940	–	Aquarium	Tropics
Verbascum densiflorum	–	1985	–	Garden	Eurasia
Verbascum nigrum	1890	1913	1960	Accidental or natural	Eurasia
Verbascum virgatum	–	1988	–	Accidental	Europe
Veronica filiformis	1950	1950s	1944	Garden	C.Asia
Veronica peregrina	1870	1865	1909	Horticultural seed	N. & S.America
Veronica persica	1847	1872	1907	Agricultural seed	S.W.Asia
Vinca major	1962	1946	1837	Garden	Eurasia
Vinca minor	1962	1812	1858	Garden	Eurasia
Viscum album	1983	1877	1950s	Garden	Europe
Vitis vinifera	–	1942	–	Accidental	Europe
Vulpia myuros	1898	1933	1969	Grain/natural spread	Widespread

Accidental		Crop, Amenity Introductions, etc.				Garden		
Unknown or accidental	*Impurity with timber, bird seed, grain, agricultural seed, ballast, etc.*	*Amenity*	*Forestry*	*General Agriculture*	*Grassland*	*Garden*	*Medicinal and culinary herbs*	*Aquarium*
10%	9%	7%	1%	3%	1%	60%	5%	3%
Total 19%		Total 12%				Total 68%		

Table 19.12. Means of introduction for naturalised alien species in Cheshire and Lancashire (VC58-60).

Region of Origin Vice-county	*Cultivated Accidental Unknown*	*Widespread Europe, N.Africa, W. & C.Asia, India*	*N. & S.America*	*S.Africa Australasia*	*E.Asia China Japan*
A. *Naturalised aliens* (VC58–60)	8%	68%	15%	3%	6%
B. *All aliens* Ex. Travis' Flora S.Lancs VC59	15%	71%	11%	2%	1%

Table 19.13. Origin of naturalised alien species in Cheshire and Lancashire (VC58–60).

of Liverpool Botanic Garden started by A.K. Bulley in the early 20th century (McLean 1997), and in the last 30 years the development of numerous garden centres and water gardens throughout the region.

Table 19.13 attempts to show the geographical origination of naturalised species. Botanists and horticulturists have explored the globe in search of suitable species for cultivation, but to prosper they must be suitable for the climatic conditions of the region. Not surprisingly some 68% of the taxa have a widespread European, Mediterranean or West Asian distribution. However, a significant number originate in the Americas.

Conclusion

The title of this paper refers to a 'game of chance'. Whilst the element of chance is present, what has and is happening to the floristic composition of the flora is not particularly chancy. Indeed there seems to be a great deal of logic about it.

Human intervention through urbanisation, industrialisation and agriculture has reduced, and in some cases almost to extinction, many natural and semi-natural habitats. Where the loss is total or almost so, then the native species have suffered badly. However, the same processes have created many new habitats and, if suitable, the native flora is not slow to take advantage of these man-made habitats (Greenwood & Gemmell 1978; Day *et al.* 1982). It is also suggested that at least some southern species are extending their range northwards, a good example being Southern Marsh-orchid

(*Dactylorhriza praetermissa*), probably as a consequence of climatic amelioration.

However, by far the most important factor influencing the composition of the flora of the region was, and continues to be, the horticultural industry. This industry flourished at least as early as the 18th century, and as the population grew and urbanisation of the region took place people of all social groupings had a desire to cultivate plants. These plants were grown for different reasons, but something like 197 of the 290 or so naturalised aliens owe their introduction to this industry. Yet it is an industry that directly leaves little noticeable mark on the landscape unlike that left by modern agriculture, extractive industries, communications systems and the built landscape itself.

Overall, therefore, human intervention has been a major cause of change to the floristic composition of the region's flora so that it is now more diverse than at any time in post-glacial history. Globally, human intervention reduces species diversity catastrophically, but the findings reported here confirm that in certain circumstances and over limited areas the reverse is true, and species diversity is increased (McNeeley *et al.* 1995). Whilst chance plays its part, change is perhaps more predictable than chancy. Furthermore, it may well be possible to predict changes more accurately in the future, given a knowledge of climatic and habitat changes and a knowledge of the species being introduced to gardens, for amenity plantings and agriculture. It is likely that the trends noted here will continue except that, as global and therefore local warming continues (United Kingdom

Climates Change Impacts Review Group 1996), coastal communities may be subject to increased erosion.

Acknowledgements

In compiling the data for this paper I am grateful to all the botanists who have sent me information over the years. However, I am especially grateful to Peter Gateley (BSBI recorder for VC59, South Lancaster) for providing recent data, for South Lancaster and commenting on an early draft, and to Professor A.D. Bradshaw for his comments and encouragement.

References

Allen, D. (1996). The earliest records of Danish Scurvy-grass, *Cochlearia danica*, on inland railway tracks. *BSBI News*, **72**, 24.

Anon. [1991]. *Lancashire, A Green Audit*. Lancashire County Council, Preston.

Anon. [1993]. *The State of Cheshire's Environment*. A broadsheet published by Cheshire County Council, Chester.

Berry, W.G. (1967). Saltmarsh development in the Ribble Estuary. *Liverpool Essays in Geography, A Jubilee Collection* (eds R.W. Steel & R. Lawton), pp. 121–35. Longmans, London.

Braithwaite, M. (1992). BSBI Monitoring Scheme 1987–88. Change at a local level – VC81, Berwickshire. *BSBI News*, **61**, 16–19.

Buxton, R. (1849). *Botanical Guide to the Flowering-Plants, etc. found indigenous within 16 miles of Manchester*. London.

Church, J. (ed.) (1996). *Regional Trends*. Office for National Statistics, HMSO, London.

Clement, E.J. & Foster, M.C. (1994). *Alien Plants of the British Isles*. Botanical Society of the British Isles, London.

Day, P., Deadman, A.J., Greenwood, B.D. & Greenwood, E.F. (1982). A floristic appraisal of marl pits in parts of north-western England and northern Wales. *Watsonia*, **14**, 153–65.

de Smidt, J.T. (1995). The imminent destruction of north west European heaths due to atmospheric nitrogen deposition. *Heaths and Moorland: Cultural Landscapes* (eds D.B.A. Thompson, A.J. Hester & M.B. Usher), pp. 206–17. Scottish National Heritage, HMSO, London.

de Tabley, Lord (1899). *The Flora of Cheshire*. Longmans, Green and Co., London.

Freeman, T.W. (1962). The Manchester Conurbation. *Manchester and its Region* (ed. C.F. Carter), pp. 47–60. Manchester University Press, Manchester.

Graham, G.G. (1988). *The Flora and Vegetation of County Durham*. The Durham Flora Committee and the Durham County Conservation Trust, Wallsend.

Gray, A.J. & Scott, R. (1987). Saltmarshes. *Morecambe Bay, An Assessment of Present Ecological Knowledge* (eds N.A.

Robinson & D.W. Pringle), pp. 97–117. Centre for North-West Regional Studies in conjunction with the Morecambe Bay Study Group, Lancaster.

Greenwood, E.F. & Gemmell, R.P. (1978). Derelict industrial land as a habitat for rare plants in S. Lancs. (VC59) and W. Lancs. (VC60). *Watsonia*, **12**, 33–40.

Hall, T.B. [1839]. *A Flora of Liverpool*, Whitaker & Co., London.

Lancashire Polytechnic (1989). *Annual Report of the Director of the Observatories*. Lancashire Polytechnic, Preston.

Lawton, R. (1982). From the Port of Liverpool to the Conurbation of Merseyside. *The Resources of Merseyside* (eds W.T.S. Gould & A.G. Hodgkiss), pp. 1–13. Liverpool University Press, Liverpool.

McLean, B. (1997). *A pioneering plantsman A.K. Bulley and the great plant hunters*. The Stationery Office, London.

McNeeley, J.A., Gadgil, M., Leveque, C., Padoch, C. & Redford, K. (1995). Human Influences on Biodiversity. *Global Biodiversity Assessment* (ed. V.H. Heywood), pp. 711–821. Cambridge University Press, Cambridge.

Newton, A. (1971). *Flora of Cheshire*, Cheshire Community Council, Chester.

Newton, A. (1990). *Supplement to Flora of Cheshire*. A. Newton, Leamington Spa.

Palmer, M.A. & Bratton, J.H. (eds) (1995). *A sample survey of the flora of Britain and Ireland*. U.K. Nature Conservation No. 8, Joint Nature Conservation Committee, Peterborough.

Perring, F.H. & Walters, S.M. (1962). *Atlas of the British Flora*. Botanical Society of the British Isles, London.

Peter, D. (1994). *In and around Silverdale*. Barry Ayre, Carnforth.

Pinkess, L.H. (1996). Danish Scurvy grass on a Worcestershire road. *BSBI News*, **72**, 25.

Pringle, W.A. (1987). Physical processes shaping the intertidal and subtidal zones. *Morecambe Bay, an assessment of present ecological knowledge* (eds N.A. Robinson & D.W. Pringle), pp. 51–73. Centre for North-West Regional Studies in conjunction with the Morecambe Bay Study Group, Lancaster.

Rich, T.C.G. & Woodruff, E.R. (1990), *BSBI Monitoring Scheme 1987–1988*. 2 vols. A report for the Nature Conservancy Council.

Savidge, J.P., Heywood, V.H. & Gordon, V. (1963). *Travis's Flora of South Lancashire*. Liverpool Botanical Society, Liverpool.

Secord, A. (1994a). Corresponding interests: artisan and gentlemen in nineteenth-century natural history, *The British Journal for the History of Science*, **27**, 383–408.

Secord, A. (1994b). Science in the pub: artisan botanists in early nineteenth-century Lancashire, *History of Science*, **32**, 269–315.

Swann, G.A. (1993). *Flora of Northumberland*. The Natural History Society of Northumbria, Newcastle upon Tyne.

United Kingdom Climate Change Impacts Review Group (1996). *Review of the Potential Effects of Climate Change in the United Kingdom. Second Report*. Prepared at the request of the Department of the Environment. HMSO, London.

Wheldon, J.A. & Wilson, A (1907). *The Flora of West Lancashire*, Liverpool.

The Changing Fauna

B.N.K. DAVIS

This last group of three chapters covers changes in the records of invertebrates, amphibians, lizards, mammals and birds in north-west England over the past 15,000 years. At the start of this period, Britain contained species characteristic of the last glacial period such as Woolly Mammoth and Giant Deer, which have since become totally extinct, while others such as Lemmings and Reindeer have retreated to Scandinavia and Siberia. Before the Industrial Revolution the wilder north of England became a haven for the larger mammals, such as Wolves and Bears, when such species had been driven out of southern and eastern England at a much earlier date. Other groups of animals have enterprisingly exploited the new range of habitats created through forest clearance, land drainage and agriculture, and the even more disturbed urban-industrial landscape. In this century, however, there has been a retraction again, particularly among the birds, butterflies and other well recorded groups as woodlands and wetlands have shrunk in extent, and land management has intensified. Forecasts of global warming over the next 50 years suggest changes as rapid as those that took place around 10,000 BP. Already there are suggestions that some invertebrates are responding to the warmer conditions. These records of past faunistic changes and present status, therefore, are not just of local significance, but may serve as markers for determining and analysing future changes at a regional and national scale.

CHAPTER TWENTY

Changes in the land and freshwater invertebrate fauna

S. JUDD

Introduction

The distribution of invertebrate animals has fluctuated continually since the retreat of the ice age from north-western England some 12,000 years ago. Post-glacial changes have been driven by changes in climate and increasingly by human influence.

There are four main categories into which the changes shown by invertebrates can be grouped: extinctions, additions, changes in distribution and changes in diversity (Foster 1992), all of which will be discussed in this chapter. Together these can indicate general trends and illustrate major changes in the fauna as a whole. Long-term monitoring work suggests that there are both short- and long-term fluctuations in insect diversity, and in the distribution and abundance of individual species. However, the features influencing change are complex, and separating the effects of climate, land-use and natural population variability are not fully understood (Luff & Woiwod 1995).

This account provides a synthesis of information for Watsonian vice-counties 58 and 59, Cheshire and South Lancaster. Examples are drawn from a wide variety of invertebrate taxa. Nomenclature and national categories for species status are taken from Ball (1992).

Invertebrate faunal studies

The strong tradition of invertebrate faunal studies in north-west England began with the publication of Byerley's *Fauna of Liverpool* in 1854. It was assisted by the foundation of the Lancashire and Cheshire Entomological Society in 1877. Initial studies, mainly on the Coleoptera and Lepidoptera (Ellis 1889, 1890; Sharp 1908), culminated in the first, and unfortunately, the only full faunal check-list for Lancashire and Cheshire, to be published by the influential Fauna Committee (Lawson 1930). This was prefaced with the proud boast that 'the fauna had been more systematically investigated and was more completely known than that of any other two counties in the whole of the British Isles'.

Subsequent notable regional contributions included work on Coleoptera, Diptera, Lepidoptera and Odonata (Mansbridge 1940; Smith 1948; Ford 1953; Kidd &

Brindle 1959; Johnson 1962; Rutherford 1983, 1994; Sumner 1985), whilst the innovative study of Cheshire dragonflies by Gabb & Kitching (1992) provided, for the first time, distribution maps, location and habitat data, and information on behaviour and developmental stages. A limited number of check-lists for orders and families have also been published in the last 10 years (Hull 1987, 1990; Judd 1987; Cross 1989; Chandler 1991; Felton 1991; Garland & Appleton 1994). These indicate that 50–60% of the British invertebrate fauna of *c*.28,000 species are present in Lancashire and Cheshire.

Sites of importance for the conservation of invertebrates in Cheshire, Lancashire, Merseyside and Greater Manchester were evaluated by Parsons (1987 a,b,c). Examples of detailed site surveys include those for Dunham Park (Johnson, Robinson & Stubbs 1977) and Hilbre Island (Wallace & Wallace 1982). Subsequently, Atkinson & Houston (1993) collated information for the Sefton Coast.

Relic faunas

The quaternary migrations of many species of insects in response to complex, and often rapid glacial/interglacial climatic oscillations, bear witness to the astonishing capacity of insect populations to track changing environments (Lawton 1995). Nevertheless, some populations became trapped in one, or a few locations in the British Isles, and are likely to have been isolated from continental populations for at least 10,000 years. These represent important biological capital in the context of conservation (Hammond 1974) and may be important during a northerly spread of species resulting from temperature increase. However, southern species with northern refugia must be distinguished from thermophilous species that make frequent sorties into northern Britain (Foster 1992).

Four relic species are recorded from single sites in the region and nowhere else in Britain. The most intriguing of these is the distinctive and flightless, parthenogenetic, *Red Data Book 1 (RDB1)* leaf beetle *Bromius obscurus,* which is associated with willow-herbs, especially Rosebay Willowherb (*Chamerion angustifolium*). It was rediscovered in 1979 on a disused railway line adjacent to the River Dane at Hugbridge, Cheshire (Kendall 1982). Its prior absence had been puzzling, because it is a common Scandinavian insect and was common in Britain during the climatic mild phase c.12,000-11,000 BP.

Two others, both spiders, are restricted to the small, isolated 'schwingmoor' at Wybunbury Moss National Nature Reserve (Felton & Judd 1997). The fourth, another relic mossland species, is the pselaphid beetle *Plectophloeus erichsoni*, which occurs on what remains of Chat Moss (Johnson & Eccles 1983).

The presence of the money spider *Carorita limnaea* and the gnaphosid spider *Gnaphosa nigerrima* at Wybunbury Moss, is a biogeographical mystery. Both must have been present in Britain, though not necessarily at Wybunbury, during the Boreal period, before the severance of the land connection with continental Europe (c.8,300 BP). It is most unlikely that they originate from re-immigrant stock. They could have been widely distributed throughout Britain during the extension of ombrogenous mires in the Atlantic Period (c.6,000 BP), their distributions then contracting due to subsequent destruction and fragmentation of this habitat. Interestingly, the Arctic-Alpine, nationally scarce Manchester Treble Bar moth (*Carsia sororiata*), is also recorded from Wybunbury.

Other rare invertebrates, once more widely distributed but now restricted to favourable sites, include the coastal *RDB* moths the Sandhill Rustic (*Luperina nickerlii*) and the Belted Beauty (*Lycia zonaria*), which are both possibly glacial-phase survivors. However, most are saproxylic species which include relic elements of a fauna associated with unmanaged forest cover in the first half of the Holocene (Harding & Alexander 1993).

Today it is almost only in pasture woodlands that the overmature tree/dead wood component of the primeval forest ecosystem survive, in lowland Britain (Harding 1979). Surprisingly, in a region with under half the national average area of woodland (chapter ten, this volume), three sites in the Mersey Basin are considered nationally important for their saproxylic fauna (Harding & Alexander 1993). They are the pasture woodland of the 16th-century Deer Park at Dunham Massey, Greater Manchester; the 18th-century Stockton's Wood, a plantation on former pasture woodland on the National Trust's Speke Hall estate, Merseyside and the ancient woodland of the lower River Weaver woodlands in Cheshire.

Dunham Park has attracted coleopterists since the 1860s. It is the only site in north-western England with an appreciable number of old trees, and supports 34 indicator species of ancient woodland (Johnson, Robinson & Stubbs 1977). Three *RDB* beetles and one *RDB* fly are recorded, together with nearly 100 nationally scarce species. It is also the only known north western site for 27 species, and the northernmost site for 12 species.

Pollution from an adjacent factory and fire damage killed many trees at Stockton's Wood in the 1980s. These dead trees have yielded a remarkable assemblage of beetles, of which 23 are indicators of ancient woodland (Eccles 1987). They are probably present throughout the wood, but the burnt trees sustain an abnormally high population. Their presence, in a secondary woodland, is something of an enigma, but it is thought that they transferred from earlier pasture woodland on this site. Alternatively, they might have originated from ancient woodland along the edge of the Mersey Estuary, which probably supported a fauna similar to the contiguous Weaver Valley, 16km distant.

Cyclical changes

Stability is the exception rather than the rule in the natural environment, and change is a normal condition in ecological ecosystems (Wyatt 1992). Four butterfly species (Table 20.1) became extinct in Merseyside and Cheshire during contractions in range, only to reappear during an expansive phase. The fluctuating presence of species at particular sites is illustrated by a study of butterflies along the Heald Green Railway Cutting in Cheshire between 1984 and 1993. Of the 21 species recorded during this period, two became established, four were seen with greater regularity and two were lost (Shaw 1993).

Britain's largest dragonfly, the Emperor (*Anax imperator*), was first recorded at Ainsdale, in the hot summer of 1976, 130km from its nearest regular breeding colony in the West Midlands. Its population peaked in 1983, when 40 were seen. Subsequently, numbers declined markedly and the loss of this isolated population must be a distinct possibility, unless more hot summers like that of 1995 re-occur (Hall & Smith 1991).

Long-term decline and extinction

Species hardly ever occupy their full ranges. Outlying records for many predominantly southern species, such as the nettle feeding seedbug *Heterogaster urticae*, indicate that they probably occurred further north than now when temperatures were 2–3°C higher. Local extinctions of species are most likely to occur at the limits of their ranges; with a few notable exceptions, most of the native terrestrial biota of the British Isles have suffered range contraction and fragmentation due to habitat loss (Vincent 1990).

Unlike the native regional flora, which has declined by 10% (chapter nineteen, this volume), extinction rates for most invertebrates are unknown. However, 18 of the 65 scarab species have been lost in Lancashire and Cheshire since 1908, and only two gained – a decline of 22% (Johnson 1962). For eight of these extinct species Lancashire and Cheshire are near the limit of their northern distribution, and it is quite possible that industrialisation has tipped the balance against their survival. Another six are coastal, where their habitat has been restricted by urban development.

The decline of Britain's butterflies has been particularly severe (Thomas 1991). Nationally, nearly half of the 59 resident species have experienced major contractions in range (Warren 1992), while ten species have disap-

Extinctions (Last record)	Cyclical extinctions & re-colonisation	Major contraction	Fluctuating	Relatively stable	Extending
Grizzled Skipper (1971)	Small Skipper	Small Pearl Bordered Fritillary	Large Skipper	Large White	Brimstone
Wood White (1850s)	Comma	Dark Green Fritillary	Dingy Skipper	Small White	
Small Blue (1850s)	Ringlet	Small Heath	Orange Tip	Green-veined White	
Silver-studded Blue (1920s)*			White-letter Hairstreak?	Green Hairstreak	
Duke of Burgundy (1850s)			Holly Blue	Purple Hairstreak	
Large Tortoiseshell (1876)			Peacock	Small Copper	
Pearl-bordered Fritillary (1941)			Speckled Wood	Common Blue?	
High Brown Fritillary (1922)			Wall Brown	Small Tortoiseshell	
Marsh Fritillary (1882)			Gatekeeper	Grayling	
Large Heath (1929)			Meadow Brown		

Vagrants: Silver-washed Fritillary, Marbled White
Migrants: Clouded Yellow, Pale Clouded Yellow, Long-tailed Blue, Red Admiral, Painted Lady, Camberwell Beauty
* Re-introduced in 1994

Table 20.1. The changing butterfly fauna of Cheshire and Merseyside since 1854. Based on Rutherford (1983), Anon (1991), Creaser (1992), Shaw (1994, 1995).

peared from Cheshire and Merseyside (Table 20.1). Excluding vagrants and migrants, this gives a local extinction rate, since 1854, of 28%, all but one before 1946.

For the well-studied macro-Lepidoptera, 62 of the 568 Cheshire species, have not been recorded since 1960 (Rutherford 1994). However, over half are either immigrants, vagrants or of dubious veracity which would suggest that no more than 5% have become extinct in the last 30 years. This is similar to the less well studied spiders, which have shown a 7% decline (25/367 species) in Lancashire and Cheshire since 1930 (Felton 1991). In contrast, a surprisingly high proportion (28/82) of ants and solitary wasps listed by Garland & Appleton (1994) have not been recorded in the last 30 years. This 34% decline contrasts with the national extinction rate for aculeates of 5% (Falk 1991), and may reflect a lack of recording as well as habitat loss.

At least 10 species recorded from the region have become extinct nationally. Of these, the strange story of the Manchester Tinea moth (*Euclemensia woodiella*) is worthy of repetition. This was discovered new to science by Robert Cribb in 1829 when c.30 specimens were taken from a hollow tree on Kersall Moor, Greater Manchester. It has never been recorded here, or anywhere else since, but there are three surviving museum specimens (Brindle 1952).

The heathland habitat of the Silver-studded Blue butterfly (*Plebejus argus*) at Bidston Hill, Merseyside, where this species was abundant in the 1850s, was planted as woodland during the latter half of the 19th century, and this species was last recorded here around 1885 and at Delamere Forest in 1921. The Marsh Fritillary butterfly (*Eurodryas aurinia*) has declined nationally due to the 'improvement' and drainage of damp grassland. It was evidently abundant in the middle of the last century in Cheshire, but had disappeared from the county by 1882. The Large Heath butterfly (*Coenonympha tullia*) occurred in profusion on Simonswood Moss in the 1850s. Forty years later it was still common on all Lancashire mosses (now mostly in the counties of Greater Manchester and Merseyside), but by 1913 it was scarce or extinct in Lancashire. The major cause of its decline was the drainage of its habitat for the development of industry and agriculture (Emmet & Heath 1989). Similarly, the nationally scarce Variable Damselfly (*Coenagrion pulchellum*) disappeared from Wirral with the infilling of ponds (Judd 1986).

Such gross changes in land-use clearly affect the structure and composition of invertebrate communities, but relatively minor changes in land use occur more often and can also have marked effects (Usher 1995). For instance, four of the extinct butterfly species in Merseyside and Cheshire – the Wood White (*Leptidea sinapis*), Duke of Burgundy Fritillary (*Hamearis lucina*), Pearl-bordered Fritillary (*Boloria euphrosyne*) and High Brown Fritillary (*Argynnis adippe*) and one declining species, the Small Pearl-bordered Fritillary (*Boloria selene*), are all woodland species whose declines are linked nationally to a reduction in coppicing (Warren 1992).

The White-letter Hairstreak butterfly (*Satyrium*

w-album) has declined due to the loss of Elm, its food-plant, from Dutch Elm disease, a pathogen introduced into the region in 1973 (chapter ten, this volume). However, the loss of the Grizzled Skipper (*Pyrgus malvae*) is more puzzling. It was recorded in the early 1970s at Childer Thornton, Cheshire and although its foodplant, Wild Strawberry (*Fragaria vesca*), was still present in the late 1970s, there was no sign of the butterfly (Rutherford 1983).

Addition and expansion

Overlooked species

The discovery of species that have always been present in the region, illustrates how poorly understood are many groups. For example, the recent discovery of the cerylonid beetle, *Anommatus diecki*, in the River Weaver woodlands was also the first British record (Eccles & Bowestead 1987). Similarly, a grass midge new to science, *Sitodiplosis phalaridis*, was found infesting the inflorescence of Reed Canary-grass (*Phalaris arundinacea*), at Fletcher's Moss, Didsbury, Greater Manchester in 1982 (Abbas 1986). Others, such as the widespread Svensson's Copper Underwing moth (*Amphipyra berbera svenssoni*) are newly recognised.

Natural expansions

Seventeen macro-Lepidoptera species have been added to the Cheshire list since 1960 through natural expansion. They include Blair's Shoulder-knot moth (*Lithomoia leautieri*), which became resident in Britain 40 years ago and which first reached Cheshire in 1989 (Rutherford 1994). Other species, such as the generally distributed Red Underwing moth (*Catocala nupta*), are worthy of mention, because they were very rare 20 years ago.

Species such as the Broad-bodied Chaser (*Libellula depressa*), are more transitory, colonising newly created water bodies, then disappearing after a few years. Another dragonfly, the Ruddy Darter (*Sympetrum sanguineum*) which was first recorded in Cheshire in 1985, is now well established.

The changing distribution over time of the seed bug, *Ischnodemus sabuleti* provides a classic example of range expansion (Figure 20.1). This species was confined to two locations in south-eastern England last century, but it is now found as far north as Cheshire and Yorkshire and is one of the more common seed bugs in Britain, occurring in 114 10 x 10km squares (Judd & Hodkinson 1998).

Four butterfly species, the Holly Blue (*Celastrina argiolus*), Speckled Wood (*Pararge aegeria*), Gatekeeper (*Pyronia tithonus*), and Small Skipper (*Thymelicus*

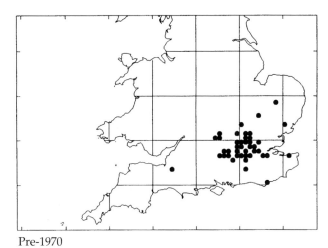

Pre-1900

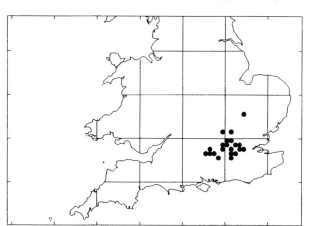

Pre-1950

Pre-1970

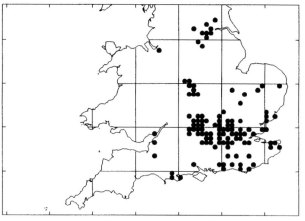

Total records

Figure 20.1. The changing distribution of *Ischnodemus sabuleti* in England over time. (From Judd & Hodkinson 1998)

sylvestris), all inhabiting different biotopes, have crossed wide expanses of unsuitable terrain to become established in previously vacant habitats in northern Cheshire and Greater Manchester since 1989 (Hardy, Hind & Dennis 1993). Other species are spreading in a north or north-westerly direction to reach sites in the Mersey valley, thus expanding their range and density of distribution.

Accidental expansions and deliberate introductions

Some native species, such as the spider *Pholcus phalangioides* which was first recorded in the region in 1992, have been accidentally spread by man (Felton 1992a,b). The first north-western record of the Spitting Spider (*Scytodes thoracica*) was probably introduced to a Liverpool Museum store amongst packing cases.

Only two species have deliberately and successfully been translocated. A colony of Silver-studded Blue butterflies was reintroduced into Cheshire, VC58, at Thurstaston Common in 1994. Nearby on the Wirral Way at West Kirby, an introduced but forgotten colony of Scarlet Tiger moths (*Panaxia dominula*) was rediscovered in 1989. This had been founded in 1961 with a stock of 13,000 larvae and has since been used for studies on gene frequency (Clarke 1993).

Alien species

Most successful colonisers are synanthropic species or those introduced with their host plant. Other accidentally introduced alien species do not normally survive. A notable exception is the scarab beetle *Psammodius caelatus*, which originated from the western seaboard of North America, and, since its discovery in 1972, has spread along the Sefton coast (Johnson 1976; Eccles 1993).

Changes in the transport and packaging of fruit and vegetables have significantly reduced the number of 'stowaway' species that used to be imported through the Liverpool docks, such as the 25 orthopteroid species recorded by Ford (1972). Large theraphosid bird-eating spiders are also rarely recorded now, although the dangerous Black Widow spider (*Latrodectus mactans*) still turns up. A specimen of the dark and sinister spider *Badumna insignis*, a well known synanthropic species in Western Australia, was recently found at Widnes amongst imported *Eucalyptus* spp. timber from Perth. None of these is likely to become established here.

Garland & Appleton (1994) listed 10 introduced or vagrant species of ants found in Greater Manchester, the most successful being the Pharaoh ant (*Monomorium pharaonis*) which occurs in heated buildings in many parts of Britain. Recently established species in Merseyside include the gnaphosid spider *Urozelotes rusticus*, which was found in a house in Kirkby, and the house centipede (*Scutigera coleoptrata*), which was twice found close to the docks (Felton 1995, 1996). Other species, such as the 'cricket on the hearth' (*Acheta domestica*), which may have been present in Britain since the Crusades, have declined this century due to improving standards of hygiene. This last species can be found free-living in man-made habitats where there is constant warmth from fermenting organic matter (Marshall & Haes 1988), e.g., Bidston tip, Merseyside.

Most of the nine alien Mollusca species recorded from the region are found in canals (McMillan 1990). The Ramshorn snail *Menetus dilatatus*, which was discovered in 1869, was introduced with the discharging waste from cotton mills (Owen *et al.* 1962). Some, such as the Zebra mussel (*Dreissena polymorpha*), which was first recorded in 1844, and the *RDB* operculate water snail *Marstoniopsis scholtzi*, established since 1900, were once abundant but are now rare. A tiny freshwater limpet *Ferrissia wautieri*, is the most recent arrival and was first recorded in a Wirral pond in 1985.

Introduced plants are sometimes accompanied by host-specific insects. An extensively naturalised plant in the Mersey Basin is *Rhododendron ponticum*, which was brought to Britain as seed in 1763. Among the five alien species recorded from it in Britain, two, a leaf hopper *Graphocephala fennahi* and a lace bug *Stephanitis rhododendri*, occur in Cheshire and Merseyside, although the latter has not been recorded for many years (Judd & Rotherham 1992). A further 31 common polyphagous species (15 on Merseyside) are associated with it nationally.

Other notable colonisers include the Knopper Gall wasp (*Andricus quercuscalicis*) which has spread north and west across Europe, following the introduction of one of its principal hosts, Turkey Oak (*Quercus cerris*) (Lawton 1995).

The mirid bug *Alloeotomus gothicus*, which was first discovered in Britain in 1951, was found in 1985 at Nunsmere, Cheshire, probably originating from an adjacent Forestry Commission nursery (Judd 1987). The introduced pines at Ainsdale support an assemblage of beetles which were recorded by Eccles (1993). He also commented on the prettily marked Asparagus beetle (*Crioceris asparagi*), with its slug-like larvae, which may be found at this site, on naturalised Garden Asparagus (*Asparagus officinalis* ssp. *officinalis*).

This review of alien species is by no means complete and many other examples could be cited, e.g., the worm-eating flatworm *Australoplana sanguinea* and the money spider *Ostearius melanopygius*, both of which originate from New Zealand.

Fauna of natural habitats

Ainsdale Sand Dunes National Nature Reserve and Dunham Park with eleven and four *RDB* species respectively, are the two nationally important 'hotspots' for invertebrate conservation in the region, and are classified by Parsons (1987 a,b,c) as grade A sites. A further 14 Grade B sites are regional 'hotspots'. All but Stockton's Wood are natural or semi-natural habitats, and five of these top sites, supporting 23 (21%) of the region's *RDB* species, are on the Sefton coast.

The study of rarities is sometimes thought of as élitist,

but this has great value. These organisms may be very subtle indicators of environmental change, apart from being worthy of conservation in their own right (Mellanby 1974). Of the 110 *RDB* species recorded from the study region, 10 (9%) are now extinct nationally and a further 73 (66%) have not been recorded since 1970, many of which are probably regionally extinct.

The destruction of natural and semi-natural habitats, which support most of the region's *RDB* species, is the major threat to them. Another key threat to invertebrates in the changing British landscape is the fragmentation into small, isolated patches of habitats frequented by specialist species with poor dispersal powers (Luff & Woiwod 1995). This is a particular problem for isolated peat mosses in the region which support 18 (16%) of the region's *RDB* species.

Fortunately eight of the regional Grade A and B sites are statutory or private nature reserves or National Trust properties. The most important regional conservation achievement was the creation of an almost continuous strip of protected land along the Sefton coast. However, even on 'safe sites' the opportunity to manage change occurs all too rarely. Usually, management of the consequences of change is the only option, and nature conservation is forced to be reactive (Langslow 1992).

Man-made habitats

There are now more man-made habitats in Britain than ever before, providing greater scope for invasion. Likewise, the increase in disturbance to natural and semi-natural vegetation, by an increasingly mobile and affluent human population, together with the gradual loss of indigenous species through land use change, pollution, pesticides, etc., may collectively make the British countryside increasingly more accessible for alien species to become established (Eversham & Arnold 1992). In general anthropogenic features may increase variation in habitat and species richness locally, but often by the introduction or spread of more eurytopic species at the expense of native species requiring larger areas of natural habitat (Sheppard 1987).

Many species recorded in a study of the effects of current urban and past land use on beetle diversity in the Eccles area of Greater Manchester were those expected in the agricultural environment before town expansion (Terrell-Neild 1994). Species richness declines with increasing penetration into the town with a greater proportion of individuals coming from fewer species. This was most marked in areas with the longest history of human habitation.

Many studies have shown the importance of naturally colonised derelict land as a resource for wildlife (e.g., Bradshaw & Chadwick, 1980; Gemmell & Connell 1984), but there have been few on the insects of such sites. One exception is the work of Sanderson (1992a,b, 1993) who investigated the factors affecting the Hemiptera of 16, diverse, naturally colonised derelict sites, undergoing succession, in Merseyside and south Lancashire. He

found that derelict land can often form relatively undisturbed ecological habitat-islands in an otherwise less habitable urban area. Hemipteran diversity indices were significantly correlated with site area, but not with soil, pH or site age, irrespective of the site's origin or history. The vegetation is of primary importance in predicting Hemiptera species composition, which is influenced by atypical plant communities on some of the nutrient poor soils, with extremes of pH. Although many of these sites had regionally rare plants growing on them, none of the 149 Hemiptera species recorded was of *RDB* status and only one species was new to Merseyside.

The artificial lime wastes around Northwich which are of high botanical interest, support a butterfly population, comprising common rough grassland species that have colonised from the surrounding countryside (Rutherford 1983). Old sand quarries, however, support an unusually rich aculeate fauna.

Urban parks provide important habitats for some invertebrates. A new British bark beetle, *Scolytus laevis* (Atkins, O'Callaghan & Kirby 1981) and three nationally scarce hoverflies are recorded from Sefton Park, Liverpool, including *Platycheirus tarsalis*, a species associated with lush wood edges. Nearby, an astonishing 18 nationally scarce species of beetle have been recorded from Clarke Gardens (Eccles 1987), including the rove beetle, *Hydrosmectina delicatissima*, new to Britain (Allen & Eccles 1988).

A few species have invaded Britain and adjusted themselves to the garden environment to the virtual exclusion of other habitats. There are several examples, two of the best known being noctuid moths: the Varied Coronet (*Hadena compta*) associated with cultivated *Dianthus*, which reached Cheshire in 1992, and the generally distributed but not common Golden Plusia (*Polychrysia moneta*) associated with *Delphinium*, has colonised gardens throughout much of the country over the last 100 years (Owen 1978; Rutherford 1994). Likewise, a few, previously rare native species, have spread rapidly with the planting of their food plants, or closely related species, in gardens. The Juniper Shield bug (*Cythostethus tristriatus*) and Juniper Carpet moth (*Thera juniperata*) have both increased with the planting of junipers and cypresses and have recently become established in Cheshire and Merseyside.

Most of the richest and most diverse butterfly habitats in Cheshire were made by humans but then abandoned (Rutherford 1983). High on this list are disused railway lines, which provide ready-made dispersal corridors; the first modern Cheshire location for the Small Skipper butterfly (*Thymelicus sylvestris*) was a disused railway line near Malpas, from which it has spread northwards.

The Heald Green railway cutting, mentioned earlier, supports 21 butterfly species which are notably absent in the heavily grazed adjacent meadows. This is because of the abundance of nectar-bearing plants in the cutting, the absence of insecticides and the shelter from prevailing westerly winds (Shaw 1993). New Ferry railway sidings support more butterfly species than any other site on Wirral and are a butterfly reserve. Former railway cuttings are also important for other invertebrates. The third British record of the nationally scarce leafhopper, *Cosmotettix caudatus,* was from Haskayne, now a small Lancashire Wildlife Trust nature reserve (Payne 1987) where limestone ballast has raised the pH of the soil, resulting in a flora similar to chalk grassland. The leafhopper, *Rhytistylus proceps,* which is normally associated with coastal or limestone habitats, also occurs here, about 10km inland (Sanderson 1992a).

A survey of 19 stretches of disused canals in Greater Manchester recorded 74 freshwater invertebrate species, including a number of nationally scarce and local species, which enhance the known importance of these sites for scarce vascular plants (Guest 1989). Recently, the freshwater winkle, *Viviparus contectus,* was reported from the Hollingwood Branch Canal (Parsons 1987b). This is primarily a southern species associated with calcareous slow flowing or standing water.

Flooded marl pits, which were dug from the 13th century well into the 19th century provide an important national aquatic habitat for invertebrates. The Cheshire Wildlife Trust estimated that 86,000, or over 25% of all English ponds, once occurred within the county although probably only 11,000 remain (chapter fourteen, this volume). Twelve species of dragonfly were recorded from the marl pits at Churton, including the nationally scarce Variable Damselfly (Gabb & Kitching 1992) and in 1995, the *RDB*1 water beetle, *Hydrochara caraboides,* was recorded from two ponds near Chester and Winsford (Guest 1996). Earlier, between 1940 and 1957, 27 species of freshwater molluscs were recorded in a study of 172 pits in the Bromborough area (McMillan 1959).

The extraction of boulder clay, sand and gravel in Cheshire has created new and important aquatic habitats for invertebrates, particularly dragonflies, as has the pumping of natural brine, which caused the formation of flashes or small lakes (Gabb & Kitching 1992). A large number of unpolluted, man-made pools, ranging from bomb craters and golf course reservoirs to scrapes excavated during the 1970s and 1980s for conservation purposes, make Ainsdale by far the most important single site for Odonata on the Sefton coast, with 14 species recorded (Hall & Smith 1991). Some species have only been recorded since their habitats were created, such as the Variable Damselfly and Emperor Dragonfly. Other species such as the Emerald Damselfly (*Lestes sponsa*), have shown marked population increases.

In addition to these man-made habitats specialised invertebrate faunas are associated with human habitation and the work place.

Conclusion

In the past man has had a positive effect on invertebrate biodiversity and abundance within the Mersey Basin. Sub-Boreal fragmentation of blanket woodland for Neolithic agriculture created a habitat mosaic which benefited species associated with forest-edge, early successional and plagioclimax habitats. The heyday for British butterflies was probably the early medieval period with its warm, dry summers together with developing habitats maintained by traditional agricultural and forestry practices (Dennis 1992; Warren 1992).

Simplistic explanations of the pattern of invertebrate declines are misleading. However, habitat loss and fragmentation resulting from man's changing agricultural and forestry practices, industrialisation and urbanisation developments, have clearly had a major detrimental impact on invertebrate populations since the mid-19th century. Man-made areas, such as gardens, derelict land and the buildings in which humans live and work, provide new habitats, particularly for opportunist species. However, nearly all detailed investigations into the declines of British butterfly species have reached the same general conclusions: that they are ultimately caused by habitat changes or loss (Warren 1992).

Currently there are many more questions relating to the changing distribution and abundance of invertebrates in the Mersey Basin than there are answers. However, there is enormous potential for monitoring their response to environmental change and quality, in this rapidly changing and highly pressurised region.

Acknowledgements

I thank Brian Davis, Tom Eccles, Chris Felton, Eric Greenwood, Nora McMillan, Barry Shaw and Ian Wallace for their assistance.

References

Abbas, A.K. (1986). A new species of grass midge (Dipt. Cecidomyiidae) infesting the inflorescence of *Phalaris arundinacea* L. in Britain. *Entomologist's Monthly Magazine,* **122,** 65–71.

Allen, A.A. & Eccles, T.M. (1988). *Hydrosmectina delicatissima* Barnhauer (Col., Staphylinidae) new to Britain. *Entomologist's Monthly Magazine,* **124,** 215–20.

Anon. [1991]. Butterflies of Merseyside their history and status as recorded by members of the British Butterfly Conservation Society Merseyside Branch. Privately published.

Atkins, P.M., O'Callaghan, D.P. & Kirby, S.G. (1981). *Scolytus laevis* (Chapuis) (Coleoptera: Scolytidae) new to Britain. *Entomologist's Gazette,* **2,** 280.

Atkinson, D. & Houston, J. (eds) (1993). *The Sand Dunes of the Sefton coast.* National Museums & Galleries on Merseyside in conjunction with Sefton Metropolitan Borough Council, Liverpool.

Ball, S.G. (1992). *Recorder user manual.* Version 3.1, English Nature, Peterborough.

Bradshaw, A.D. & Chadwick, M.J. (1980). *The restoration of land.* Blackwell Scientific Publications, Oxford.

Brindle, A. (1952). The strange case of *Schiffemuelleria woodiella* Hub. (Lep: Oecophoridae). *Entomologist's Gazette,* **3,** 235–37.

Byerley, I. (1854). *Fauna of Liverpool.* H. Greenwood, Liverpool.

Chandler, P.J. (1991). Some corrections to the fungus gnats (Diptera, Mycetophiloidea) of Lancashire and Cheshire. *Annual Report and Proceedings of the Lancashire and Cheshire Entomological Society,* **114,** 38–53.

Clarke, C. (1993). *Panaxia dominula*, the Scarlet Tiger moth: queries arising from an artificial colony rediscovered after 28 years. *Antenna*, **17**, 177–83.

Creaser, A. (1992). The Wirral Peninsula, entomological past and present. *Annual Report and Proceedings of the Lancashire and Cheshire Entomological Society*, **115**, 72–75.

Cross, S. (1989). Provisional Lancashire and Cheshire checklists for Plecoptera, Ephemeroptera, Siphunculata and Siphonaptera. *Annual Report and Proceedings of the Lancashire and Cheshire Entomological Society*, **112**, 139–42.

Dennis, R.L.H. (1992). An evolutionary history of British butterflies. *The ecology of butterflies in Britain* (ed. R.L.H. Dennis), pp. 217–45. Oxford University Press, Oxford.

Eccles, T.M. (1987). Speke Hall Wood or Stockton Heath Wood. *Review of invertebrate sites in England, Merseyside and Greater Manchester* (ed. M. Parsons), (not paginated). Invertebrate site register 96. Nature Conservancy Council, Peterborough.

Eccles, T.M. (1993). Beetles (Coleoptera). *The Sand Dunes of the Sefton Coast* (eds D. Atkinson & J. Houston), pp. 99–103. National Museums & Galleries on Merseyside in conjunction with Sefton Metropolitan Borough Council, Liverpool.

Eccles, T.M. & Bowestead, S. (1987). *Anommatus diecki* Reitter (Coleoptera: Cerylonidae) new to Britain. *Entomologist's Gazette*, **38**, 225–27.

Ellis, J.W. (1889). *The coleopterous fauna of the Liverpool District*. Turner, Routledge, Liverpool.

Ellis, J.W. (1890). *The lepidopterous fauna of Lancashire and Cheshire*. McCorquodale, Leeds.

Emmet, A. & Heath, J. (1989). *Hesperidae – Nymphalidae, the butterflies. The moths and butterflies of Great Britain and Ireland*, 7:1. Harley Books, Colchester.

Eversham, B.C. & Arnold, H.R. (1992). Introductions and their place in British wildlife. *Biological recording of changes in British wildlife* (ed. P.T. Harding), pp. 44–64. Institute of Terrestrial Ecology Symposium, No. 26. HMSO, London.

Falk, S. (1991). *A review of the scarce and threatened bees, wasps and ants of Great Britain*. Research and survey in nature conservation, No. 35. Nature Conservancy Council, Peterborough.

Felton, C. (1991). Checklist of Lancashire and Cheshire spiders. *Annual Report and Proceedings of the Lancashire and Cheshire Entomological Society*, **114**, 57–70.

Felton, C. (1992a). Spider notes from Lancashire. *British Arachnological Society Spider Recording Scheme Newsletter*, **13**, 4.

Felton, C. (1992b). Foreigners in Cheshire. *British Arachnological Society Spider Recording Scheme Newsletter*, **13**, 4.

Felton, C. (1995). Urozelotes rusticus (L. Koch). (Gnaphosidae) new to Lancashire. *British Arachnological Society Spider Recording Scheme Newsletter*, **22**, 1–2.

Felton, C. (1996). The house centipede *Scutigera coleoptrata* (L.) in Lancashire. *British Myriapod Group Newsletter*, **24**, 1.

Felton, C. & Judd, S. (1997). *Carorita limnaea* (Araneae: Linyphiidae) and other Araneae at Wybunbury Moss, Cheshire – A unique refuge for two relict species of spider in Britain. *Bulletin of the British Arachnological Society*, **10**, 298–302.

Ford, W.K. (1953). Lancashire and Cheshire Odonata (a preliminary list). *The North Western Naturalist*, **6**, 227–33. New Series No. 2.

Ford, W.K. (1972). Orthopteroid stowaways. *Lancashire and Cheshire Fauna Society*, **61**, 17–21.

Foster, G.N. (1992). The effects of changes in land use on water beetles. *Biological recording of changes in British wildlife* (ed. P.T. Harding), pp. 27–30. Institute of Terrestrial Ecology Symposium, No. 26. HMSO, London.

Gabb, R. & Kitching, D. (1992). *The dragonflies and damselflies of Cheshire*. National Museums & Galleries on Merseyside, Liverpool.

Garland, S. & Appleton, T. (1994). Solitary wasps and ants (Sphecidae and Formicidae) of Lancashire and Cheshire.

Lancashire Wildlife Journal, **4**, 1–25.

Gemmell, R.P. & Connell, R.K. (1984). Conservation and creation of wildlife habitats on industrial land in Greater Manchester. *Landscape Planning*, **11**, 175–86.

Guest, J. (1989). *Freshwater invertebrates in the canals of Greater Manchester*. Unpublished survey for the Nature Conservancy Council. Copy at Environmental Data Bank, Liverpool Museum, William Brown Street, Liverpool L3 8EN.

Guest, J. (1996). *Hydrochara caraboides* (Linnaeus) (Hydrophilidae) in Cheshire. *The Coleopterist*, **5**, 19.

Hall, R.A. & Smith, P.H. (1991). Dragonflies of the Sefton coast sand-dune system, Merseyside. *Lancashire Wildlife Journal*, **1**, 22–34.

Hammond, P.M. (1974). Changes in the British coleopterous fauna. *The changing flora and fauna of Britain* (ed. D.L. Hawksworth), pp. 323–69. The Systematics Association Special Volume No. 6. Academic Press, London.

Harding, P.T. (1979). *A survey of the trees at Dunham Massey Park, Greater Manchester*. Institute of Terrestrial Ecology Project No. 405, Huntingdon.

Harding, P.T. & Alexander, K.N.A. (1993). The saproxylic invertebrates of historic parklands: progress and problems. *Dead wood matters: the ecology and conservation of saproxylic invertebrates in Britain* (eds K.J. Kirby, & C.M. Drake), pp. 58–73. English Nature Science, No. 7. English Nature, Peterborough.

Hardy, P.B., Hind, S.H. & Dennis, R.L.H. (1993). Range extension and distribution-infilling among selected butterfly species in north-west England: evidence for inter-habitat movements. *Entomologist's Gazette*, **44**, 247–55.

Hull, M. (1987). Check-list of the Neuropteroidea of Lancashire and Cheshire. *Annual Report and Proceedings of the Lancashire and Cheshire Entomological Society*, **110**, 66.

Hull, M. (1990). Check-list of the sawflies of Lancashire and Cheshire. *Annual Report and Proceedings of the Lancashire and Cheshire Entomological Society*, **113**, 38-53.

Johnson, C. (1962). The scarabaeoid (Coleoptera) fauna of Lancashire and Cheshire and its apparent changes over the last 100 years. *The Entomologist*, **95**, 153–65.

Johnson, C. (1976). Nine species of Coleoptera new to Britain. *Entomologist's Monthly Magazine*, **111**, 177–83.

Johnson, C., Robinson, N.A. & Stubbs, A.E. (1977). *Dunham Park a conservation report on a parkland of high entomological interest*. Chief Scientist's Team Notes, No. 5, Nature Conservancy Council, London.

Johnson, C. & Eccles, T.M. (1983). *Plectophloeus erichsoni occidentalis* Besuchet (Coleoptera: Pselaphidae) new to Britain. *Entomologist's Gazette*, **34**, 267–69.

Judd, S. (1986). The past and present status of the damselfly *Coenagrion pulchellum* (Van Der Linden) (Odonata: Coenagriidae) – in Cheshire and parts of its adjacent counties, corresponding to the 100km square SJ3–3—. *Entomologists' Record and Journal of Variation*, **98**, 57–61.

Judd, S. (1987). A checklist of Lancashire and Cheshire Heteroptera. *Annual Report and Proceedings of the Lancashire and Cheshire Entomological Society*, **110**, 60–65.

Judd, S. & Rotherham, I.D. (1992). The phytophagous insect fauna of *Rhododendron ponticum* L. in Britain. *The Entomologist*, **111**, 134–50.

Judd, S. & Hodkinson, I.D. (1998). The biogeography and regional biodiversity of the British seed bugs (Hemiptera: Lygaeidae). *Journal of Biogeography*, **25**, 227–49.

Kendall, P. (1982). *Bromius obscurus* (L.) in Britain (Col., Chrysomelidae). *Entomologist's Monthly Magazine*, **117**, 233–34.

Kidd, L.N. & Brindle, A. (1959). The Diptera of Lancashire and Cheshire, Part 1. *Lancashire and Cheshire Fauna Committee*, **33**, 1–136.

Langslow, D.R. (1992). Legislation and policies – Managing the changes in British wildlife: the effects of nature conservation. *Biological recording of changes in British wildlife* (ed. P.T. Harding), pp. 65–70. Institute of Terrestrial Ecology Symposium, No. 26. HMSO, London.

Lawson, A.K. (1930). *A check list of the fauna of Lancashire and Cheshire. Part 1.* Lancashire and Cheshire Fauna Committee, Buncle, Arbroath.

Lawton. J.H. (1995). The response of insects to environmental change. *Insects in a changing environment* (eds R. Harrington & N.E. Stork), pp. 1–26. 17th Symposium of the Royal Entomological Society of London. Academic Press, London.

Luff, M.L. & Woiwod, I.P. (1995). Insects as indicators of land-use change: A European perspective, focusing on moths and ground beetles. *Insects in a changing environment* (eds R. Harrington & N.E. Stork), pp. 399–422. 17th Symposium of the Royal Entomological Society of London. Academic Press, London.

Mansbridge, W. (1940). *The lepidopterous fauna of Lancashire and Cheshire* (Revised Ellis list). Lancashire and Cheshire Entomological Society, Liverpool.

Marshall, J.A & Haes, E.C.M. (1988). *Grasshoppers and allied insects of Great Britain and Ireland.* Harley Books, Colchester.

Mellanby, K. (1974). Summing up. *The changing flora and fauna of Britain* (ed. D.L. Hawksworth), pp. 419–23. The Systematics Association Special Volume No. 6. Academic Press, London.

McMillan, N.F. (1959). The Mollusca of some Cheshire marl-pits: a study in colonization. *Journal of Conchology*, **24**, 299–315.

McMillan, N.F. (1990). The history of alien freshwater Mollusca in North-West England. *Naturalist*, **115**, 123–132.

Owen, D.E., Deyd, E.L., Smith, S.G., Brindle, A. *et al.* (1962). *Fauna of the Manchester area.* Manchester University Press, Manchester.

Owen, D.F. (1978). Insect diversity in an English suburban garden. *Perspectives in urban entomology* (eds G.W. Frankie, & C.S. Koehler), pp. 13–29. Academic Press, New York, USA.

Parsons, M. (1987a). *Review of invertebrate sites in England, Cheshire.* Invertebrate site register 95. Nature Conservancy Council, Peterborough.

Parsons, M. (1987b). *Review of invertebrate sites in England, Merseyside and Greater Manchester.* Invertebrate site register 96. Nature Conservancy Council, Peterborough.

Parsons, M. (1987c). *Review of invertebrate sites in England, Lancashire.* Invertebrate site register 98. Nature Conservancy Council, Peterborough.

Payne, K. (1987). Haskayne Cutting. *Review of invertebrate sites in England, Lancashire 98* (ed. M. Parsons), (not paginated). Nature Conservancy Council, Peterborough.

Rutherford, C.I. (1983). *Butterflies in Cheshire 1961–1982.* Supplement to the *1981–1982 Proceedings of the Lancashire and Cheshire Entomological Society.*

Rutherford, C.I. (1994). *Macro-moths in Cheshire 1961–1993.* Lancashire and Cheshire Entomological Society.

Sanderson, R.A. (1992a). Hemiptera of naturally vegetated derelict land in north-west England. *Entomologist's Gazette*, **43**, 221–26.

Sanderson, R.A. (1992b). Diversity and evenness of Hemiptera communities on naturally vegetated land in NW England. *Ecography*, **15**, 154–60.

Sanderson, R.A. (1993). Factors affecting the Hemiptera of naturally colonised derelict land in North West England. *The Entomologist*, **112**, 10–16.

Sharp, W.E. (1908). *The Coleoptera of Lancashire and Cheshire.* Lancashire and Cheshire Entomological Society. Gibbs and Bamforth Ltd, St. Albans.

Shaw, B.T. (1993). Results of a butterfly survey conducted along the Heald Green railway cutting (1984-1993 inclusive). Unpublished pamphlet – copy deposited in Liverpool Museum entomology library.

Shaw, B.T. (1994). *Cheshire butterfly recording scheme 1994 annual report.* Unpublished – copy deposited in Liverpool Museum entomology library.

Shaw, B.T. (1995). *1995 Cheshire butterfly report.* Unpublished – copy deposited in Liverpool Museum entomology library.

Smith, S.G. (1948). The butterflies and moths found in the counties of Cheshire, Flintshire, Denbighshire, Caernarvonshire, Anglesey and Merionethshire. *Proceedings of the Chester Society of Natural Science, Literature and Art*, **2**, 1–251.

Sumner, D.P. (1985). The geographical and seasonal distribution of the dragonflies of Lancashire and Cheshire. *Annual Report and Proceedings of the Lancashire and Cheshire Entomological Society*, **108**, 177–94.

Terrell-Neild, C. (1994). Beetle diversity and land use in the Eccles area of Greater Manchester. *Lancashire Wildlife Journal*, **4**, 35–55.

Thomas, J.A. (1991). Rare species conservation: case studies of European butterflies. *The scientific management of temperate communities for conservation* (eds I.F. Spellerburg, F.B. Goldsmith & M.G. Moriss), pp. 149–97. Blackwell Scientific Publications, Oxford.

Usher, M.B. (1995). A world of change: land-use patterns and arthropod communities. *Insects in a changing environment* (eds R. Harrington & N.E. Stork), pp. 371–422. 17th Symposium of the Royal Entomological Society of London. Academic Press, London.

Vincent, P. (1990). *The biogeography of the British Isles – an introduction.* Routledge, London.

Wallace, I.D. & Wallace, B. (1982). Land invertebrates, excluding Mollusca, spiders and harvestmen. *Hilbre the Cheshire Island – its history and natural history* (ed. J.D. Craggs), pp. 87–116. Liverpool University Press, Liverpool.

Warren, M.S. (1992). The conservation of British butterflies. *The ecology of butterflies in Britain* (ed. R.L.H. Dennis), pp. 246–74. Oxford University Press, Oxford.

Wyatt, B.K. (1992). Resources for documenting changes in species and habitats. *Biological recording of changes in British wildlife* (ed. P.T. Harding), pp. 20–26. Institute of Terrestrial Ecology Symposium, No. 26. HMSO, London.

CHAPTER TWENTY-ONE

'All the birds of the air' – indicators of the environment

D. NORMAN

Introduction

Birds are now probably the best known part of the fauna, having been greatly studied, with reliable identification, for almost two centuries, and avifaunas of all parts of the Mersey Basin were published in the 19th century. These were based on relatively limited information, however, depending mainly on the authors' network of correspondents. It is only in the last few decades that truly comprehensive information has become available, mainly through surveys organised by the British Trust for Ornithology (BTO), largely carried out by amateur participants who travel throughout the region. The most notable surveys are the national atlases, of the distribution of breeding birds in 1968–72 (Sharrock 1976), of the distribution and numbers of wintering birds in 1981–84 (Lack 1986), and of the distribution and numbers of breeding birds in 1988–91 (Gibbons, Reid & Chapman 1993). The local ornithological societies have organised atlases of the breeding distribution of birds on a 2 x 2km square (tetrad) basis in Greater Manchester (Holland, Spence & Sutton 1984) and Cheshire and Wirral (Guest et al. 1992) but, regrettably, not in Merseyside or Lancashire. Because most publications concentrate on descriptions of distribution rather than numbers, it is often difficult to tell if a species has increased or decreased if it has not also changed its range: nevertheless, birds are probably the only taxon for which some quantitative data exist.

This chapter considers various factors that have affected the birds of the Mersey Basin. Examples are given of species in each category, but these are illustrative rather than comprehensive, because of limited space, and I concentrate on the breeding season, with occasional discussion of wintering birds.

Key factors

Natural effects

Climatic change in Britain

Several species which used to breed in Cheshire and Lancashire no longer do so, including Woodlark (*Lullula arborea*), last recorded around the 1860s, Nightingale (*Luscinia megarhynchos*), Nightjar (*Caprimulgus*

europaeus), Red-backed Shrike (*Lanius collurio*), which nested at Bootle in the 19th century, and Wryneck (*Jynx torquilla*), the last two now being almost extinct in Britain. All are insect-eaters of southerly distribution, probably affected by climatic change (Burton 1995).

The Mersey Basin has long been on the edge of the distribution of the Turtle Dove (*Streptopelia turtur*), so its status in this area has fluctuated considerably. It underwent a northerly and westerly expansion in the 19th century, colonising Cheshire in the 1870s and first breeding in Lancashire in 1904, steadily increasing to 1930, possibly assisted by agricultural recession and the availability of more weed seeds. Southern Lancashire continued to be the most north-westerly regular breeding area until the late 1960s, when Turtle Doves again started to retreat south-eastwards. A comparison between the two BTO Atlases (Sharrock 1976 and Gibbons et al. 1993) shows a dramatic shift out of north-west England, and this species is now close to extinction in the region.

The Reed Warbler (*Acrocephalus scirpaceus*) is another species close to its north-western limit in the Mersey Basin. It expanded its range northward to reach Lancashire in about 1850, probably assisted by climatic amelioration, and by 1900, bred as far north as Morecambe and east to the Yorkshire border. By the 1930s the species had retreated towards the south – rather against the climatic trend – only being found near the coast in Lancashire and not in Wirral or the eastern hills. It has spread again in the last 20 or 30 years, with odd breeding sites now occupied along the coast almost as far as the Scottish border. The extensive drainage of the lowlands cannot have helped Reed Warblers, but many of the region's *Phragmites* reedbeds, their required breeding habitat, are now in the control of nature conservation organisations and are well managed and protected. Cuckoos (*Cuculus canorus*) in the area mainly use Dunnock (*Prunella modularis*) or Meadow Pipit (*Anthus pratensis*) hosts, but have extended to cuckold Reed Warblers in Cheshire recently: there were single records in 1934 and 1977, then annually from 1988 onwards at Rostherne and Woolston (Calvert 1989; Smith & Norman 1989).

Two species, Barn Owl (*Tyto alba*) and Song Thrush

(*Turdus philomelos*), have occasioned much concern for their decline over the last 30 or 40 years. Several factors are implicated, including agricultural habitat changes and pesticides, and climatic factors must be playing a part. They are both southern species which here reach close to their northern limit.

The Yellow Wagtail (*Motacilla flava*) used to be widespread in the lowlands here, but must have been severely hit by loss of habitat from drainage, intensification of agriculture and replacement of grasslands with cereals. However, when Stuart Smith studied Yellow Wagtails breeding alongside the River Mersey in suburban Manchester over the seven years 1939–45, he wrote in his New Naturalist Monograph (1950) 'there is little doubt that it is increasing'. Since then there has clearly been a contraction in range, south-eastwards, with some local declines or extinctions.

Climatic change outside Britain

The reducing rainfall in western Africa has drastically affected the populations of several of our summer visitors. The Sedge Warbler (*Acrocephalus schoenobaenus*) used to be regarded as common throughout the Mersey Basin, except for the eastern hills, and Oakes (1953) declared it to be '*the* warbler of the plains' of old Lancashire. Its population appeared stable from 1800 to 1968 although it had disappeared from areas where the habitat had been drained or developed. Then, along with other species – notably the Whitethroat (*Sylvia communis*) (Winstanley, Spencer & Williamson 1974) – the population crashed in 1969 and again in 1984, conclusively linked to west African drought (Peach, Baillie & Underhill 1991). Although their status in the Mersey Basin region has been poorly recorded until recently, my ringing studies of Sand Martins (*Riparia riparia*) breeding in mid-Cheshire quarries show that the survival of adults varies annually between about 8% and 50%, which correlates strongly with west African rainfall (Norman & Peach, unpublished). The local populations of Tree Pipit (*Anthus trivialis*), Whinchat (*Saxicola rubetra*), Grasshopper Warbler (*Locustella naevia*) and Redstart (*Phoenicurus phoenicurus*) have not been so well studied, but they are all probably mainly determined by African climate changes.

Severe winter weather

The breeding numbers of the Grey Heron (*Ardea cinerea*) are particularly well known, from the BTO censuses from 1928 onwards, the longest running census of any bird species. There are about 500 nests in the Mersey Basin, the great majority in Cheshire. Although pesticides can reduce the survival of adults, the most important factor affecting their numbers is severe winters, when their food becomes locked under ice.

Populations of insectivorous residents such as Goldcrests (*Regulus regulus*), Stonechats (*Saxicola torquata*) and Green Woodpeckers (*Picus viridis*) fluctuate with expansions and local extinctions, mainly correlated with hard winters.

Range expansion

The Little Ringed Plover (*Charadrius dubius*) first bred in the Mersey Basin (Cheshire) in 1954, and there are now regularly up to 40 or 50 pairs, mostly in sand quarries or lime beds. Collared Doves (*Streptopelia decaocto*) have spectacularly spread from the Balkans in the 1930s to reach the Mersey Basin in 1960, finding an available niche as a medium-sized grain-eater tolerant of man, and now it is one of our commoner species.

Black Redstarts (*Phoenicurus ochruros*) and Bearded Tits (*Panurus biarmicus*) bred here in the 1970s or early 1980s, but have failed to establish themselves. One of the most extraordinary events of recent years was the nesting of Marsh Warbler (*Acrocephalus palustris*) at Woolston in 1991, their most northerly breeding in Britain this century (Norman 1994). Regrettably this has not been sustained. Single pairs of Black-necked Grebe (*Podiceps nigricollis*) bred several times in Cheshire between 1938 and 1953 and then during the 1980s, and from the late 1980s onwards several pairs have bred each year.

Shelducks (*Tadorna tadorna*) and Oystercatchers (*Haematopus ostralegus*) built up during this century to a considerable breeding population on the coast and near the River Mersey, and are now moving inland more frequently to breed (Gibbons *et al.* 1993). Similarly the Redshank (*Tringa totanus*), previously a rare coastal breeder, completely changed its status during the late-19th and 20th century to become a common inland breeder (Oakes 1953), although usually close to the major rivers.

Unintentional human impacts
Agricultural changes

Species such as Skylark (*Alauda arvensis*) and Lapwing (*Vanellus vanellus*) must have been scarce up to about 5,000 BC, perhaps limited to the coastal marshes and the Pennine fringes, but clearance of trees for land cultivation would have allowed them to spread and become characteristic species of open farmland.

Skylarks largely shifted from natural pastures and grassland to arable fields, apparently with little change in distribution or numbers, until about 1980, since when their population has almost halved. Good habitat used to be provided by short-term leys of grass or clover, but the shift to autumn-sown cereals leaves fewer seeds for winter feeding, and by April the crops are too tall and dense for nesting. They have declined in my study area at Frodsham Marsh, however, where little change in habitat or farming practice has occurred, and I suspect that there may be other influences such as the recent prevalence of cold, wet springs.

Lapwings are still common in the region, but like Skylarks are declining fast with agricultural changes, mainly the transition from hay to silage, and autumn-sown cereals that grow too tall in spring for the species to nest in such fields.

Corncrakes (*Crex crex*) used to be quite common in north-west England, and their rapid decline from about

1880 onwards is a classic tale of inimical agricultural changes. Their habitat of tall grass and herbs, especially hay meadows, became much less common as farmers turned to making silage. What hay remained was grown more rapidly with inorganic fertilisers, leading to a season too short for Corncrakes to rear their young before the onslaught of mechanical harvesting. The species almost disappeared from the Mersey Basin during the Second World War, and has bred only sporadically since then.

The lowland mosses used to be grouse moors; Carrington Moss until 1886 and Chat Moss until the mid-1930s, but they were drained for agriculture, and Red Grouse (*Lagopus lagopus*) is now confined to the heather-dominated eastern hills. Their population has roughly halved in the last 50 years, as moorlands have been lost to pastoral farming and subsidies have encouraged heavy sheep grazing (chapter eleven, this volume). Twite (*Carduelis flavirostris*) and Short-eared Owls (*Asio flammeus*) also used to breed regularly on the mosses, but are now restricted to breeding in heather moorland.

The Corn Bunting's (*Miliaria calandra*) population declined in the agricultural depression of 1870–1930 with the much reduced area of cereals. It recovered from 1940–1970, but then dropped greatly again as winter survival suffered through the loss of stubble. A reduction in farmland invertebrates, taken as food for their young, is also thought to be a contributory factor. Despite this gloomy picture, the lowlands of west Lancashire are one of the most densely occupied areas of all Britain (Gibbons *et al.* 1993).

Tree Sparrows (*Passer montanus*) have shown puzzling, irregular cycles in numbers, being high from the 1880s to 1930s, dropping to a low about 1950 then peaking again in 1960–78. They are now in a major decline phase undoubtedly caused by the reduction of weeds through the use of herbicides. House Sparrows (*Passer domesticus*) increased substantially, along with the human population in the 19th century, and major attempts to destroy them had little general effect. Their population showed little change until the 1970s, but has decreased markedly since then, probably because of a decline in both invertebrates and weed seeds. Greenfinches (*Carduelis chloris*) and Linnets (*Carduelis cannabina*) survived the depredations of bird-catchers in the 19th century, but their population fluctuations in the 20th century, now sharply downwards, are apparently linked to the availability of seeds on agricultural land.

The Yellowhammer (*Emberiza citrinella*) used to be common through most of the region, but has undergone a serious contraction in range since 1930–50, especially noticeable in the east of the Mersey Basin (Holland *et al.* 1984). They are very dependent on hedgerows, mainly as breeding sites and song-posts. However, as much as 60% of Cheshire hedges have gone in this century, for agricultural production, urban and industrial growth, mineral extraction and road-building. Lesser Whitethroats (*Sylvia curruca*) became less common during the Second World War, which Boyd (1950) ascribed to the practice of trimming hedges lower to allow cultivation closer to the field edges.

The story of the disastrous effects of the organochlorine seed-dressings used in the 1950s and early 1960s is well known: the poisons were ingested by seed-eating birds, and populations of Stock Dove (*Columba oenas*) and many farmland finches were badly affected. The pesticide residues also accumulated up the food chain to devastate the populations of many bird-eating raptors. The effects were partly direct, with many birds killed outright, but also more insidious, with a reduction in breeding success caused by thinner egg-shells. Sparrowhawks (*Accipiter nisus*) were wiped out from much of England, but persisted in the region, probably because of the lower acreage of arable land. The Peregrine (*Falco peregrinus*) was also badly affected, but was not a common bird and certainly did not breed in the Mersey Basin at that time.

Poisoning by agricultural chemicals is not just a recent phenomenon: in the first half of the 19th century, Grey Partridges (*Perdix perdix*) were poisoned in large numbers through feeding on seedcorn that had been steeped in arsenic to kill wireworms.

In the last few years, pesticides have become still more efficient and now kill many of the farmland invertebrates, probably affecting species such as Rook (*Corvus frugilegus*) and Starling (*Sturnus vulgaris*). The Rook has always been common through most of the Mersey Basin, although it is missing from much of the lowlands north of the River Mersey because of the shortage of nesting trees. There used to be some sport shooting and collection of nestlings for food, but they were not persecuted as much as the other corvids because of the early (mid-19th century) recognition of its role in eating agricultural pests. Breeding Starlings, which also eat tremendous numbers of soil invertebrates, declined in the 18th century, but then increased greatly in the 19th century, perhaps due to climatic changes. They have decreased sharply in recent years, from 1980 on, although still attaining a very high density north of the River Mersey (Gibbons *et al.* 1993).

Finally, one species which has adapted its behaviour recently to take advantage of agricultural produce is the winter-visiting Fieldfare (*Turdus pilaris*). Their habit of feeding on orchard apples helps them to survive when the ground is frozen (Norman 1994).

Drainage of wetlands

The Bittern (*Botaurus stellaris*) became extinct in the Mersey Basin by 1850. Initially its decline was due to drainage of its reedbed habitat, but numbers were reduced further by shooting for the table and for its feathers to make flies for fishing; then, as they became rare, collectors of skins and eggs hastened the decline. Another reedbed specialist, the Marsh Harrier (*Circus aeruginosus*), probably bred until about 1820 in the Mersey valley and last nested in the region about 1860 around Martin Mere (Holloway 1995). Drainage and enclosure through the 18th and 19th centuries caused a widespread decline in Mallard (*Anas platyrhynchos*) and Teal (*Anas crecca*) breeding, and although the former is still common and widespread, Teal is still dwindling as a breeding species. Snipe (*Gallinago gallinago*) used to nest abundantly on the mosslands of Lancashire (Oakes 1953); it was common on the plains and very common on the moors, but has now gone from almost all the lowland areas.

Many water birds such as Little Grebe (*Tachybaptus ruficollis*) and Reed Bunting (*Emberiza schoeniclus*) have taken advantage of the patchwork of ponds created by marling since the 13th century and formerly such a feature of the Mersey Basin (chapter fourteen, this volume). However, the use of fertilisers and piping of water for stock to drink meant that the two reasons for marl pits to remain no longer applied, and many were filled in or left to become overgrown with scrub.

Urban and industrial development

Most rural species have suffered, often with local extinctions, when their habitats have been taken for development, and it would be tedious to list examples. There are, on the other hand, several examples where development has benefited birds. Some species now depend almost exclusively on man-made structures for their nest sites, including Swifts (*Apus apus*) and House Martins (*Delichon urbica*): when they had to find cliffs or tree-holes they must have been scarce in the Mersey Basin. Swallows (*Hirundo rustica*) must also have changed their breeding sites since historical time, almost all now nesting in buildings, but even during the Roman occupation commensal breeding was common. From about 1960 the vast majority of the Sand Martin population has nested in sand quarries. The Little Gulls (*Larus minutus*) that now congregate on spring passage at the mouth of the River Mersey benefit from the lagoons in the docks at Seaforth for feeding, bathing and resting (Smith 1995).

Conurbations also have a 'heat island' effect, whereby the mean temperature may be several degrees above that of surrounding areas, making a substantial difference to birds' overnight survival, especially in winter. The massive urban roosts of immigrant Starlings certainly take advantage of this effect.

The former coastal colony of up to 300 pairs of Common Terns (*Sterna hirundo*) at Ainsdale and some Arctic Terns (*Sterna paradisaea*) declined in the 1930s with adjacent house-building, and subsequent war-time activities – establishment of a naval station on the dunes – finished them off. The period between the wars saw a widespread decline in Ringed Plover (*Charadrius hiaticula*), as there was a dramatic increase in human use of the coasts, for recreation, holiday homes and sea defences, with consequent disturbance and loss of habitat. The species' salvation was a shift in habit, moving to breed inland, sporadically from 1946 and regularly for about the last 20 years. The Little Tern (*Sterna albifrons*), which used to nest on south Lancashire shores (Sefton coast), has been similarly affected, but has suffered much more since it has not moved inland to breed in this country as it commonly does in continental Europe.

Afforestation

The maturing conifer plantations are assisting Crossbills (*Loxia curvirostra*) and Siskins (*Carduelis spinus*) to breed in greater numbers. The Redpoll (*Carduelis flammea*) used to be scarce but has flourished, especially since the early 1960s, benefiting particularly from Birch (*Betula* spp.) on the mosslands and those planted around the new towns. Coal Tits (*Parus ater*) have increased substantially since the 19th century but are still scarce away from trees. Chiffchaffs (*Phylloscopus collybita*) expanded their range northwards in the 19th century, and are mainly associated with wooded areas. Although still present at low densities north of the River Mersey, they have increased and spread in recent years.

Several species that depend on woodland for breeding, notably Grey Heron, Sparrowhawk, Nuthatch (*Sitta europaea*), Wood Warbler (*Phylloscopus sibilatrix*) and Goldcrest showed an obvious gap in their distribution in the 1968–72 breeding Atlas north of the River Mersey in Merseyside and Greater Manchester, which had largely been filled in by the time of the 1988–91 Atlas. Other species are still missing from that area, including Woodcock (*Scolopax rusticola*), Green Woodpecker, Lesser Spotted Woodpecker (*Dendrocopos minor*), Rook, Jackdaw (*Corvus monedula*), Marsh Tit (*Parus palustris*), Redstart, Garden Warbler (*Sylvia borin*) and Tree Pipit. On the other hand the Willow Tit (*Parus montanus*) is more abundant in the Mersey valley and mosslands than anywhere else in the region (Gibbons *et al.* 1993). Also, the thousands of Bramblings (*Fringilla montifringilla*) that visited Liverpool in the winter of 1980–81 took advantage of the mature Beeches (*Fagus sylvatica*), planted in suburban avenues around 1810, by feeding avidly on the mast.

Water and air quality

The almost relentless decline of the Kingfisher (*Alcedo atthis*) in the 19th century was the product of persecution, hard winters and pollution of watercourses. There has been a general recovery in the 20th century following a reduction in persecution, and milder winters, but canalisation of streams and rivers has reduced the numbers of nesting sites and fishing perches. Dippers (*Cinclus cinclus*) and Grey Wagtails (*Motacilla cinerea*) are found in limited numbers in the hill-streams throughout the uplands, but the latter species also spreads into urban areas and is less sensitive to pollution or acidification of its watercourses because they take a wider range of insect prey.

The waterfowl wintering on the Mersey Estuary, especially Pintail (*Anas acuta*), Teal, Wigeon (*Anas penelope*), Shelduck, Dunlin (*Calidris alpina*) and Redshank, have undoubtedly benefited greatly from the improvement in water quality from an almost lifeless river before the 1970s (Thomason & Norman 1995).

The aerial pollution of the Industrial Revolution might

have suppressed flying insects and their predators such as House Martins. Both Long-tailed Tits (*Aegithalos caudatus*) and Chaffinches (*Fringilla coelebs*) were scarce in the Mersey valley and to the north of it, perhaps because of poor air quality limiting the lichens with which they camouflage their nests. Long-tailed Tits have recently spread into the Liverpool area, and Chaffinches might now be increasing (Marchant *et al.* 1990).

Recreational disturbance

Several scarce moorland breeders are endangered by the increasing numbers of walkers in the hills, with the very small populations of Golden Plover (*Pluvialis apricaria*) (Yalden & Yalden 1989), Common Sandpiper (*Actitis hypoleucos*) (Holland *et al.* 1982), Dunlin and Ring Ousel (*Turdus torquatus*) especially at risk.

Mute Swans (*Cygnus olor*) have been fairly common in the region for centuries, although often ignored as a domesticated species. They underwent a catastrophic decline between 1950 and 1985, attributed to poisoning through ingestion of lead weights discarded by anglers. The species reached its nadir around 1985 when it was one of the scarcest breeding birds in Cheshire (13 pairs), but since the phasing out of lead weights, the population has rapidly recovered to reach at least 70 pairs only ten years later.

Deliberate human impacts
Legislation

Laws such as the Sea Birds Preservation Act (1869), the Lapwing Protection Act (1926) and the Wildlife and Countryside Act (1981) have had major effects on conservation of some species, but affect the whole of the UK and are not specific to the Mersey Basin.

Provision of food and protection

Greenfinches and Siskins have benefited greatly from their recently-acquired habit of feeding on peanuts in gardens, which allows them to attain good condition at a time of year when natural seeds are scarce. Many tits likewise owe their over-winter survival to food put out by householders. Blackcaps (*Sylvia atricapilla*) appeared as a wintering species in the Mersey Basin from about 1964 onwards, as birds of southern German and Austrian breeding stock discovered the benefits of a shorter migration to an area with reasonably mild winters and ample food supplies: they favour well-stocked gardens, and perhaps 100 individuals are now recorded in the region each winter.

In the 19th century the Pied Flycatcher (*Ficedula hypoleuca*) was not present locally although it was common to the west (Wales) and north (Cumbria and northern Yorkshire). From 1939 onwards they nested sporadically in the Mersey Basin, with increasing numbers especially in the last decade or so as more nest-boxes were provided. The only Common Tern breeding records in our region since the Second World War have been in the last two decades on man-made nesting sites in Merseyside and Greater Manchester. Goldeneyes

(*Bucephala clangula*) have bred recently in nest-boxes in Lancashire.

Reserves like Martin Mere have allowed hundreds of wild Bewick's Swan (*Cygnus columbianus*) and Whooper Swans (*Cygnus cygnus*) – previously very scarce birds – to winter here. The Ribble National Nature Reserve hosts tens of thousands of Pink-footed Geese (*Anser brachyrhynchus*) and Wigeon, manifold increases on the figures of 50 years ago.

It is probably protection, or at least the decline of shooting, that has allowed several duck species to breed from early this century, e.g., Shoveler (*Anas clypeata*) and Tufted Duck (*Aythya fuligula*), with other species, e.g., Gadwall (*Anas strepera*) and Pochard (*Aythya ferina*), colonising only in the last 20 years.

Persecution

During the 19th century, most wildlife was seen as competitors with humans for food, and widespread systematic persecution of almost any species of bird took place. The large estates employed many gamekeepers who killed anything they thought, rightly or wrongly, might threaten their sport. The effect was most noticeable when the numbers of keepers dropped markedly during both World Wars, and many birds took advantage. Sparrowhawks used to be heavily persecuted throughout the region, but their secretive breeding habits and high reproductive rate ensured they survived, even in heavily keepered woods. Some nests are still robbed in Merseyside, apparently by amateur would-be falconers. Red Kites (*Milvus milvus*) were persecuted into extinction around 1800, and Hen Harriers (*Circus cyaneus*) and Buzzards (*Buteo buteo*) a little later. Buzzards are now making a welcome resurgence, having bred in Cheshire from 1980 onwards. Peregrines, having survived the depredations of falconers for hundreds of years, mainly by nesting in inaccessible and dangerous nesting places, have also nested at several Cheshire sites in the last five years, as have Ravens (*Corvus corax*).

Jays (*Garrulus glandarius*) and Magpies (*Pica pica*) declined during the 19th century due to persecution, but increased during the First World War and afterwards, again increasing noticeably during the Second World War so that Boyd (1950) thought Magpies 'far too abundant'. The species has now changed its habits to tolerate man and to nest in hedgerows and in conurbations. Carrion Crows (*Corvus corone*), hated for so long, also increased during the Great War and afterwards, with another rise during the Second World War.

Kestrels (*Falco tinnunculus*) and Tawny Owls (*Strix aluco*) were uncommon in the 19th century, through direct persecution, but the former also through reduction of old corvid nests to take over. Both species have greatly increased during this century, particularly between 1928 and 1953, and have become common in towns, especially feeding on House Sparrows in Manchester.

The Bullfinch's (*Pyrrhula pyrrhula*) habit of eating

fruit-tree buds resulted in a bounty of a penny a head in the mid-17th century, about 50 pence at today's prices, but their population increased after the Wild Birds Protection Acts of the 1880s and 1890s.

There are modern-day calls to reduce numbers of some species. The Cormorant (*Phalacrocorax carbo*) is an abundant non-breeding visitor to the region and is spreading inland to gorge on artificially stocked fishing waters, much to the disgust of anglers who, dubbing it the 'black plague', are lobbying hard for licences to shoot it. Many of the region's migrant species are hunted in southern Europe, but there is little evidence on whether or not this actually affects their populations. Wildfowling is extensively practised in the Mersey Basin: properly exercised, with appropriate adherence to closed seasons and bag limits, it appears to have no effect on the populations of ducks or geese.

Collection for food
Large numbers of Lapwing eggs used to be collected for food, and the species decreased greatly in the 19th century, with loss of habitat as well. A rapid resurgence in fortunes occurred with the Lapwing Protection Act of 1926. Black-headed Gulls (*Larus ridibundus*) bred in the Delamere area of Cheshire from at least 1617, but a steady population decline throughout Britain during the 19th century, probably through large numbers of eggs being harvested from most colonies, brought them close to extinction. The Delamere colony survived, being recorded in 1860, and the Sea Birds Preservation Act (1869) helped the species to thrive, although they declined again after 1958, possibly due to changes in water-level caused by increased sand extraction in the area, and have not bred there since 1965.

Collection for fashion
The fashion for skins and feathers increased in the 19th century so that by 1860 the national population of the Great Crested Grebe (*Podiceps cristatus*) was reduced to about 42 breeding pairs, of which 20 were in Cheshire. Following legal protection and the formation of the Royal Society for the Protection of Birds (RSPB), the species increased so that a survey in 1931 found 78 pairs breeding in Cheshire. Green Woodpeckers and Jays were also shot for their feathers to make fishing flies and for women's fashions, and it is difficult now to comprehend that in the 1890s Robins (*Erithacus rubecula*) were popular in millinery, and that countless individuals of Britain's national bird were killed to adorn women's hats.

Collection as specimens
In the 19th century the Goldfinch (*Carduelis carduelis*) was a very popular cage-bird, and so easy to catch that it was rare in the area until its inclusion in the Wild Birds Protection Acts of 1880 and 1881 allowed slow increases in population and range. An increase in weeds, especially thistles, in the agricultural depression of the late 19th and early 20th centuries also helped them, but all publications up to 1950 indicate exceptional scarcity in

Greater Manchester. They are now widespread but not numerous across the Mersey Basin.

Many Victorian drawing rooms used to include a stuffed specimen, but no birds have been collected for this purpose for many years. The oologists of the 19th and the first half of the 20th centuries took many complete clutches, often from the rarer species, and must have affected their populations, before egg-collection was outlawed by various Acts of Parliament.

Introductions
Introductions have almost always benefited the species concerned, allowing an expansion of range beyond that naturally occupied, but many introductions have had deleterious effects on the indigenous species, either by competition or displacement. Pheasants (*Phasianus colchicus*) were introduced about 1,000 years ago but remained uncommon until the rise of modern game-keeping, and are now widespread and common, with extensive management and releasing. The Ruddy Duck (*Oxyura jamaicensis*) bred in Cheshire from 1968 and in Greater Manchester since 1981, but migrating individuals may now be endangering the closely related scarce White-headed Duck (*Oxyura leucocephala*) in Spain through hybridisation. At the end of the last century, the Canada Goose (*Branta canadensis*) was rare, but it has spread rapidly in recent decades, the Cheshire population trebling from 1953 to 1976 and still increasing, though they have bred in Greater Manchester only since 1974. The Little Owl (*Athene noctua*) was introduced to England in the 19th century, and first bred in several parts of the region around 1920. The population rocketed until about 1960, since when it has become less common, perhaps with reductions in prey from pesticides and tidier farm practices.

Introduced mammals can affect birds. Rabbits (*Oryctolagus cuniculus*), introduced probably around 1100, have shaped parts of the countryside so that some birds depend on the habitat they produce. Indirectly the decline of Wheatears (*Oenanthe oenanthe*) after the Second World War was accentuated by the catastrophic drop in rabbit populations from myxomatosis, since their burrows are the Wheatear's favourite nest site, and the rabbits graze the grass to the short sward the birds need. The Wigeon that winter on the Mersey Estuary themselves graze almost exclusively on the areas that have been cropped, first by summer cattle and then by rabbits. American Mink (*Mustela vison*), escaped or deliberately released from farms (chapter twenty-two, this volume), are now major predators on waterfowl in some areas, although their population appears to be cyclical and they have been checked by organised trapping.

Species showing little change
As well as those species discussed above that have changed their status, there are several whose range or numbers have kept stable, often because their catholic habits have enabled them to adapt to the changing environment. Pied Wagtails (*Motacilla alba*), Dunnocks,

Robins, Blue Tits (*Parus caeruleus*) and Great Tits (*Parus major*) have been common for at least two centuries, and have experienced no widespread changes, although local or temporary declines have occurred following severe winters. Similarly, the Wren (*Troglodytes troglodytes*) has long been abundant throughout the region, although less so north of the River Mersey (Gibbons *et al.* 1993); being mainly resident, and certainly not leaving the British Isles, and although insectivorous all year round, thus suffering huge losses in hard winters it can bounce back to become, in some years, Britain's most numerous bird.

Future prospects

Prophecy can be dangerous, but I feel there is room for cautious optimism. Most of the destructive actions of the past were taken in ignorance of their likely effect, whereas now there is a reasonable understanding of the way in which birds and habitats interact. Indeed, in recent years there is a greater awareness of the interdependence of all the components of our 'green and pleasant land'. Governments have enthusiastically embraced the concept of biodiversity, and action plans are devised for species and habitats. My main concern is that there is too much emphasis on special sites and species, and that the overview of the wider countryside could still be lost. Bitterns and Black-necked Grebes are important and must be encouraged and protected, but we must not concentrate on the rarities so much that we overlook agricultural policies that could drive Skylarks and Lapwings to the same state of rarity.

References

Included here are a number of publications used in compiling this chapter but not necessarily cited in the text.

Boyd, A.W. (1946). *The Country Diary of a Cheshire Man*. Collins, London.

Boyd, A.W. (1950). *A Country Parish*. The New Naturalist Series. Collins, London.

Burton, J.F. (1995). *Birds and Climate Change*. Christopher Helm, London.

Calvert, M. [1989]. Cheshire Reed Warblers as Cuckoo Hosts. *Cheshire & Wirral Bird Report*, **1988**, 87.

Cheshire County Council (1992). *Cheshire State of the Environment Project*. Cheshire County Council, Chester.

Coward, T.A. & Oldham, C. (1900). *The Birds of Cheshire*. Sherratt & Hughes, Manchester.

Coward, T.A. (1910). *A Vertebrate Fauna of Cheshire and Liverpool Bay*. Witherby, London.

Flegg, J.J.M. (1975). Bird Population and Distribution Changes and the Impact of Man. *Bird Study*, **22**, 191–202.

Gibbons, D.W., Reid, J.B. & Chapman, R.A. (1993). *The New Atlas of Breeding Birds in Britain and Ireland 1988–1991*. T. & A.D. Poyser, London.

Guest, J.P., Elphick, D., Hunter, J.S.A. & Norman, D. (1992). *The Breeding Bird Atlas of Cheshire and Wirral*. Cheshire & Wirral Ornithological Society.

Hardy, E. (1941). *Birds of the Liverpool Area*. Buncle, Arbroath.

Holland, P., Robson, J.E. & Yalden, D.W. (1982). The status and distribution of the Common Sandpiper (*Actitis hypoleucos*) in the Peak District. *Naturalist*, **107**, 77–86.

Holland, P., Spence, I. & Sutton, T. (1984). *Breeding Birds in Greater Manchester*. Manchester Ornithological Society.

Holloway, S. (1995). *The Historical Atlas of Breeding Birds in Britain and Ireland 1875–1900*. T. & A.D. Poyser, London.

Lack, P. (1986). *The Atlas of Wintering Birds in Britain and Ireland*. T. & A.D. Poyser, Calton.

Latham, F.A. (ed.) (1991). *Delamere: The History of a Cheshire Parish*. Local History Group, Tarporley.

Lever, C. (1977). *The Naturalized Animals of the British Isles*. Hutchinson, London.

Marchant, J.H., Hudson, R., Carter, S.P. & Whittington, P. (1990). *Population Trends in British Breeding Birds*. BTO, Tring.

Mitchell, F.S. (1892). *The Birds of Lancashire*. Gurney and Jackson, London.

Norman, D. [1994]. First Breeding of Marsh Warbler in Cheshire: Woolston 1991. *Cheshire & Wirral Bird Report*, **1993**, 94–96.

Norman, D. (1994). *The Fieldfare*. Hamlyn, London.

Oakes, C. (1953). *The Birds of Lancashire*. Oliver & Boyd, Edinburgh.

Peach, W.J., Baillie, S.R. & Underhill, L. (1991). Survival of British Sedge Warblers *Acrocephalus schoenobaenus* in relation to west African rainfall. *Ibis*, **133**, 300–05.

Sharrock, J.T.R. (1976). *The Atlas of Breeding Birds in Britain and Ireland*. T. & A.D. Poyser, Berkhamsted.

Smith, P.H. (1995). Gulls and Terns of the Mersey Estuary. *The Mersey Estuary – Naturally Ours* (eds M.S. Curtis, D. Norman, & I.D. Wallace), pp. 60–68. Mersey Estuary Conservation Group, Warrington.

Smith, M.G. & Norman, D. [1989]. Cuckoos at Woolston 1988. *Cheshire & Wirral Bird Report*, **1988**, 88.

Smith, S.E. (1950). *The Yellow Wagtail*. New Naturalist Monograph. Collins, London.

Thomason, G. & Norman, D. (1995). Wildfowl and Waders of the Mersey Estuary. *The Mersey Estuary – Naturally Ours* (eds M.S. Curtis, D. Norman, & I.D. Wallace), pp. 33–59. Mersey Estuary Conservation Group, Warrington.

Williamson, K. (1975). Birds and climatic Change. *Bird Study*, **22**, 143–64.

Winstanley, D., Spencer, R. & Williamson, K. (1974). Where have all the Whitethroats gone? *Bird Study*, **21**, 1–14.

Yalden, D.W. & Yalden, P.E. (1989). The sensitivity of breeding Golden Plovers (*Pluvialis apricaria*) to human intruders. *Bird Study*, **36**, 49–55.

A much extended version of this manuscript is available on request from the author.

CHAPTER TWENTY-TWO

The lost ark: changes in the tetrapod fauna of the Mersey Basin since 15,000 BP

C.T. FISHER AND D.W. YALDEN

Introduction

In discussing the history of amphibians, reptiles and mammals in the Mersey Basin, some historical licence is necessary. There are some instances where tetrapod species are known from local records, many more where tetrapods are only known certainly as records elsewhere on a national basis but which are assumed also to have occurred locally; yet others can only be surmised as existing here by analogy with what happened in other places and with other organisms. In trying to present a coherent story from fragmentary evidence, it is necessary to combine these elements.

15,000 BP

At the beginning of the record, the ice sheet of the last glaciation (Devensian) was just retreating, and the mammal fauna of Britain included Woolly Mammoth (*Mammuthus primigenius*), Woolly Rhinoceros (*Coelodonta antiquitatis*), Reindeer (*Rangifer tarandus*), Giant Deer (*Megaloceros giganteus*), lemmings (*Dicrostonyx torquatus* and *Lemmus lemmus*), Wolf (*Canis lupus*) and Brown Bear (*Ursus arctos*).

Direct evidence for the existence of these species locally comes from caves in the uplands surrounding the Mersey Basin (in the Peak District, Furness and in North Wales), and from more local peat and gravel deposits. The exact dating of such records is uncertain, and some remains (especially of Woolly Mammoth and Woolly Rhinoceros) may belong to earlier periods in the Devensian, but most are late-glacial (Figure 22.1; Table 22.1).

A Mammoth from Cae Gwynn is dated at 18,000 BP, so is certainly full-glacial in age, but the recently discovered mammoths from Condover, Shropshire, dated 12,400 BP (Lister 1991) are among the latest known from Britain, and confirm their presence here until the late glacial.

Reynolds (1933, after Dawkins 1875) comments that the abundant remains of Reindeer (and Bison (*Bos bonasus*), which is not discussed here) at Windy Knoll, just west of Castleton in Derbyshire, strongly indicate that this pass was a regular migration route from the

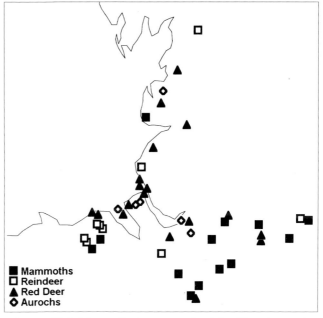

Figure 22.1. Mersey Basin mammals of the Glacial period and later Mesolithic – distribution of Mammoths, Reindeer, Red Deer and Aurochs.

Derwent Valley through to the plains of Lancashire and Cheshire.

Mersey Basin records of Giant Deer are few, but include those at Freshfield and Wallasey (Table 22.1). There is also Giant Deer material from Helsfell Point (near Kendal, Cumbria) in the collections at Liverpool Museum, National Museum & Galleries on Merseyside. Lemmings of both genera are recorded from the area (Table 22.1).

According to Rackham (1986) there can be little doubt that the larger mammals of this era died out as a result of human activity, including hunting; this may be an over-simplified view.

14,000 to 11,000 BP

During this period the climate became warmer and Birch scrub spread into the area. The most notable tetrapod record from these times is the specimen of an Elk (*Alces*

	North Wales	Merseyside	Lancashire	Cheshire	Derbyshire
Mammoth	Cefn Caves[1,7] Tremeirchion[1,7] Plas Heaton[GMC] Cae Gwynn[4]		Blackpool[3]	Adlington[2,6] Coppenhall[2,6] Marbury[2,6] Mere Hall[2,6] Sandbach[2,6] Wrenbury[2,6] Northwich[1] Bolesworth[7]	Buxton[1] Castleton[1]
Woolly Rhinoceros	Ffynnon Beuno[7] Cae Gwynn[7,GMC] Gop Cave[7] Gwaenysgor[7]				
Reindeer	Cefn Caves[7,10] Bont Newydd[7,10] Plas Heaton[7,10] Galltfaenan[7,10] Ffynnon Beuno[7,10] Cae Gwynn[7,10] Gop Cave[7,10] Gwaenysgor[7,10] Lynx Cave[JB]	Freshfield[10]	Whittington Hall[10]	Chester[10]	Castleton[3,10] Ravencliffe[3,10] Wirksworth[3,10]
Giant Deer	Cefn Caves[7,9] Ffynnon Beuno[7,9] Gwaenysgor[7,9]	Wallasey Pool[3,5,8]	Ravensbarrow Hole[12]		
Lemmings (D. torquatus, L. lemmus)	Gwaenysgor[7] (D.t) Lynx[JB] (D.t, L.l)		Wharton Crags[11] (D.t, L.l)		Dowel Cave[11] (D.t, L.l) Fox Hole[11] (D.t) Etches Cave[11] (L.l)

Key
1 = Adams (1877–79)
2 = Coward & Oldham (1910a)
3 = Fisher Card Index
4 = Lister (1991)
5 = Moore (1858)
6 = Morton (1898)
7 = Neaverson (1940–43)
8 = Rance (1875)
9 = Reynolds (1929)
10 = Reynolds (1933)
11 = Sutcliffe & Kowalski (1976)
12 = Whitehead (1964)
GMC = Collections of the Grosvenor Museum, Chester
JB = John Blore, pers. comm.

Table 22.1. Sites with fossil remains from the glacial period. Mammoth, Woolly Rhinoceros, Reindeer, Giant Deer and Lemmings.

alces) found at High Furlong near Blackpool, which was dated at 12,400 BP from analysis of bone from the skeleton (Hallam et al. 1973; Clutton-Brock 1991). This provides the earliest direct evidence locally of human hunting; from lesions on the long bones the Elk appears to have been attacked about the legs with barbed points about two weeks before death, and the subsequent multiple wounds – caused by both barbed points (found with the skeleton) and probably axes – show how the animal finally succumbed.

The large collections of Giant Deer from Ballybetagh Bog near Dublin date from this period (Barnosky 1985), and it is probable that they also still roamed the Mersey Basin at this time. Indeed many of the imprecisely dated late-Pleistocene tetrapod records cited above could belong to either the earlier or this later period.

11,000 to 10,000 BP

Between 11,000 and 10,000 BP the climate cooled again,

and this cold spell probably wiped out Giant Deer for good. However, during this time Horse (Equus caballus), Reindeer and lemmings returned to Britain. The nearest certain record of Reindeer dated to this period is that of 10,600 BP from Ossom's Cave, Manifold Valley, Staffordshire. Elsewhere, there are numerous records for this period (e.g., Chelm's Coombe and Gough's New Cave, Somerset; Currant 1991). Lemmings are referred to by Yalden (1982), and ^{14}C dates were compiled for Reindeer and Horse by Housley (1991).

10,200 BP to the present

The post-glacial or Flandrian period in which we live began rather abruptly at about 10,200 BP. From both records of beetle faunas and direct climatological evidence, the warming of about 8°C in summer temperature took only 50 years (Yalden 1982; Dansgaard, White & Johnsen 1989, and see also chapter one, this volume). The Reindeer may have lingered on in the uplands for the next 1,000 years, but the improved climate led to an equally rapid change in the fauna. Red Deer (*Cervus elaphas*), Roe Deer (*Capreolus capreolus*), Elk (Moose), Aurochs (*Bos primigenius*) and Wild Boar (*Sus scrofa*) became the predominant prey of Mesolithic hunters. There seem to be no local sites with bones of these species that are contemporary with the famous Star Carr, Yorkshire and Thatcham, Berkshire camps. However, there are numerous undated records of Red Deer and Wild Boar from the Mersey Basin flood plain which could belong to this period, and numerous records of Mesolithic flints (chapter four, this volume) which imply that hunters were active in the area.

One very interesting set of fossils from this period is the early amphibian fauna from the Whitemoor Channel, near Bosley in east Cheshire, dated to 10,000–8,800 BP, which includes Palmate Newt (*Triturus helveticus*), Common Newt (*T. vulgaris*), Common Toad (*Bufo bufo*), Natterjack (*B. calamita*) and Common Frog (*Rana temporaria*) (Holman & Stuart 1991). There has been some debate over the years about how the isolated Merseyside coastal populations of Natterjack and Sand Lizard (*Lacerta agilis*) reached this area, when their main British populations are on the heathlands of southern England (Beebee 1978, 1980; Yalden 1980a, 1980b). The Whitemoor fossil fauna strongly suggests that they did indeed migrate into the north-west of England, reaching the north Solway coast at an early post-glacial date before the spread of woodland made the English midlands too inhospitable for them (as argued by Yalden 1980a) – rather than after the Neolithic tree-clearance.

During the well-wooded Mesolithic and on into the Neolithic period, the Mersey Basin mammal fauna must have been dominated by the large ungulates (Figure 22.1; Table 22.2). An interesting picture of this fauna is provided by the footprints exposed at low tide from time to time, but especially in the 1990s, on the foreshore at Formby and Hightown (see chapter two, this volume). These include Aurochs, Red Deer and Humans (*Homo sapiens*), the prints often appearing to hint at a state of pursuit – although which species is in flight is open to conjecture (C.T. Fisher, personal observation). Remains of large sub-fossil Red Deer have been recorded from several localities in the Mersey Basin (see Table 22.2).

Leigh (1700) included the location of most of the Lancashire Red Deer specimens in the following passage:

> … in which five Yards within the Marle I saw the Skeleton of a Buck standing upon his Feet, and his Horns on its Head, which are yet preserv'd at Ellel-Grange near Lancaster … eight Yards within Marle in Larbrick near Preston in Lancashire, was found the entire Head of a Stag, with the Vertebrae of the Neck whole… In a Place in Lancashire call'd the Meales, under the Moss four Yards within Marle was found an exotic Head … the Brow-Antlers were bigger than usually the Arm of a man is … the Beams were near 2 Yards in height …

Roe Deer sub-fossil bones seem less well recorded, but occur in several river deposits (see Table 22.2). There are many records of Aurochs from the Mersey Basin.

The Neolithic is marked faunally by the arrival of domestic Sheep (*Ovis aries*) and Goats (*Capra hircus*) with the new farmers from the Middle East, and the level of human interference in the ecology of the Basin increases steadily from this point to the present. Direct evidence is, however, poor compared with the south of England, until the Roman period. House Mice (*Mus musculus*) arrived in Iron Age times in southern England, and it was probably the Romans who brought Black Rats (*Rattus rattus*). Both House Mice and Yellow-necked Mice (*Apodemus flavicollis*) were recorded from Roman

	North Wales	Merseyside	Lancashire	Cheshire
Red Deer	Grognant Beach, Prestatyn[NMGM]	Blundellsands Shore[NMGM] Rimrose[7] Formby Bank[2] Dove Point, Leasowe[1] Meols[8] West Kirby[NMGM] New Brighton Shore[NMGM]	Ellel[3] Larbreck[3] Meols Hall[3] Preston[8]	Macclesfield[10] Norton[1] Tytherington[10] Combermere[8] Rostherne Mere[1] Thornton le Moors[NMGM]
Roe Deer	Ffynnon Beuno[8]	Meols[8]	Warton Crags[8]	Rushton[8]
Aurochs		Dove Point, Leasowe[NMGM,9] Moreton[NMGM] Hilbre Island[6] Wallasey Pool[4,5]	Preston[9] Pilling Moss[9]	River Weaver at Kingsley[NMGM] River Dee at Chester[NMGM] Northwich[9] Runcorn[1]

Key
1 = Coward & Oldham (1910a)
2 = Edwards & Trotter (1954)
3 = Leigh (1700)
4 = Moore (1858)
5 = Neaverson (1940–43)
6 = Newstead
7 = Reade (1872)
8 = Reynolds (1933)
9 = Reynolds (1939)
10 = Sainter (1878)
NMGM = Collections of National Museums & Galleries on Merseyside

Table 22.2. Sites with Mesolithic sub-fossil remains (from 10,200 BP). Red Deer, Roe Deer and Aurochs.

Manchester (Yalden 1984); the latter is now confined to southern Britain.

From Roman Chester come the remains of Red Deer, Roe Deer, Wild Boar and Wolf. Leigh's plate (1700) of 'The horn of the Rane-Deer found under the altar at Chester' actually depicts a Roebuck antler. Anglo-Saxon archaeological evidence is locally absent, but a clue to the fauna then comes from place-name evidence; the Beaver (*Castor fiber*) is recalled in two lost Cheshire names, Beuer'feld and Buernes, while the Wolf is remembered in at least twenty names in Cheshire and seven in Lancashire (Aybes & Yalden 1995). Wildboar Cloughs in Cheshire and Derbyshire, Wildboar Farm in Lancashire and Yeverleye (eofor – leah = boar clearing) in Cheshire record that species, and the Wildcat (*Felis silvestris*) is represented by Wildcathishevede, Cheshire (Aybes & Yalden 1995).

By Norman times, the English fauna had probably already lost Beaver, Brown Bear and Aurochs as a consequence of hunting, but species diversity was restored somewhat by the introduction of Fallow Deer (*Dama dama*) and Rabbit (*Oryctolagus cuniculus*) by the Normans (Corbet & Harris 1991). Rabbits were initially kept and protected in warrens as a source of meat and fur, while Fallow Deer, also originally from the Mediterranean, were kept in deer parks, being better adapted than the native woodland deer to grazing.

The herd-dwelling Red Deer were also imparked, as indeed were Wild Boar. Delamere Forest is obliquely mentioned in the Domesday Book of 1086, along with *haiae capreolorum* at Kingsley and Weaverham; there are in all 99 hays detailed in the Cheshire Domesday volume, and they were apparently enclosures for harvesting woodland game. Roe Deer would have been the most common of these, though other species were probably taken (Yalden 1987). One other intriguing entry appears in the Domesday Book for Cheshire: in the records for the county town, it is reported that the city paid in revenue £45 and 3 timbers of marten skins. A timber was a bulk quantity of 40–60 skins and presumably the martens were Pine Martens (*Martes martes*).

Deer played an increasingly important part in the economy and in the social standing of the ruling classes. Forests were originally designated by the king for deer hunting (Rackham 1986), and later by others of the nobility under licence from him. Many were in wooded areas, but others were largely treeless; the forest of High Peak, on the Pennine slopes of the Mersey Basin, was re-established in Elizabethan times, and contained a herd of Red Deer which increased under enhanced protection from under 30 in 1579, to over 120 in 1586, before finally being disestablished in 1650 (Shimwell 1977). Parks were created out of large estates, and the remnant Red Deer from the forest of High Peak were reputedly enclosed in Lyme Park. Other major parks were on the outskirts of Liverpool, at Croxteth, Knowsley and Toxteth (Harrison 1902), and one is still clearly marked in Sefton Park on the Ordnance Survey map of south Liverpool for 1908. Toxteth ('Stochestede', or 'woody place') was originally enclosed, with Smithdown, to form the park for Liverpool Castle in 1204. In 1337 Ranulf de Dacre, parson of Prescot, was brought before the assize for poaching deer in Toxteth Park; Robert de Barton, a priest from Aigburth, was also arrested for encouraging him. Toxteth's status as a forest was removed in 1596 (Griffiths 1907).

The Elizabethan period also marked the start of formal vermin control, with the infamous act of 1564 'for the preservation of Grayne'; this empowered churchwardens to pay bounties for various pest species, not only rats and mice but also many predators. Out of this antipathy seems to have developed the increasing persecution which led to the virtual extinction of many carnivores locally. The Pine Marten was last recorded in Cheshire in the 1880s (Coward & Oldham 1910; Forrest 1918) but survived in northern Lancashire until the 1950s (Ellison 1959). The Polecat (*Mustela putorius*) died out in Cheshire in 1900 and in Lancashire was gone by 1910. The Wildcat was extinct in Cheshire before 1800 and gone from Lancashire by 1825 (Langley & Yalden 1977).

The same story of persecution involves the Otter (*Lutra lutra*), which Lomax hunted in the north-west of England from 1829 to 1871; he caught 21 animals from the rivers Ribble and Hodder alone from 1830 to 1833 (Lomax 1892). Otters must once have been common in the undrained areas between Southport and Liverpool; the last record from the Blackpool area was at Marton Mere in 1955 (Ellison 1959, 1963). Otterspool, a creek off the Mersey in south Liverpool, clearly indicates their historical presence on the banks of this now most industrial of rivers. The last period when Otters appear to have been relatively common was the 1930s; for instance, two cubs were shot 'above Formby' in March 1933, according to the *Proceedings of the Liverpool Naturalists' Field Club* for 1934. Looking through those cards in the North West Biological Field Data Bank at the Liverpool Museum, which record historical occurrences of Otters in Lancashire and Cheshire (also see Ellison 1959), it is apparent how many were killed by trains, as the tracks were often laid across wetlands. Otters held on in north Lancashire, and by the mid-1990s appeared to be spreading south again. However, polluted waters still hinder their re-establishment in the Mersey Basin (Jefferies 1989).

During this period one other significant addition was made to the fauna; the Brown Rat (*Rattus norvegicus*) arrived, with human assistance, from Russia in the late 1720s and rapidly displaced the Black Rat. Brown Rats are now the major economic vertebrate pest in this country, while the Black Rat is almost extinct. Liverpool docks and the city centre were one of the last strongholds of this rather attractive species, but sadly it has not been recorded there for several years (Twigg 1992).

By Victorian times, when the predators were being enthusiastically exterminated, other species were being introduced for their 'attractive appearance'. The American Grey Squirrel (*Sciurus carolinensis*) was introduced to Cheshire at Henbury in 1876, the earliest introduction documented by Shorten (1954). From there it spread throughout the county and displaced the native Red Squirrel (*Sciurus vulgaris*), as it has done in many places in Britain. Red Squirrels survived in Wirral until the 1960s but they are now extinct there. The River Mersey and its associated industrial belt acted for a long time as a barrier to the Grey Squirrel but during the 1980s they spread north through Greater Manchester, and also spread westwards down the River Ribble from Yorkshire in a pincer movement. Red Squirrels still occur in some numbers along the Ainsdale – Formby pinewoods, but there are repeated claims that this population derives in part from German stock released in the early 1940s (Ellison 1959). Ecological research on this population is currently under way to determine the beneficial or other effect of all the supplementary food (peanuts) given to them by humans, and genetic research based at the Institute of Zoology in London is investigating the origins of this population. In view of the demise of the species elsewhere, this population should be regarded as of considerable conservation importance. Small populations of Red Squirrel occur in Lancashire and more widely further north, but they are endangered everywhere.

Other introductions include the Sika Deer (*Cervus nippon*) population in the Bowland Forest, released near Gisburn in 1907 (Ellison 1959). The Muntjac (*Muntiacus reevesi*) was accidentally released in Cheshire in 1989, although it has been present in southern England since the 1940s (Chapman, Harris & Stanford 1994). The American Mink (*Mustela vison*) has spread through the area since the 1950s after accidental escapes from fur farms in the north of Lancashire (Tapper 1992), although the oldest Mink record from north-west England is from Farnworth, Bolton in 1956 (Bolton Museum and Art Gallery collections).

The present

Currently, the pattern of change is as dynamic as ever. Deer are increasing in numbers and range in the Mersey Basin; Fallow Deer occur in a few parks (Dunham, Lyme and Knowsley) and small populations occur in small woodlands, but they are still scarce compared with their status elsewhere in England. Roe Deer are spreading back down the Pennines; *c.*100 Red Deer are established, but vulnerable to poaching, in the Cheshire Hills (albeit from mainly re-introduced stock), and the survivors of the Peak Forest herd remain in Lyme Park. Muntjac and Sika Deer are also increasing in range.

Among the carnivores, there is currently evidence of Otters and Polecats spreading back into Cheshire from Wales. On the other hand, Mink appear to be responsible for the final demise of many local Water Vole (*Arvicola terrestris*) populations, though this animal has been declining all this century in many parts of the country as a consequence of habitat change (Strachan & Jefferies 1993). Red Squirrels remain precariously poised in the face of the threat from Grey Squirrels, and the small Sand

Lizard population is being sustained in part by captive breeding. The Natterjack also seemed doomed to extinction during the drought years in the 1970s, but conservation efforts and improved rainfall have rescued them.

As for the bats, more than ten years of dedicated, organised recording by the county bat groups have started to show accurately the effect of human impact on members of this order. Daubenton's Bat (*Myotis daubentonii*), a species associated with water and thus vulnerable to the effects of pollution, is probably much less common than it was in the 19th century. Nevertheless, it holds on in some surprising places – such as in a river tunnel running underneath a factory which uses a vast variety of chemicals – and may now be increasing again. The woodland-loving Bechstein's Bat (*Myotis bechsteinii*) probably occurred historically in north-west England, though it has not been definitely recorded here – the species was once common in other parts of England, as is indicated by the large assemblages of bones in Grimes Graves in Norfolk (Corbet & Harris 1991). There are now only about 1,500 individuals in Britain (Morris 1993). Conversely, the Pipistrelle (at present alluded to as *Pipistrellus pipistrellus*, but actually two distinct species), although declining by an estimated 60% in Britain as a whole between 1978 and 1986, has found many new roost sites in houses, particularly those built since 1960, with easier access to roof spaces and wall cavities (Morris 1993).

Conclusion

Thus the mammal, particularly the large mammal, fauna of the Mersey Basin has been almost entirely shaped by humans. The original herds of large game were hunted and, together with partly man-induced habitat change, eventually led to actual or near-extinction. With the arrival of Neolithic farmers came domestic species, and a different attitude to harvesting animals. The large predators – like Wolf and Bear – were eliminated as a threat to man or livestock, and were hunted for their fur, along with Beavers and other species. The ungulates and other food animals (like the Rabbit and Hare) were carefully guarded as beasts of the chase, or even more strictly controlled in parks. The survival of deer in this region clearly depended on their inclusion in parks, and with the waning political correctness of hunting – and increased opportunities for the economy offered by forestry – they are establishing themselves again as part of the regional fauna.

Currently, however, the mammal fauna is dominated by some 250,000 dairy cattle, 330,000 pigs and over a million sheep on the surrounding uplands (Government Statistical Service: *Digest of Agricultural Census Statistics*, UK 1991). In addition, there are a large number of cats, dogs and other domestic animals. The human population of north-west England also occupies much of the land. Not much space or habitat is left for other species; recent discussions about re-introducing the Dormouse (*Muscardinus avellanarius*) to Cheshire almost foundered because there was scarcely anywhere suitable. Yet the willingness to consider such actions, along with the efforts to encourage species such as the Otter and Natterjack, offer the hope that a tetrapod fauna in the next millennium will be at least as rich as that of the 20th century. Changes in fashion in human food consumption away from meat and dairy products – whether temporary or permanent – may mean more land returns to habitats suitable for wild species. Attempts to re-establish the fauna of pre-Roman times are premature, if not impossible to achieve, but at least we can dream of the Lost Ark.

Acknowledgements

We are grateful for assistance from staff in the Local History Library and Municipal Research Departments of Liverpool City Libraries.

We would also like to thank Brian Davis and Peter Jones for their very pertinent remarks on, and improvements to, this manuscript. Dr A.J. Morton's DMAP programme produced Figure 22.1.

References

Adams, A.L. (1877–1879). *Monograph on the British fossil elephants.* Part 1. *Dentition and osteology of* Elephas antiquus *(Falconer).* Part 2. *Dentition and osteology of* Elephas primigenius *(Blumenbach).* The Palaeontographical Society, London.

Aybes, C. & Yalden, D.W. (1995). Place-name evidence for the former distribution and status of Wolves and Beavers in Britain. *Mammal Review, 25,* 201–26.

Barnosky, A.D. (1985). Taphonomy and herd structure of the extinct Irish Elk *Megaloceros giganteus. Science, 228,* 340–44.

Beebee, T.J.C. (1978). An attempt to explain the distributions of the rare herptiles *Bufo calamita, Lacerta agilis* and *Coronella austriaca* in Britain. *British Journal of Herpetology, 5,* 763–70.

Beebee, T.J.C. (1980). Historical aspects of British herpetofauna distribution. *British Journal of Herpetology, 6,* 105.

Cantor, L.M. (1982). *The English medieval landscape.* Croom Helm, London.

Cantor, L.M. (1983). *The medieval parks of England: a gazetteer.* Loughborough University of Technology, Loughborough.

Chapman, N., Harris, S. & Stanford, A. (1994). Reeves' Muntjac *Muntiacus reevesi* in Britain: their history, spread, habitat selection, and the role of human intervention in accelerating their dispersal. *Mammal Review, 24,* 113–60.

Clutton-Brock, J. (1991). Extinct species. *The Handbook of British Mammals* (eds G.B. Corbett & S. Harris), pp. 571–75. The Mammal Society and Blackwell Scientific Publications, Oxford.

Corbet, G.B. & Harris, S. (1991), *The handbook of British mammals,* 3rd ed. The Mammal Society & Blackwell Scientific Publications, Oxford.

Coward, T.A. & Oldham, C. (1910). *The Vertebrate fauna of Cheshire and Liverpool Bay* Vol. 1. *The mammals and birds of Cheshire.* Part 2. *The reptiles and amphibians of Cheshire,* (ed. T.A. Coward). Witherby & Co., London.

Currant, A. (1991). A Late Glacial Interstadial mammal fauna from Gough's Cave, Somerset, England. *The Late Glacial in north-west Europe: human adaptation and environmental change at the end of the Pleistocene* (eds N. Barton, A.J. Roberts & D.A. Roe), pp. 48–50. Council for British Archaeology Research Report 77, London.

Dansgaard, W., White, J.W.C. & Johnsen, S.J. (1989). The abrupt termination of the Younger Dryas climate event. *Nature, 339,* 532–34.

Dawkins, W. Boyd. (1875). On the Mammalia found at Windy Knoll. *Quarterly Journal of the Geological Society, 31,* 248.

de Rance, C.E. (1875). On the relative age of some valleys in the north and south of England, and of the various and post-glacial deposits occurring in them. *Proceedings of the Geological Association, 4,* 221–53.

Edwards, W. & Trotter, F.M. (1954). *British Regional Geology (Geological Survey & Museum): the Pennines and adjacent areas, based on previous editions by D.A. Wray.* 3rd ed. British Geological Survey, HMSO, London.

Ellison, N.F. (1959). *A checklist of the fauna of Lancashire and Cheshire. Mammalia, reptilia, amphibia. Revised to the end of 1956.* In *Thirty-first report of the recorders.* Lancashire & Cheshire Fauna Committee, Buncle & Co., Arbroath.

Ellison, N.F. (1963). *Report on the mammals, reptiles and amphibians to the end of 1962.* In *Thirty-third report of the recorders.* Lancashire & Cheshire Fauna Committee, Buncle & Co., Arbroath.

Fisher, J.M.Mc. 'card index': refers to manuscript reference cards and bibliography relating to all vertebrate post-glacial records in Britain; index now amongst Fisher Collection Archive in the Archives Section of the Natural History Museum, London.

Forrest, H.E. (1918). Pine Marten in Shropshire. *The Naturalist, 738,* 231.

Government Statistical Service, (1991). *The Digest of Agricultural Census Statistics, United Kingdom 1991.* HMSO, London.

Griffiths, R. (1907). *The history of the royal and ancient park of Toxteth, Liverpool.* Privately printed; copies in Liverpool City Library.

Hallam, J.S., Edwards, B.J.N., Barnes, B. & Stuart, A.J. (1973). The remains of a Late Glacial Elk associated with barbed points from High Furlong, near Blackpool, Lancashire. *Proceedings of the Prehistoric Society, 39,* 115–21.

Harrison, W. (1902). Ancient forests, chases and deer parks in Lancashire. *Transactions of the Lancashire & Cheshire Antiquarian Society, 19* (for 1901), 1–37.

Holman, J.A. & Stuart, A.J. (1991). Amphibians of the Whitemoor Channel early Flandrian site near Bosley, East Cheshire; with remarks on the fossil distribution of *Bufo calamita* in Britain. *Herpetological Journal, 1,* 568–73.

Housley, R.A. (1991). AMS dates from the Late Glacial and early Postglacial in north-west Europe: a review. *The Late Glacial in north-west Europe: human adaptation and environmental change at the end of the Pleistocene* (eds N. Barton, A.J. Roberts & D.A. Roe), pp. 25–39. Council for British Archaeology Research Report 77, London.

Jefferies, D.J. (1989). The changing otter population of Britain 1700–1989. *Biological Journal of the Linnean Society, 38,* 61–69.

Langley, P.J.W. & Yalden, D.W. (1977). The decline of the rarer carnivores in Great Britain during the nineteenth century. *Mammal Review, 7,* 95–116.

Leigh, C. (1700). *The natural history of Lancashire, Cheshire and the Peak, in Derbyshire.* Oxford.

Lister, A. (1991). Late Glacial mammoths in Britain. *The Late Glacial in north-west Europe: human adaptation and environmental change at the end of the Pleistocene* (eds N. Barton, A.J. Roberts & D.A. Roe), pp. 51–59. Council for British Archaeology Research Report 77, London.

Lomax, J. (1892). *Diary of otter hunting from A.D. 1829 to 1871.* Henry Young & Sons, Liverpool.

Moore, T.J. (1858). Notice of mammalian remains discovered in the excavations at Wallasey for the Birkenhead New Docks. *Transactions of the Historical Society of Lancashire & Cheshire, 1857–1858,* 265–68.

Morris, P.A. (1993). *A red data book for British mammals.* The Mammal Society, London.

Morton, G.H. (1898). The elephant in Cheshire. *Transactions of Liverpool Biological Society, 12,* 155–58.

Neaverson, E. (1940–43). A summary of the records of Pleistocene and Postglacial Mammalia from North Wales and Merseyside. *Proceedings of Liverpool Geological Society*, **18**, 70–85.

Rackham. O. (1986). *The history of the countryside*. Dent, London.

Reade, T.M. (1872). The geology and physics of the post-glacial period as shewn in the deposits and organic remains in Lancashire and Cheshire. *Abstract and Proceedings Liverpool Geological Society*, **1871–1872**, 36–88.

Reynolds, S.H. (1929). *A monograph on the British Pleistocene Mammalia* Vol. III, part III. *The Giant Deer*. The Palaeontographical Society, London.

Reynolds, S.H. (1933). *A monograph on the British Pleistocene Mammalia* Vol. III, part IV. *The Red Deer, Reindeer and Roe*. The Palaeontographical Society, London.

Reynolds, S.H. (1939). *A monograph on the British Pleistocene Mammalia* Vol. III, part VI. *The Bovidae*. The Palaeontological Society, London.

Sainter, J.D. (1878). *Scientific rambles round Macclesfield*. Swinnerton & Brown, Macclesfield.

Shimwell, D.J. (1977). Studies in the history of the Peak District landscape: I Pollen analysis of some podzolic soils on the Limestone Plateau. *University of Manchester School of Geography Research Papers*, **3**, 1–54.

Shorten, M. (1954). *Squirrels*. New Naturalist Monograph. Collins, London.

Strachan, R. & Jefferies, D.J. (1993). *The Water Vole Arvicola terrestris in Britain 1989–1990: its distribution and changing status*. Vincent Wildlife Trust, London.

Sutcliffe, A.J. & Kowalski, K. (1976). Pleistocene rodents of the British Isles. *Bulletin of the British Museum (Natural History), Geology*, **27**, 31–147.

Tapper, S. (1992). *Game Heritage*. Game Conservancy, Fordingbridge.

Twigg, G. (1992). The Black Rat *Rattus rattus* in the United Kingdom in 1989. *Mammal Review*, 22: 33–42.

Whitehead, G.K. (1964). *The deer of Great Britain and Ireland*. Routledge & Kegan Paul, London.

Yalden, D.W. (1980a). An alternative explanation of the distribution of the rare herptiles in Britain. *British Journal of Herpetology*, **6**, 37–40.

Yalden, D.W. (1980b). Historical aspects of British herpetofauna distribution: a reply. *British Journal of Herpetology*, **6**, 105–06.

Yalden, D.W. (1982). When did the mammal fauna of the British Isles arrive? *Mammal Review*, **12**, 1–57.

Yalden, D.W. (1984). The Yellow-necked mouse *Apodemus flavicollis*, in Roman Manchester. *Journal of Zoology*, London, **203**, 285–88.

Yalden, D.W. (1987). The natural history of Domesday Cheshire. *Naturalist*, **112**, 125–31.

Conclusion

Figure 23.1. The south prospect of Prescot, engraving of 1743 by W. Winstanley. (Philpott 1988 p. 29).

CHAPTER TWENTY-THREE

The consequences of landscape change: principles and practice; problems and opportunities

J.F. HANDLEY AND R.W.S. WOOD

Introduction

The Industrial Revolution began in Britain with north-west England, and the Mersey Basin in particular, in the forefront. Consequently, it was one of the first regions in the world to experience the full force of industrialisation and urbanisation. The advent of a new millennium coincides with the end of that first industrial era. Indeed for the past two decades, the region has been undergoing a painful process of economic restructuring associated with the transition from a manufacturing to a knowledge-based economy. It is therefore timely to take stock, to assess the scale of change and to review the consequences for biodiversity.

Ecology and Landscape Development: A History of the Mersey Basin takes a long time frame from the last glaciation to the present day and beyond. Throughout this time there have been two major drivers of landscape change: climate and human influence. During the past two centuries these processes have become interlinked as the burning of fossil fuels has contributed to global warming and accelerated climate change. This chapter examines the causes and consequences of landscape change, draws out some underlying ecological principles and shows how these can be applied to overcoming problems and realising opportunities within the Mersey Basin. The study of the ecology of the Mersey Basin is interesting and worthwhile in its own right. However, it has a wider significance of providing lessons which can help guide environmental planners and managers, both here and elsewhere, along a new development path.

The Industrial Revolution replaced a rural economy which had lasted for a thousand years and which had itself profoundly changed the landscape, fauna and flora of the Mersey Basin. The publication of the Strategic Plan for the North West (SPNW) in 1973 (North West Joint Planning Team 1973) is an important bench-mark providing, for the first time, a systematic review of the state of the region's environment. This coincided with the end of two centuries of rapid population growth in the Mersey Basin and associated urbanisation. That development had taken place with little regard for environmental capacity and this was reflected in the poor quality of air, land and water throughout the Mersey

Basin. There were clear signs by the 1970s that a once prosperous region was undergoing structural economic decline and SPNW broke new ground by recognising that improving the state of the environment was fundamental to addressing that decline. The subsequent period, from the mid-1970s to the present day, has been one of 'stabilisation and renewal'.

Progress in improving environmental quality has been uneven, with the reduction of smoke and sulphur dioxide in the urban centres perhaps the outstanding achievement. It might be expected, given the scale and intensity of industrialisation and urbanisation, and the limited progress in solving inherited problems from that era, that biodiversity in the Mersey Basin would have suffered grievously. The actual position is much more complex and the processes of change have had positive as well as negative consequences. There are important lessons to be learnt.

The tide of change

Woodland clearance and habitat modification

The tide of change has been flowing strongly since the end of the most recent glaciation around 10,000 years ago. Human influence was there to see even in the late-glacial (12,000 BP with the skeleton of an Elk (*Alces alces*) showing clear evidence of human predation (chapter twenty-two, this volume).

Fossil footprints of a later date on the shores of Liverpool Bay show Mesolithic hunters sharing the habitat with Aurochs (*Bos primigenius*) and Red Deer (*Cervus elaphus*) (Figure 2.3, chapter two, this volume). There is evidence even from these early times of significant habitat modification in the form of burning by Mesolithic hunters in the woodlands flanking the Pennine uplands (chapter eleven, this volume). The influence of human induced disturbance through fire, stock grazing and soil erosion in prehistoric times is well documented by Innes *et al.* (chapter four, this volume). Indeed it seems that many of the supposed 'natural environments', especially the heathlands and moorlands, of the Mersey Basin are in fact plagioclimax communities created by human intervention long before the Roman

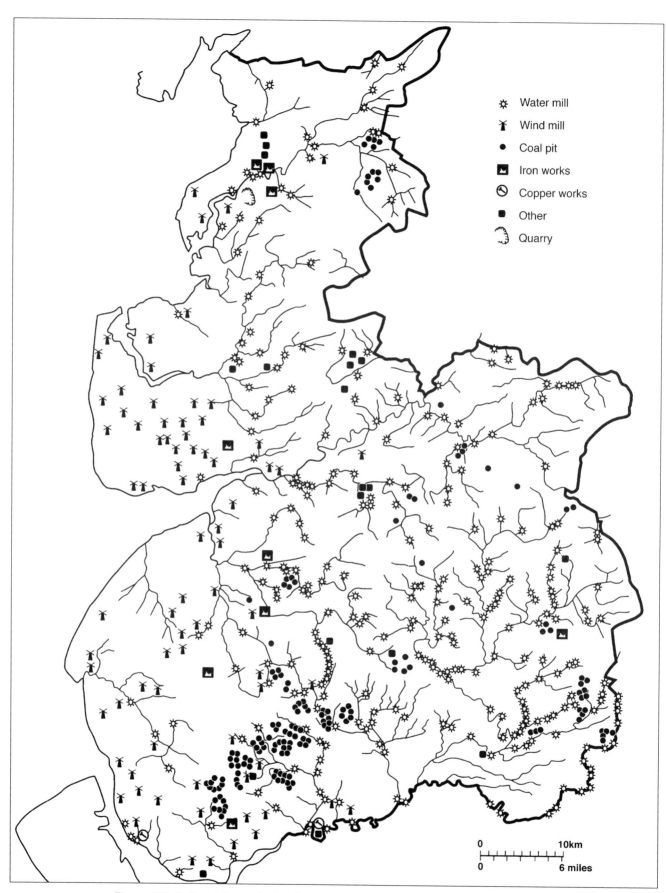

Figure 23.2. Industrial sites depicted on Yates' 1786 map of Lancashire. (Harley 1968).

conquest (chapters four and eleven, this volume).

Cowell (chapter four, this volume) concludes that

> the landscape pattern for approximately 9,000 years before the Domesday Survey was one rooted in woodland, fen, swamp, peat bog and long sweeps of coast. Early human intervention in this environment, due to low population pressure and the sustainable nature of human activity, meant that the effects were limited both in scale and distribution.

Nevertheless, by the time of the 1086 Domesday Survey the nature and extent of woodland cover had been profoundly altered (chapter three, this volume); the woodland matrix, which once stretched from the coast to the highest hills, had been permanently opened up and fragmented. For example, two estimates for Cheshire, based on a Domesday Survey, suggest only a 25–27% woodland cover (Rackham 1986; Yalden 1987). Forest clearance proceeded rapidly in subsequent centuries to fuel the demands of a burgeoning population. Climate change and disease in the 14th century caused social and economic dislocation and with it a switch to pastoral farming. As population growth resumed after the plagues which swept Europe, urban centres began to take shape within an increasingly enclosed pastoral landscape (chapter five, this volume; Philpott 1988). By the mid-17th century, Lancashire was described as a 'close county full of ditches and hedges' (Walker 1939) and we can see this tightly connected landscape in the 1745 tythe map for Newton-in-Makerfield (Figure 5.4, chapter five, this volume) and in the early 18th-century engraving of Prescot (Figure 23.1).

The Industrial Revolution

The second great shaping of the landscape begins with the Industrial Revolution. This profound change in the social and economic fortunes of the region had its roots in textiles and other putative industries of Elizabethan times, but gained pace rapidly in the 18th century (Walton 1987). We have seen the enclosure landscape of pre-industrial England in Figure 23.1, but the signs of change are already evident here. Coal mining is under way in the foreground and in the town itself we can see the first smoke stacks from large-scale industrial processes. But the real action was taking place in the south where in the 1750s and 1760s, as Walton (1987 p. 74) recounts

> a clear tendency was emerging for large-scale new industries to concentrate in Liverpool and close to the Mersey and its associated waterways. Brass, copper, salt and sugar boiling, brewing, pottery and glass-making all followed this pattern, and the initiatives of the mid-century were to provide the basis for a new heavy industrial economy in the classic 'Industrial Revolution' years, although ultimately this lucrative harvest was reaped in St Helens, Warrington and the surrounding areas rather than in Liverpool itself'. (see also chapter six, this volume).

The subsequent development of St Helens, the archetypal industrial town, and the nature of that 'lucrative harvest' has been graphically documented by Barker & Harris (1993).

Yates' map of Lancashire from shortly after this time (1786) shows an economy in transition from renewable energy to fossil fuel (Figure 23.2). Windmills are distributed along the coastal strip and water mills like strings of beads along the streams. However, the growing importance of coal is also evident with a cluster of coal pits on the exposed coalfield of south Lancashire and scattered pits along the higher Pennine valleys. As Jarvis & Reed observe (chapter six, this volume) there is 'only a finite number of foot-pounds of work in an entire river system'. The introduction of the steam engine 'changed the rules and it was now possible to employ almost any amount of power in any location where coal was obtainable at an acceptable price'.

Coal is such a bulky item that transportation costs rather than extraction costs had a great influence on its use in the Industrial Revolution period (von Tunzelmann 1986). The development of transport infrastructure was therefore critical and began with the creation of the canal system.

Which sources of coal grew most rapidly was determined by three principal factors: '…where the demand was greatest, where connections to supply more distant markets were best developed, and where the seams themselves were most easily accessible' (Freeman, Rodgers & Kinvig 1966 p. 90). All three conditions were met in three locations: in the Douglas Valley near Wigan, using the Leeds & Liverpool Canal (completed in 1816); St Helens using the Sankey Canal of 1757, providing a connection to the Mersey Estuary west of Warrington and via the Weaver navigation into the saltworks in Cheshire (which exhausted local wood supplies by the middle of the 17th century); and the area to the north and east of Manchester using the Ashton and Rochdale canals, supplementing local production at Ardwick and Pendleton. This was a profoundly symbiotic relationship where the growth of coal mining stimulated the growth of the textile industry and vice versa.

The pattern was further reinforced by the creation of the railway network (Figures 23.3 and 23.4). The rapid establishment of a net of lines covering the Mersey Basin was substantially completed by 1850, enabling strong inter- and intra-regional connections. 'The main effect of the railways was to open up much of Lancashire to a far wider market than had been known before and especially to stimulate the growth of Liverpool as a centre of both overseas and local trade' (Freeman *et al.* 1966 p. 89–90). Indeed Widnes and Crewe owed their existence to the railway and, more generally, the network of lines served further to strengthen local specialisation and contributed to industrial locational inertia. The pattern of industrial development which was substantiated and rapidly expanded in these early years of the railway largely persisted until the precipitous industrial decline of the middle and late 20th century.

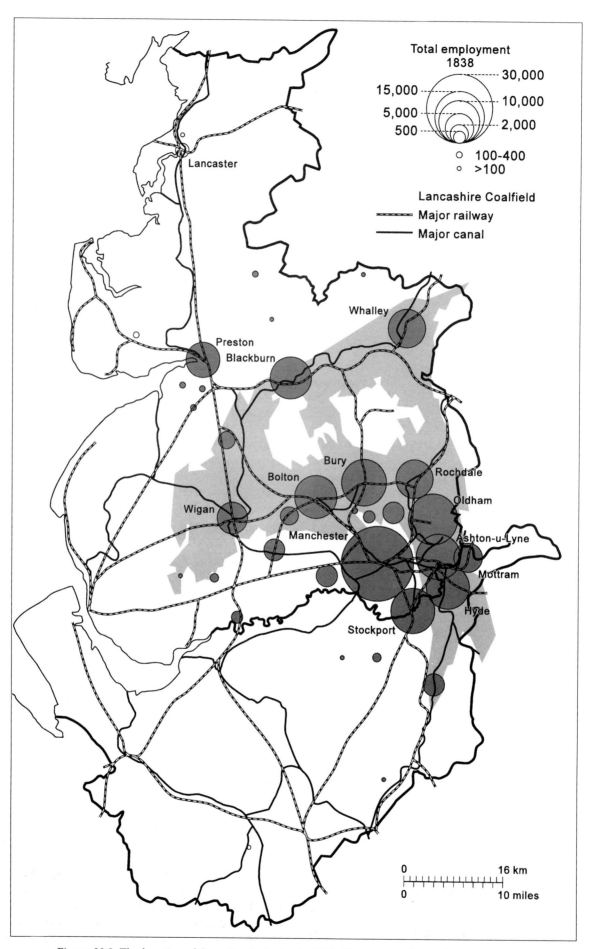

Figure 23.3. The location of the cotton industry in 1838. (Freeman *et al.* 1966 p. 91 and p. 96).

Figure 23.4. The Stockport railway viaduct 1986 with modern office block and 19th century mills by the River Mersey
(*Photo: John Davis*).

The Mersey Basin had by 1851 become the world's greatest manufacturing region. Its wealth was based primarily upon textile manufacture (Figure 23.5), accounting for some 63% of the British textile industry which itself yielded over 50% of the total value of the nation's exports. Industrial and commercial activity stretched throughout a network of industrial towns united by a complex web of waterways and latterly railways. The two great commercial centres of Liverpool and Manchester dominated the scene, the focus for the activities of a number of smaller towns which by now specialised in aspects of the textile, mining, chemical, and engineering industries. Together, these formed a unique industrial heartland and the model for the development of industrial regions throughout the world.

One of its more notable, and alarming, features was the rapidity with which this textile-based economy had expanded. Quite simply 'the growth of this single great staple industry had reshaped both the economy and the landscape of Lancashire at a speed unmatched in any other region of Great Britain' (Freeman *et al.* 1966 p. 103).

Despite the massive dominance of the cotton industry, regional industrial specialisation was nevertheless very important, and St Helens exemplifies the importance of local industrial tradition in setting a course for subsequent development induced by the supply of cheaper fuel and the availability of growing markets both at home and in the colonies. The glass, chemical and copper refining industries have a long history in St Helens. Glass-making in the area was founded upon and largely tied to a silica-rich source of Shirdley Hill sand and soda ash, derived from the local salt refining industry (the origins of Shirdley Hill sand are discussed in chapter two, this volume). The glass-making process was revolutionised in 1773 by the construction of the first large factory, and a developing supply of cheap coal led to the rapid growth of this and the other power-hungry processes of chemical manufacture and copper refining (Barker & Harris 1993).

The urbanisation of the Mersey Basin

In the mid-19th century, some 68% of the population lived in urban areas at a time when the rest of Britain, let alone the world, had a primarily rural-based population. Though the exact definition of what constituted an urban area was by this time not yet substantiated, a dense settlement pattern had evolved on the Lancashire coalfield and along the line of the Mersey Valley, exploiting the natural resources of water and coal, and progressively forming a closely interconnected web which, even in 1851, would have appeared to form a great metropolis. Thus in 1801 Liverpool had some 83,000 inhabitants

Figure 23.5. Ashton-under-Lyne, 1985, with 19th-century cotton mills and houses as well as more recent housing. The Pennine hills are in the background. (*Photo: John Davis*).

which swelled to 375,000 by 1851; similarly, Manchester and Salford grew from some 93,000 to around 388,000 over this period; the influx of migrants from rural areas, other regions and countries was the key to the speed and scale of growth. Immigration from Ireland in the wake of the potato famine was a particularly important source, together with migrants from eastern Europe en route to the Americas who stayed to make a permanent home in north-west England.

As well as being fundamental to industrial growth and development, the railways also provided the first practical means of extensive suburban development and the consequent urbanisation of the hitherto largely agricultural hinterlands of the two dominant growth poles (Figure 23.6). The fine Victorian suburbs of Liverpool and Manchester such as Wirral and south Manchester were the result. The level of pollution in the city centres provided a further stimulus to the exodus from the increasingly commercial cores of both cities.

By the mid-19th century a complex settlement pattern was emerging, centred on the cities of Liverpool and Manchester. Long-established textile towns such as Burnley, Rochdale and Oldham were becoming increasingly linked to the commercial centre of Manchester, just as Liverpool was serving as the focus for satellite towns

such as Runcorn and St Helens. By 1851, the Mersey Basin was a predominantly urban area with the newly opened railways acting as a stimulus to growth and the increasing coalescence of urban areas.

The environmental legacy of industrialisation and urbanisation

The flowering of the economic fortunes of the Liverpool–Manchester urban-industrial area was co-extensive with the greatest pressures upon biota, many of which continue into the present: industrial and urban development creating air and water pollution, and the extinguishing and fragmentation of natural habitats on an unprecedented scale. The contemporary accounts of those such as Frederic Engels (1844) of the conditions of the urban dwellers in Manchester give a startling picture of the unregulated environmental degradation which inevitably accompanied such growth; we can only begin to imagine the scale and intensity of the environmental pollution which accompanied so many to an early grave. Charles Dickens' evocative description of 'Coketown' in *Hard Times* – probably based on Preston, but nevertheless good for most of the region's industrial towns – gives us a flavour, albeit lyrical, of the fetid conditions which prevailed:

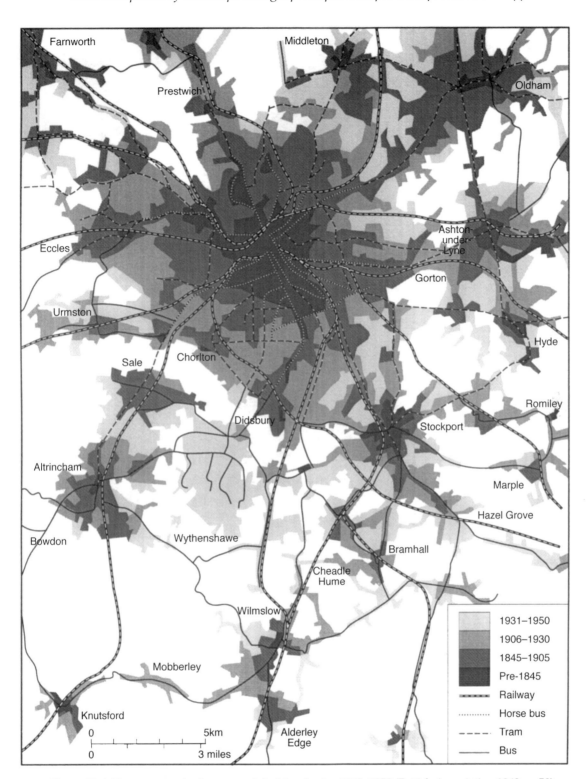

Figure 23.6. The pattern of urban growth in Manchester 1845–1950 (British Association 1962, p. 50).

It was a town of red brick, or of brick that would have been red if the smoke and ashes had allowed it; but as matters stood it was a town of unnatural red and black like the painted face of a savage. It was a town of machinery and tall chimneys, out of which interminable serpents of smoke trailed themselves for ever and ever, and never got uncoiled. It had a canal in it, and a river that ran purple with ill-smelling dye, and vast piles of buildings full of windows where there was a rattling and a trembling all day long, and where the piston of the steam engine worked monotonously up and down like the head of an elephant in a state of melancholy madness. (Dickens 1854.)

Plant and animal communities thus faced the challenges of extinction, marginalisation and adaptation as the tide of industrial and concomitant urban development swept the Mersey Basin. Natural habitats such as the peat mosses were, by 1851, largely reclaimed for intensive agricultural production to supply the ever more demanding urban areas, and those same areas spewed forth choking atmospheric pollution from coal burning and poisonous discharges into the watercourses, the legacy of which is still being tackled today. In St Helens the alkali industry created 'one scene of desolation' in the surrounding countryside, with leafless trees and lifeless hedgerows whilst the nearby stream was 'an open sepulchre full of pestiferous odours' (Barker & Harris 1993, p. 350 and p. 416 respectively). A more extended account of the alkali industry, its environmental impact and the emergence of pollution control regulation is provided by Jarvis & Reed (chapter six, this volume).

Whilst the industrial heartland of the Mersey Basin formed the centrepiece of national economic boom in the period 1850 to 1875, the agricultural hinterland of Cheshire, though to an extent industrialised through the operations of the salt industry centred on Northwich, remained outside the immediate effects of the stimuli of town growth and improved communications. The broad pattern of dairy farming and market gardening observable today was established in this period.

There is much anecdotal evidence from these times for the impact of historic pollution on biodiversity in the Mersey Basin (see chapters six and eleven, this volume), but this topic deserves more systematic research. As this volume demonstrates, the consequences have in many respects been lasting, and include acute damage to biota from water pollution in rivers and canals (chapter thirteen, this volume) and the Mersey Estuary (chapter fifteen, this volume); air pollution damage to bryophytes in the uplands (chapter eighteen, this volume) and lichens in the lowlands (chapter eighteen, this volume). Nevertheless, it seems likely that urbanisation has had a more permanent impact through direct habitat loss, habitat fragmentation and indirectly through habitat modification (see chapter seven, this volume). Michael Dower (1965, p. 5) set the scene in graphic terms:

> Three great waves have broken across the face of Britain since 1800. First, the sudden growth of the dark industrial towns. Second, the thrusting movement along far flung railways. Third, the sprawl of car-based suburbs. Now we see under the guise of a modest word, the surge of a fourth wave which could be more powerful than all the others. The modest word is leisure.

We have had the tools to control the pattern of urban development since the Town & Country Planning Act of 1947. However, habitat loss to development continued unabated in the post-war years, and the impact of cumulative development has often gone unnoticed (Handley 1984). Taking but one example, the loss of precious sand dune habitat on the Sefton coast has been well documented (Figure 23.7), together with the consequences for vulnerable animal species such as the Sand Lizard (*Lacerta agilis*).

The first period of urbanisation in this case coincided with the construction of the Lancashire & Yorkshire Railway from Liverpool to Southport. The next wave of

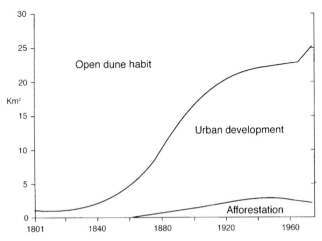

Figure 23.7. Loss of sand dune habitat to urbanisation and afforestation (Handley 1984, p. 229, after Jackson 1979).

development (post 1945) coincided with the expansion of car-based suburbs. The dune coastline itself bore the full brunt of Michael Dower's 'fourth wave' as the sand dunes were destabilised by intensive recreational use threatening habitat integrity and tidal inundation (Handley 1982). The Sefton Coast Management Scheme has been conspicuously successful in addressing these problems as we shall see later. However, it is only in the past two decades that the importance of protecting what remains of the Mersey Basin's critical natural capital has been properly recognised and, even today, primary habitats are still being lost to meet society's demand for transport infrastructure, waste disposal and minerals.

Stabilisation and renewal

Unsurprisingly, for an economy so heavily dependent upon one staple industry, adaptation to the changing demands of the world economy proved to be extremely difficult. Precipitous industrial decline began in the 1920s and gathered pace in the 1930s, reaching its peak during the 1950s and 1960s, where the high-wage textile industry of north-west England simply could not compete with the newly industrialising countries on the same terms. The industrial structure of the Mersey Basin did adapt, refocusing into high investment industries such as electrical engineering, car and aircraft manufacture, chemical production and oil refining, which grew in areas away from the traditional mining and textile centres. In turn, however, these 'new' industries have experienced decline in the face of fierce global competition.

Population growth in the North West Region (most of which is accounted for by the Mersey Basin) continued until 1971 but then declined and finally stabilised at around 6.4 million (Figure 23.8). Urban expansion continued during the post-war period (see for example Figure 23.10 for Merseyside). Between Liverpool and Manchester significant new development also took place in the New Towns of Runcorn and Warrington. Here service and distributive industries were able to take full

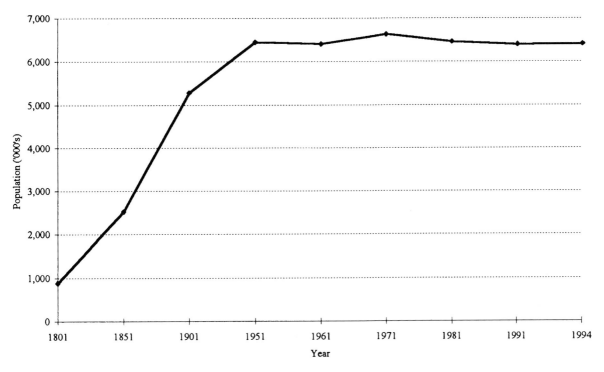

Figure 23.8. Regional population change 1801–1994 (Department of Environment *Longterm Population Distribution in Great Britain – a Study*. HMSO, London. 1971 Table 1.4, p. x; Office for National Statistics *Regional Trends*, 1996 HMSO, London. Table 3.1, p. 46).

advantage of their central location not only in the region but also in the UK. This reinforced the trend of population migration from the core of the older urban centres to new housing estates on the periphery.

The *Strategic Plan for the North West* [SPNW] (North West Joint Planning Team 1973) marked a very significant shift in policy with a new emphasis on urban renewal and a proposal to concentrate new development along 'growth corridors' between Liverpool and Manchester. In particular, 'the corridors should not be seen mainly as fringes of new development attached to the present urban areas. The concept implies action along corridors extending outwards from the conurbation centres. They should play their part in structuring renewal and rehabilitation of existing urbanisation as well as that of new development beyond existing urban boundaries.' (SPNW 1973, p. 241.)

One important policy tool was the creation of extensive tracts of Green Belt within the Mersey Basin to constrain and shape the future development pattern (Figure 23.9).

The Green Belt is an effective and well understood planning tool for containing development, but it has little or no influence on the quality of the countryside within it, which is all too often severely degraded in urban fringe locations (Elson 1986). Today, the physical integrity of the Green Belt is also under threat. Whilst the population has stabilised in the Mersey Basin, there is a continuing process of counter-urbanisation. This is being reinforced by an increased rate of household formation due to changes within society such as greater longevity, more one-person households, etc. (Office for National

Figure 23.9. Green belt and the community forests in the Mersey Basin (Wood *et al.* 1996, p. 40).

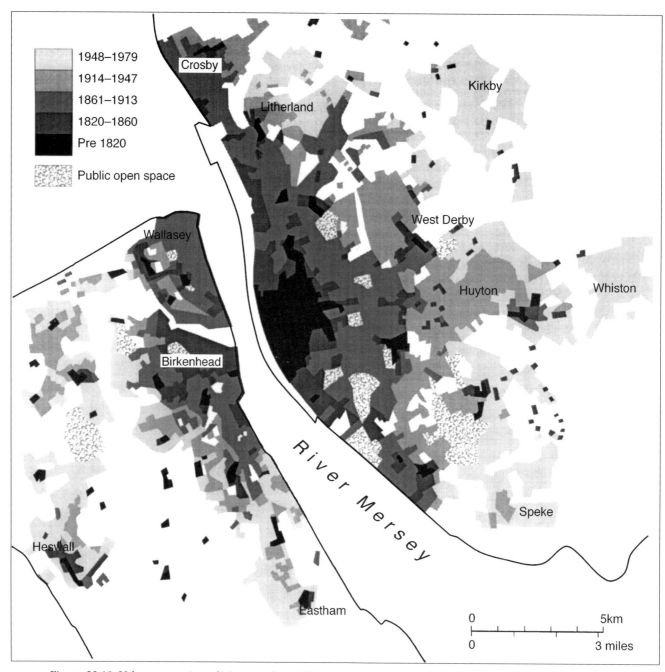

Figure 23.10. Urban expansion of Merseyside to 1979, (contrast this map of Merseyside with Yates' map of 1786 (Figure 5.5, p. 53, this volume) (Lawton 1982, p. 8).

Statistics 1996). The most recent policy statement – Regional Planning Guidance for the North West (RPG 13) (Government Office for the North West/ Government Office for Merseyside 1996) – anticipates that projected development can be accommodated without encroaching on the Green Belt, but given the continuing 'flight from the cities' this policy stance will be difficult to maintain.

SPNW also recognised that poor environmental quality was a contributing factor in regional decline, and that in this respect 'the North West is significantly worse than other regions'. The Report highlighted three priorities for action: river pollution, derelict land and air pollution. The subsequent progress on these fronts has

been mixed. For some time, despite an upturn in expenditure on infrastructure, water quality in the rivers continued to deteriorate. However, thanks to major investment during the past 15 years, there are now signs of significant progress both in river and canal systems (Figure 23.11) and in the Mersey Estuary (National Rivers Authority 1995; Mersey Basin Campaign, 1997 and chapter fifteen, this volume).

The reduction in urban concentrations of smoke and sulphur dioxide which resulted from smoke control programmes (combined with changing patterns of fuel consumption) is a remarkable achievement. These pollutants have now fallen below World Health Organisation guidelines in both towns and cities (Figure 23.13) (see

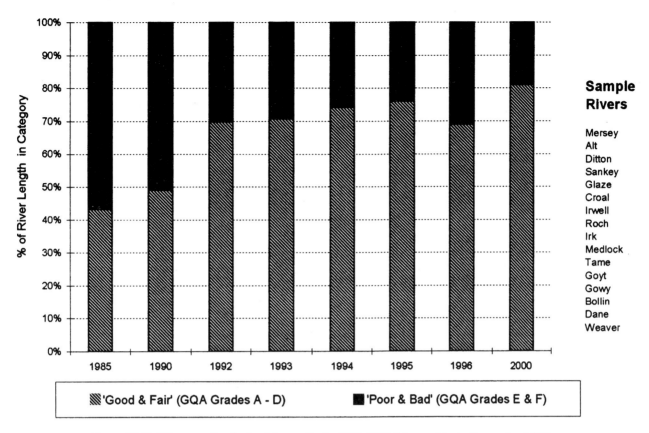

Figure 23.11. Water quality in the Mersey Basin 1985–2000 (Mersey Basin Campaign 1997).

chapter eighteen, Figure 18.2, this volume).

The position with regard to derelict land is less satisfactory. SPNW predicted that if current levels of investment in land reclamation were projected forward, the region's stock of derelict land, much of which had been inherited by the demise of the old wealth creating industries, would be finally cleared in 1994. In fact, despite huge investment, the stock of derelict land has actually increased (Figure 23.12). This is because the rate at which new dereliction has been created by industrial obsolescence has more than matched the rate of derelict land clearance. This situation presents one of the sternest future challenges; not only must the 'engine of dereliction' be reversed, but the legacy of derelict land has to be used wisely as part of a wider strategy for the promotion of biodiversity and conservation.

Industrialisation, urbanisation and biodiversity

The Mersey Basin was at the forefront of the Industrial Revolution and is one of the first 'regions' in the world to cope with the socially painful process of adjustment

Figure 23.12. Derelict land in north-west England (ha) – Components of Change (Wood *et al.* 1996, p. 106, derived from DoE 1991b, 1995).

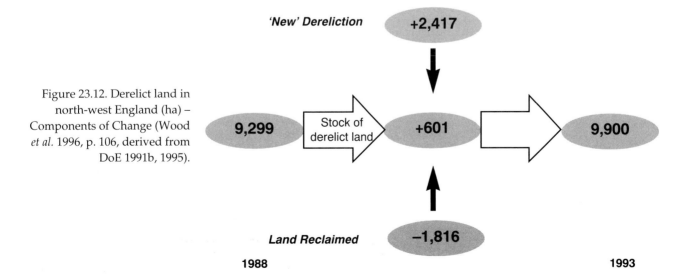

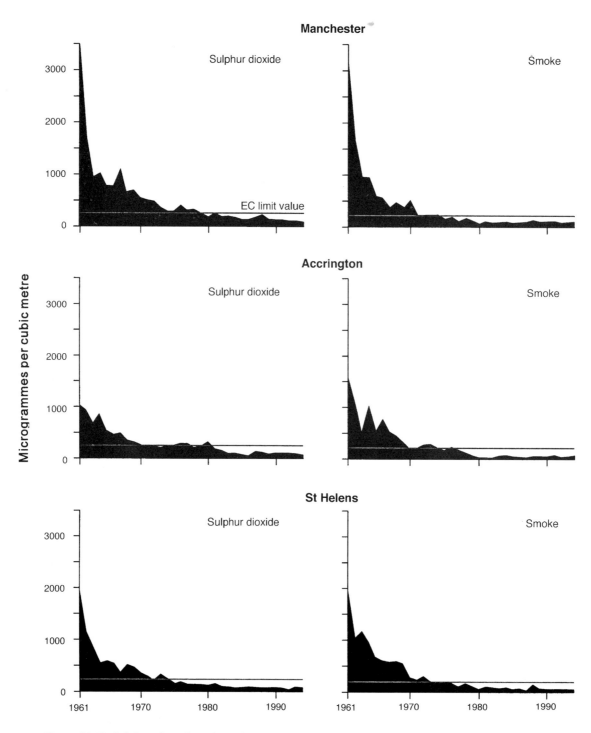

Figure 23.13. Sulphur dioxide and smoke trends in north-west England 1961–1994 (annual peak values).
(AEA Technology 1995).

to a post-industrial economy. As this volume demonstrates, it is also one of the best documented parts of Britain, and therefore the world, in terms of its natural history. Here we explore some of the key processes which drive landscape change, the consequences for flora and fauna, and any lessons learnt which may be transferable to other regions experiencing a similar transition.

We can identify two fundamental cultural drivers of landscape change in modern times: obsolescence and dysfunction.

Obsolescence/Loss of function

Obsolescence, or loss of function, is associated with changing patterns of land-use. This is a potent factor in the countryside of the Mersey Basin where cessation of traditional management practices has released successional processes in plagioclimax communities and initiated landscape change which threatens habitat integrity. Examples cited in this volume include scrub development following cessation or reduction of grazing on sand dunes (chapter sixteen, this volume) heathland (chapter eight, this volume), and the virtual disappearance of

| | **Number of Woods** | | |
Use	*Wirral*	*North Merseyside*	*Total*
Major Uses			
Environmental protection	0	22	22
Nature conservation	10	5	15
Game management	19	67	86
Grazing	28	21	49
Timber production	0	17	17
Recreation	67	72	139
Visual amenity	26	70	96
No use	25	54	79
Ancillary uses			
Vandalism to trees	11	29	40
Refuse disposal	73	52	125

Table 23.1. The utilisation of woodlands in Merseyside 1974–1976. Note that many woods have two or three uses in either part or the whole of the wood (Berry & Pullen 1982 p. 112).

traditional woodland management systems such as coppicing (chapter ten, this volume). A survey of woodland uses in Merseyside in the 1970s highlighted the poverty of woodland management (Table 23.1), and for all the commendable efforts of the Mersey and Red Rose Forests, the woodland management situation is little better today. The situation in the farmed countryside is a matter of particular concern where valuable (but often obsolete) features, such as hedgerows and marl pits, are especially at risk (see chapters seven and fourteen, this volume).

On a more positive note, obsolescence in the urban economy is a source of great biological opportunity (Handley 1996). The abandoned 'disturbance corridors' of the former transport system can quickly become 'regeneration corridors', providing new opportunities for people and wildlife (Foreman 1995 and chapters eight and twelve, this volume). Similarly, the abandoned spoil heaps and subsidence flashes of the post-industrial landscape provide a wide variety of starting points for natural recovery (see Figures 23.14 and 23.15 and chapter twelve, this volume). Derelict land is inherently variable, providing a wide range of starting points for natural succession with variation in pH, fertility, water relations, slope, aspect, stability and toxicity.

Those plant communities and their associated fauna which develop, represent the interaction between this inherent physical and chemical variability (often a product of industrial history) and the biological potential of the area (Ash 1991). It is no accident that in the more heavily urbanised parts of the region we find a strong match between designation of sites for nature conservation and industrial disturbance (Morrish 1996).

We can even see natural recovery at work in industrial water bodies notably the docks of Salford Quays ('fresh' water) and Liverpool (salt water) (see chapter fifteen, this volume). The recovery of the Liverpool Docks has been relatively self-sustaining since the initial dredging operation and installation of lock gates to re-

establish permanent water. However, by contrast, the establishment of an aquatic ecosystem at Salford Quays has faced a series of biological challenges which have been overcome by some imaginative ecological engineering (Hendry *et al.* 1993; Struthers 1997).

Dysfunction

Dysfunction refers to the disruption caused by a mismatch between the type and intensity of land-use and the character of the receiving landscape. Philip Grime recognised three fundamental challenges to existence for plants, and perhaps other components of the ecosystem (Grime 1979, 1986). These are:

Figure 23.14. Colonisation of a colliery spoil heap at Bold Moss, St Helens: heathland.

Figure 23.15. Colonisation of a colliery spoil heap at Bold Moss, St Helens: wetland.

1. competition – where conditions for growth are favourable;

2. stress – where some factor in the environment depresses growth potential; and

3. disturbance – the physical removal of biomass by processes such as grazing, burning and trampling.

In heavily urbanised areas, competition may be intensified by nutrient enrichment leading to a loss of biodiversity; examples in this volume are algal blooms in dockland ecosystems (chapter thirteen, this volume), and worries about enrichment of heathland by atmospheric pollution (chapter eight, this volume). Intensive agricultural production with heavy use of nitrogen fertilizers in the lowlands of Lancashire and Cheshire is leading to enrichment of lakes and streams (chapter thirteen, this volume) and ponds (chapter fourteen, this volume). There is also concern about groundwater pollution by nitrates. Part of Cheshire has been designated as a Nitrate Sensitive Area (Wood *et al.* 1996).

Stress is typically associated with environmental pollution of air, water and land. Figure 23.16 shows how the consequences of stress may be felt at different levels from the individual to the ecosystem. Examples are found at all levels in this account of the ecology of the Mersey Basin.

In some cases where organisms are simply not equipped to cope with new pressures, like sulphur dioxide pollution, whole taxa have been obliterated in the more polluted areas (*Sphagnum* spp. on blanket peat, chapter eleven and lichens in town centres, chapter eighteen, this volume) though for lichens at least we see an encouraging pattern of recovery as pollution levels have abated (see Figures 18.7 and 18.8, chapter eighteen, this volume).

Remarkable examples of both phenotypic and genotypic response of marine algae to stress are provided by Russell *et al.* (chapter seventeen, this volume). Classic examples of genotypic response to pollutants in the Mersey Basin include industrial melanism in lepidoptera (chapter twenty, this volume) and evolution of metal tolerant genotypes of both plants and animals in industrial centres such as Prescot (Bradshaw & McNeilly 1981, p. 43; Parry, Johnson & Bell 1984).

There are also good examples of the population response to opportunities created by stress such as the recent expansion of the halophyte, Danish Scurvygrass (*Cochlearia danica*) along the salt-laden road verges of the region (Figure 19.3, chapter nineteen, this volume). The influence of stress factors at the community level is illustrated by the post-industrial landscape where low fertility, extremes of pH, etc., hold back competitor species and create very distinct plant communities (Ash 1991; chapter eight, this volume). Stress conditions have even been introduced artificially by topsoil removal and mixing of industrial waste materials to produce sustainable wild flower meadows on otherwise fertile but species-poor urban grasslands (Ash, Bennet & Scott 1992).

The most significant impacts of stress are felt at the

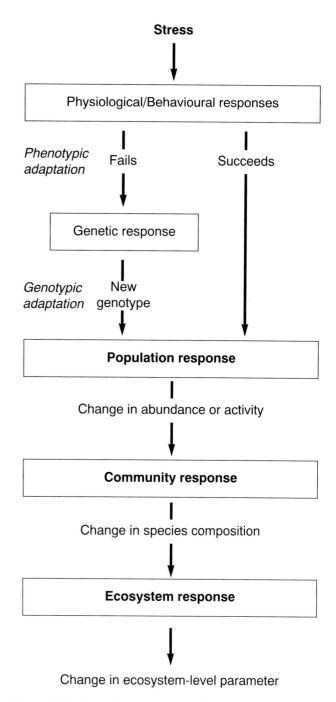

Figure 23.16. The effect of stress at different levels of biological organisation. Note: each box represents some level at which there is some buffering capacity, reducing the impact of the stress on the level above (Beeby 1993, p. 22).

level of the ecosystem. The complexity of ecosystem processes at this level of organisation is well exemplified by the recent work of Moss *et al.* on the Cheshire Meres (chapter thirteen, this volume). In the Pennine uplands, by contrast, the problem is one of soil impoverishment by acid rain. Acidification results from the long distance transport of sulphur and nitrogen from pollution sources in the lowlands which are then washed out by rainfall or captured by dry deposition in the uplands (Figure 23.17). Consequently, the Pennine uplands of the Mersey

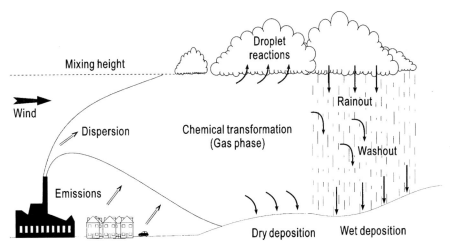

Figure 23.17. Processes which may be involved between the emission of an air pollutant and its ultimate deposition to the ground. (Harrison 1992).

Basin receive one of the highest acid pollution insults in rural Britain and this on an ecosystem of very low carrying capacity. Critical load exceedence maps point to severe problems for upland soils and water bodies (Critical Loads Advisory Group 1994; Handley & Perry 1995).

'Disturbance' is often associated with traditional vegetation management practices and we have already seen that the conservation of many relict habitats depends on the maintenance of the disturbance regime. By contrast, disturbance in the form of excessive (intensive recreation) and malignant (vandalism) use can be very damaging. Erosion problems from concentrated visitor use are a feature of heathland and sand dune habitats of the lowlands, and overused footpaths in the uplands. A systematic survey of vegetation damage on Merseyside in the late 1970s using colour infra-red aerial photography identified disturbance as a much more potent source of vegetation damage than environmental pollution (Handley 1980).

Managing over-load

In general, an effective response to dysfunction is to reduce the load from environmental stress or disturbance to within the carrying capacity of the receiving system. The Government's long-term pollution control strategy for large combustion plants will reduce levels of SO_2 and NO_x very significantly in the Pennine uplands (Critical Loads Advisory Group 1994). The same is true of water pollution to rivers, which is being addressed through the Mersey Basin Campaign. In the Mersey Basin area, over three quarters of all watercourses are now of 'good' or 'fair' water quality – that is, able to support coarse fish populations – with the ambitious target that by 2010 all watercourses will be of this standard. Just ten years ago, only some 40% of watercourses were of good or fair quality (Figure 23.11). Similar improvements are being achieved in the Mersey Estuary (chapter fifteen, this volume).

An alternative to load reduction is to increase the carrying capacity of the environment. This has been done successfully at pressure points created by concentrated visitor use in fragile dune systems of the coastal zone (Wheeler, Simpson & Houston 1991), and on eroding footpaths in the uplands (Ruff & Maddison 1994). However, effective countryside management requires more than technical solutions, and the Mersey Basin has been at the forefront of devising innovative management mechanisms which bring landowners, land users and government agencies together in a spirit of partnership (Wheeler *et al.* 1991; Countryside Commission 1976, 1993). One remarkable innovative exercise in capacity building is the creation of new community forests in the urban fringes of Manchester and Liverpool. Initiated in 1991, the two designated Community Forests cover around 170,000 hectares and are by far the largest of the twelve designated in England, and virtually encircle the conurbations (Mersey Community Forest 1995; Red Rose Community Forest 1994). Continuous woodland forming a bridge between town and country is not envisaged; rather a mosaic of planting which will give structure to the landscape in one of the least wooded areas of the country. The planting programme will extend over some 30 to 40 years and it is intended that tree cover will be increased by three to five times. The community forests cover five broad themes:

1. creating networks of wooded greenways;

2. greening transport routes;

3. returning farmland to forestry;

4. weaving woodland into new development; and

5. capitalising on woodland assets.

Rebuilding environmental capital through woodland creation and enhancement, and using that new structure intelligently, represents an overdue refocusing of environmental planning priorities towards a sensitive balance of ecology, development and amenity.

Ecology and sustainability

Towards sustainable development

If the publication of the Strategic Plan for the North West was a landmark in environmental consciousness within the region, the publication in 1990 of the Environment White Paper *This Common Inheritance* was equally significant in marking an official shift in Government policy towards a more sustainable development path. The White Paper presaged the emergence of an international consensus at the United Nations Conference on Environment and Development at Rio de Janeiro in 1992 (the Earth Summit). The UK Government responded to the Summit by approving the following commitments (Her Majesty's Government 1994a, b, c, d):

1. UK Strategy for Sustainable Development;

2. Climate Change Convention;

3. Biodiversity Convention; and

4. Sustainable Forestry.

This new national emphasis on sustainability is already being reflected in regional policy. The recently published Regional Planning Guidance for the North West (the successor to SPNW, Government Office for the North West/Government Office for Merseyside 1996) states (para. 2.9) that:

> the priority for the future is to maximise the competitiveness, prosperity and quality of life in the Region through *sustainable development*. It is envisaged that the North West can become:
>
> 1. a world class centre for the production of high quality goods and services;
>
> 2. a green and pleasant region; and
>
> 3. a region of first class links to the rest of Europe and the world.

Whatever the limitations of such rhetorical flourishes, and the underpinning motivations, land-use and economic planning are nevertheless beginning to explore how the demands of the environment, economy and society can be reconciled at the regional scale. Roberts (1994 p. 782) suggests that four principles might be used to guide sustainable regional planning:

1. the 'standard' elements of sustainable development related to the environment, futurity and equity;

2. elements related to the diversification of the regional economy, intended to make it better able to deal with adversity;

3. the question of self-sufficiency, intended to minimise environmentally and economically wasteful resource inputs or transfers; and

4. the question of territorial integration both within the individual region and between regions.

The exact interpretation of what constitutes a 'region' and hence the most appropriate scale for planning and management is problematic. The establishment of the 'city region' as a basis for sustainable planning is one approach (see for example Breheny & Rookwood 1993; Ravetz 1994; Owens 1994; Roberts & Chan 1997) of particular relevance to the Mersey Basin.

Planning to conserve and enhance biodiversity

One of the key government publications following the Earth Summit was the *National Biodiversity Action Plan* (Her Majesty's Government 1994c). The Biodiversity Action Plan aims to 'conserve and enhance biological diversity within the UK and to contribute to the conservation of global diversity through all appropriate mechanisms'. The objectives of the Strategy are to conserve, and where practicable, to enhance:

1. overall population and natural ranges of species and quality and range of wildlife habitats and ecosystems;

2. internationally important and threatened species, habitats and ecosystems;

3. species, habitats and natural and managed ecosystems that are characteristic of local areas; and

4. the biodiversity of natural and semi-natural habitats where this has been diminished over recent decades.

The government established a Biodiversity Steering Group to lead implementation of the Plan and one of their early recommendations was for the establishment of regional biodiversity action plans. It is encouraging to note that the first steps have already been taken in establishing such a plan for the North West Region centred on the Mersey Basin (Bennett *et al.* 1996). The collected papers in this volume provide an invaluable bench-mark for biodiversity planning and many important pointers for effective action.

The biodiversity balance sheet

The long history of human settlement in the Mersey Basin, culminating in the Industrial Revolution and its aftermath, has had a profound effect on landscape ecology and hence on biodiversity. Woodland clearance gradually eroded the matrix of primary habitat, replacing it in time by a fragmented countryside with isolated habitat patches of woodland, moor and heath (chapter seven, this volume). The connectivity between habitats has been reduced by the progressive removal of landscape features such as hedgerows and the downgrading of stream corridors (chapters seven and nine, this volume). Urban development has resulted in the direct loss of habitat and indirect impacts through enrichment, stress and disturbance.

However, perhaps the most remarkable feature of the history of landscape development in the Mersey Basin is the resilience of nature in the face of this massive disruption. Among the more striking positive findings are:

1. contribution to species richness within the habitat (α

diversity) by environmental stress which reduces the dominance of competitor species;

2. contribution to habitat richness (β diversity) by industrial obsolescence initiating natural succession and creative conservation; and

3. contribution to the total species complement (δ diversity) by escapes, introductions and possibly climate change.

As the chapters in this volume show, not only are the latter findings cause for optimism about the future, they also suggest that there is much we can learn about the study of ecology in a heavily urbanised region; for example the importance of disturbance corridors (including motorways) for promoting species migration within the region. The tentative findings of Greenwood (chapter nineteen, this volume) about the increasing total species complement (at least of higher plants) are most unexpected, and, together with comparable data on avifauna, lepidoptera and certain other taxa, vindicates the efforts of local recorders in documenting the biological consequences of the tide of change in this remarkable region.

Conclusions and way forward

The urban and industrial challenge to the natural integrity of the Mersey Basin has abated somewhat, but its forces are by no means spent. There will be continuing demands for development, especially housing, waste disposal and transport infrastructure which will threaten what remains of the region's critical natural capital. Similarly, as historic problems of air and water pollution are overcome, new challenges are emerging such as the primary and secondary pollutants from motor vehicles (Royal Commission on Environmental Pollution 1995; Wood *et al.* 1996). In addition, we have hardly begun to tackle the deep-seated historic problems of contaminated land and its associated ground water pollution (Royal Commission on Environmental Pollution 1996).

One critical issue is that the two principal drivers of landscape change (climatic and cultural influence) are now inextricably linked through carbon forcing (chapter one, this volume; Department of the Environment 1991a, 1996). An examination of the likely consequences of climate change for landscape and biodiversity in the region must now be a high priority. More fundamentally, we need to recognise that, despite the closure of the region's coal mines, north-west England is still essentially a fossil fuel economy. A move back toward renewable energy (so important in 1786, Figure 23.2) would bring very significant environmental impacts, not least on the Pennine uplands (wind farms) and on the Mersey Estuary, Britain's second most favourable estuary for generating tidal power (Department of Energy 1992).

Despite these concerns for the future, we can take heart from the new emphasis on sustainability emerging in national and regional policy. This will involve not only conserving what remains of the long-established biota, but actively repairing past damage and rehabilitating the environmental capital of the Mersey Basin. The Mersey and Red Rose Community Forests are beginning the long haul of revitalising the region's woodland capital (at only 4% this is one of the most impoverished landscapes in Europe). Creative conservation is a powerful force which could be used to greater effect, especially in wetland creation, as the outstanding achievements at Martin Mere demonstrate (chapter twenty-one, this volume).

There has been an exponential growth in environmental initiatives addressing these and other issues (Kidd *et al.* 1996), but the targeting of these initiatives is uneven. The landscape infrastructure of the rural lowlands seems to be especially vulnerable and, with some notable exceptions (e.g., the Cheshire Special Landscapes Project) is not well served by current policy and practice. Similarly, whilst there is now a comprehensive coverage of management schemes in the coastal zone of the Mersey Basin, they are for the most part substantially under-resourced and therefore ineffective. The Sefton Coast Management Scheme is a notable and important exception.

What is evident is that the Mersey Basin is fortunate, despite being the test-bed for the world's first industrial economy, in possessing a rich ecological legacy, and action is needed to secure what remains for future generations. The way forward, as envisaged by Adams (1996) is likely to involve:

1. maintaining diversity of landscapes and ecosystems;

2. building room for nature into economic life;

3. building connections between people and nature; and

4. allowing nature to function and creating conditions for it to do so.

The development of environmental policy from international to local levels, but especially through the new regionalism, is putting these aspirations into practice. Environmental assessment is now enabling more informed judgements to be made about the likely impact of new developments on landscape and ecology and the potential for mitigation. We are only at the start of a process which is beginning to recognise the complexities of the interconnections between ecology and development. Industrialisation and urbanisation have taken a heavy ecological toll in the past, and in some respects continue to do so, but there now exists an unparalleled opportunity to develop new understandings of the relationship between conservation and development. The timely and authoritative contributions to this volume represent the establishment of an information base which will help to realise this aspiration.

References

Adams, W. (1996). *Future Nature*. Earthscan, London.

AEA Technology (1995). *Air Pollution in the UK: 1993/4*. AEA, Stevenage.

Ash, H. (1991). Soils and Vegetation in Urban Areas. *Soils in the Urban Environment* (eds P. Bullock & P. Gregory), pp. 153–70. Blackwell Scientific Publications, Oxford.

Ash, H., Bennett, R. & Scott, R. (1992). *Flowers in the Grass*. English Nature, Peterborough.

Barker, T. & Harris, J. (1993). *A Merseyside Town in the Industrial Revolution: St Helens 1750–1900*. Cass & Co., London.

Beeby, A. (1993). *Applying Ecology*. Chapman & Hall, London.

Bennett, C., Fox, P., Marhall, I., Bruce, N. & Jepson, P. (eds) (1996). *Biodiversity North West: Co-ordinating Action for Biodiversity in North West England*. Conference Proceedings Environment Agency, Warrington.

Berry, P. & Pullen, R. (1982). The Woodland Resource: Management and Use. *The Resources of Merseyside* (eds W. Gould & A. Hodgkiss), pp. 101–18. Liverpool University Press, Liverpool.

Bradshaw, A. & McNeilly, T. (1981). *Evolution and Pollution*. Studies in Biology No. 30. Edward Arnold, London.

Breheny, M. & Rookwood, R. (1993). Planning the Sustainable City Region. *Planning for a Sustainable Environment: a Report by the Town & Country Planning Association* (ed. A. Blowers), pp. 150–89. TCPA, London.

British Association (1962). *Manchester and Its Region*. Manchester University Press, Manchester.

Countryside Commission (1976). *The Bollin Valley: A Study of Land Management in the Urban Finge*. **CCP97**. Countryside Commission, Cheltenham.

Countryside Commission (1993). *Countryside Management Projects*. **CCP403**. Countryside Commission, Cheltenham.

Critical Loads Advisory Group (1994). *Critical Loads of Acidity in the United Kingdom: Summary Report*. HMSO, London.

Department of Energy (n.d.). *Survey of Tidal Energy in the U.K.* Report No. STP 102. Department of Energy, London.

Department of Energy (1992). *Tidal Power from the River Mersey: A Feasibility Study. Stage III Report*. Department of Energy, London.

Department of the Environment (1971). *Long Term Population Distribution in Great Britain: a Study*. HMSO, London.

Department of the Environment (1991a). *The Potential Effects of Climate Change in the UK*. UK Climate Change Impacts Review Group 1st Report. HMSO, London.

Department of the Environment (1991b). *Survey of Derelict Land in England, 1988*. HMSO, London.

Department of the Environment (1995). *Survey of Derelict Land in England, 1993*. HMSO, London.

Department of the Environment (1996). *Review of the Potential Effects of Climate Change in the UK*. UK Climate Change Impacts Review Group 2nd Report. HMSO, London.

Dickens, C. (1854). *Hard Times*.

Dower, M. (1965). *The Challenge of Leisure*. Civic Trust, London.

Elson, M. (1986). *Green Belts: Conflict Mediation on the Urban Fringe*. Heinemann, London.

Engels, F. (1844). *The Condition of the Working Class in England*.

Foreman, R. (1995). *Land Mosaics: the Ecology of Landscape and Regions*. Springer Verlag, New York, USA.

Freeman, T., Rodgers, H. & Kinvig, R. (1966). *Lancashire, Cheshire and the Isle of Man*. Nelson & Sons, London.

Government Office for the North West/Government Office for Merseyside (1996). *Regional Planning Guidance for the North West: RPG 13*. HMSO, London.

Grime, J. (1979). *Plant Strategies and Vegetation Processes*. John Wiley, Chichester.

Grime, J. (1986). Manipulation of Plant Species and Communities. *Ecology and Design in Landscape* (eds A. Bradshaw, D. Goode & E. Thorpe), pp. 175–94. Blackwell Scientific Publications, Oxford.

Handley, J. (1980). The Application of Remote Sensing to Environmental Management. *International Journal of Remote Sensing*, **1**, 181–95.

Handley, J. (1982). The Land of Merseyside. *The Resources of Merseyside* (eds W. Gould & A. Hodgkiss), pp. 83–100. Liverpool University Press, Liverpool.

Handley, J. (1984). Ecological Requirement for Decison-Making Regarding Medium Scale Developments in the Urban Environment. *Planning and Ecology* (eds R. Roberts & T. Roberts), pp. 222–38. Chapman & Hall, London.

Handley, J. (1996). *The Post Industrial Landscape: a Groundwork Status Report*. Groundwork Foundation, Birmingham.

Handley, J. & Perry, D. (1995). *The Regional Environment of the Transpennine Corridor*. Occasional Paper No. 41. Department of Planning & Landscape, University of Manchester.

Harley, J. (1968). *A Map of the County of Lancashire, 1786, by William Yates*. Historic Society of Lancashire & Cheshire.

Harrison, R. (1992). *Understanding Our Environment*. Royal Society of Chemistry, London.

Hendry, K., Webb, S., White, K. & Parsons, N. (1993). Water Quality and Urban Regeneration: a Case Study of the Central Mersey Basin. *Urban Waterside Regeneration: Problems and Prospects* (eds K. White, E. Bellinger, A. Saul, M. Symes & K. Hendry), pp. 271–82. Ellis Howard, Chichester.

Her Majesty's Government (1994a). *Sustainable Development: the UK Strategy*. HMSO, London.

Her Majesty's Government (1994b). *Climate Change: the UK Programme*. HMSO, London.

Her Majesty's Government (1994c). *Biodiversity: the UK Action Plan*. HMSO, London.

Her Majesty's Government (1994d). *Sustainable Forestry: the UK Programme*. HMSO, London.

Jackson, H. (1979). The Decline of the Sand Lizard, *Lacerta Agilis* L. Population on the Sand Dunes of the Merseyside Coast, England. *Biological Conservation*, **16 (3)**, 177–93.

Kidd S., Handley, J., Wood, R. & Douglas, I. (1996). 'Greening the North West': a Regional Landscape Strategy Working Paper 2 – Strategic Environmental Initiatives. Occasional Paper No. 54. Department of Planning & Landscape, University of Manchester.

Lawton, R. (1982). From the Port of Liverpool to the Conurbation of Merseyside. *The Resources of Merseyside* (eds W. Gould & A. Hodgkiss), pp. 1–13. Liverpool University Press, Liverpool.

Mersey Basin Campaign (1997). *Mersey Basin Campaign: Mid Term Report*. Mersey Basin Campaign, Manchester.

Mersey Community Forest (1995). *The Mersey Forest: Forest Plan*. Mersey Community Forest, Warrington.

Morrish, B. (1997). *The Realisation of Ecological and Amenity Benefits through Land Reclamation Programmes in Greater Manchester*. MSc thesis, University of Manchester.

National Rivers Authority (1995). *The Mersey Estuary: a Report in Environmental Quality*. HMSO, London.

North West Joint Planning Team (1973). *Strategic Plan for the North West: SPNW Joint Planning Team Report*. HMSO, London.

Office for National Statistics (1996). *Regional Trends 1996*. HMSO, London.

Owens, S. (1994). Can Land Use Planning Produce the Ecological City? *Town & Country Planning* **63, (6) June**, 170–73.

Parry, G., Johnson, M. & Bell, R. (1984). Ecological Surveys for Metalliferous Mining Proposals. *Planning and Ecology* (eds R. Roberts & T. Roberts), pp. 40–55. Chapman & Hall, London.

Philpott, R. (1988). *Historic Towns of the Merseyside Area: a Survey of Urban Settlement to c.1800*. National Museums & Galleries on Merseyside Occasional Papers, Liverpool Museum No. 3. Liverpool Museum, Liverpool.

Rackham, O. (1986). *The History of the Countryside*. Dent, London.

Ravetz, J. (1994). Manchester 2020 – a Sustainable City Region Project. *Town & Country Planning,* **63 (6) June,** 181–85.

Red Rose Community Forest (1994). *Red Rose Forest Plan.* Red Rose Forest, Manchester.

Roberts, P. (1994). Sustainable Regional Planning. *Regional Studies,* **28 (8),**781–87.

Roberts, P. & Chan, R. (1997). A Tale of Two Regions: Strategic Planning for Sustainable Development in East and West. *International Planning Studies,* **2 (1),** 45–62.

Royal Commission on Environmental Pollution (1995). *Eighteenth Report: Transport and Environment.* HMSO, London.

Royal Commission on Environmental Pollution (1996). *Nineteenth Report: Sustainable Use of Soil.* HMSO, London.

Ruff, A. & Maddison, C. (1994). Footpath Management in the National Parks. *Landscape Research,* **19 (2),** 80–87.

Struthers, W. (1997). From Manchester Docks to Salford Quays: Ten Years of Environmental Improvements in the Mersey Basin Campaign. *Journal of the Chartered Institute of Water & Environmental Management,* **11 (2),** 1–7.

von Tunzelmann, N. (1986). Coal and Steam Power. *Atlas of Industrialising Britain 1760–1914* (eds J. Langton & R. Morris), pp. 72–79. Methuen, London.

Walker, F. (1939). Historical Geography of South West Lancashire Before the Industrial Revolution. *Chetham Society,* **103,** New Series.

Walton, J. (1987). *Lancashire: a Social History 1558–1939.* Manchester University Press, Manchester.

Wheeler, D., Simpson, D. & Houston, J. (1991). Dune Use and Management. *The Sand Dunes of the Sefton Coast.* (eds D. Atkinson & J. Houston), pp. 129–49. National Museums & Galleries on Merseyside in conjunction with Sefton Metropolitan Borough Council, Liverpool.

Wood, R., Handley, J., Douglas, I. & Kidd, S. (1996). *'Greening the North West': a Regional Landscape Strategy Working Paper 1 – Regional Landscape Assessment.* Occasional Paper No. 53. Department of Planning and Landscape, University of Manchester.

Yalden, D. (1987). The Natural History of Domesday Cheshire. *Naturalist,* **112,** 125–31.

Authors' Addresses

Allen, Dr J.R., Port Erin Marine Laboratory, School of Biological Sciences, University of Liverpool, Port Erin, Isle of Man, IM9 6JA.

Ash, Dr H.J., 5 Dearnford Avenue, Bromborough, Wirral, L62 6DX.

Atkinson, Dr D., Population Biology Research Group, School of Biological Sciences, Nicholson Building, University of Liverpool, Liverpool, L69 3BX.

Barr, C.J., Institute of Terrestrial Ecology, Merlewood Research Station, Grange-over-Sands, Cumbria, LA11 6JU.

Boothby, Dr J., Pond *Life* Project, Liverpool John Moores University, Trueman Building, 15–21 Webster Street, Liverpool, L3 2ET,

Bradshaw, Professor A.D., School of Biological Sciences, University of Liverpool, Liverpool, L69 3BX.

Butterill, G., Cheshire Wildlife Trust, Grebe House, Reaseheath, Nantwich, Cheshire, CW5 6DG.

Cowell, R.W., Liverpool Museum, National Museums & Galleries on Merseyside, William Brown Street, Liverpool, L3 8EN.

Davis, Dr B.N.K., Brook House, Easton, Huntingdon, Cambs., PE18 0TU.

Doody, Dr J.P., National Coastal Consultants, 5 Green Lane, Brampton, Huntingdon, Cambs., PE18 8RE.

Eaton, Dr J.W., School of Biological Sciences, University of Liverpool, Liverpool, L69 3BX.

Fairhurst, Dr J., Cheshire County Council, Environmental Planning, Commerce House, Hunter Street, Chester, CH1 2QP.

Fielding, Dr N.J., School of Biological Sciences, University of Liverpool, Liverpool, L69 3BX.

Fisher Dr C.T., Liverpool Museum, National Museums & Galleries on Merseyside, William Brown Street, Liverpool, L3 8EN.

Fox, Professor B.W., Tryfan, Longlands Road, New Mills, High Peak, Derbyshire, SK22 3BL.

Gonzalez, Dr S., School of Biological and Earth Sciences, Liverpool John Moores University, Byrom Street, Liverpool, L3 3AF.

Greenwood, E.F., Liverpool Museum, National Museums & Galleries on Merseyside, William Brown Street, Liverpool, L3 8EN.

Handley, Professor J.F., OBE, Department of Planning and Landscape, University of Manchester, Oxford Road, Manchester, M13 9PL.

Hawkins, Professor S.J., Biodiversity &

Ecology Division, School of Biological Sciences and Centre for Environmental Sciences, University of Southampton, Shackleton Building, Highfield, Southampton, S017 1BJ.

Holland, D.G., The Environment Agency, Mirwell, Carrington Lane, Sale, Cheshire, M33 5NL.

Huddart, Dr D., School of Education and Community Studies, Liverpool John Moores University, I.M. Marsh Campus, Barkhill Road, Liverpool, L17 6BD.

Hull, Dr A., Pond *Life* Project, Liverpool John Moores University, Trueman Building, 15–21 Webster Street, Liverpool, L3 2ET.

Huntley, Professor B., Environmental Research Centre, Department of Biological Sciences, University of Durham, Durham, DH1 3LE.

Innes, Dr J.B., Environmental Research Centre, Department of Geography, University of Durham Science Laboratories, South Road, Durham, DH1 3LE.

Jarvis, A.E., Centre for Port & Maritime History, GWR Building, Merseyside Maritime Museum, National Museums & Galleries on Merseyside, Liverpool, L3 1DG.

Jemmett, Dr A.W.L., School of Biological Sciences, University of Liverpool, Liverpool, L69 3BX.

Judd, Dr S., Liverpool Museum, National Museums & Galleries on Merseyside, William Brown Street, Liverpool, L3 8EN

Lageard, Dr J.G.A., Division of Environmental Science, Crewe and Alsager Faculty, Manchester Metropolitan University, Crewe Green Road, Crewe, Cheshire, CW1 1DU.

Lewis, Dr J.M., University of Liverpool, P.O. Box 147, Liverpool, L69 3BX.

Long, Dr A.J., Environmental Research Centre, Department of Geography, University of Durham, Science Laboratories, South Road, Durham, DH1 3LE.

Mitcham, T., Lancashire Wildlife Trust, Cuerden Park Wildlife Centre, Shady Lane, Bamber Bridge, Preston, Lancashire, PR5 6AU.

Morries, G., Lancashire County Council, P.O. Box 160, East Cliff County Offices, Preston, Lancashire, PR1 7EX.

Moss, Professor B., School of Biological Sciences, University of Liverpool, Liverpool, L69 3BX.

Nolan, P., The Environment Agency, Mirwell, Carrington Lane, Sale, Cheshire, M33 5NL.

Norman, Professor D., Rowswood Cottage, Ridding Lane, Sutton Weaver, Runcorn, Cheshire, WA7 6PF.

Oldfield, P., Environmental Services Directorate, Halton Borough Council, Grosvenor House, Halton Lea, Runcorn Cheshire, WA7 2GW.

Oldfield, Professor F., Executive Director, Past Global Changes, International Geosphere-Biosphere Programme, Bärenplatz 2, CH - 3011 Bern, Switzerland.

Philpott, Dr R.A., Liverpool Museum, National Museums & Galleries on Merseyside, William Brown Street, Liverpool, L3 8EN.

Plater, Dr A.J., Department of Geography, University of Liverpool, Liverpool, L69 3BX.

Reed, P.N., Central Services Division, National Museums & Galleries on Merseyside, P.O. Box 33, 127 Dale Street, Liverpool, L69 3LA.

Russell, Dr G., School of Biological Sciences, University of Liverpool, Liverpool, L69 3BX.

Smart, R.A., 10 Elizabeth Crescent, Queens Park, Chester, CH4 7AZ.

Stark, G.J., Institute of Terrestrial Ecology, Merlewood Research Station, Grange-over-Sands, Cumbria, LA11 6JU.

Tallis, Dr J.H., School of Biological Sciences, University of Manchester, 3.614 Stopford Building, Oxford Road, Manchester, M13 9PT.

Tooley, Professor M.J., School of Geography and Geology, Purdie Building, North Haugh, University of St Andrews, Fife, KY16 9ST.

Walker, Professor S., Regional Water Manager, Environment Agency, North West Region, P.O. Box 12, Richard Fairclough House, Knutsford Road, Warrington, Cheshire, WA4 1HG.

Wallace, Dr I.D., Liverpool Museum, National Museums & Galleries on Merseyside, William Brown Street, Liverpool, L3 8EN.

Weekes, L., Cheshire Wildlife Trust, Grebe House, Reaseheath, Nantwich, Cheshire, CW5 6DG.

Wilkinson, Dr S.B., School of Biological Sciences, University of Liverpool, Liverpool, L69 3BX.

Wood, R.W.S., Department of Planning and Landscape, University of Manchester, Oxford Road, Manchester, M13 9PL.

Yalden, Dr D.W., School of Biological Sciences, University of Manchester, 3.614 Stopford Building, Oxford Road, Manchester, M13 9PT.